AF675508

36705000002954+

179419

TJ Buck, Gordon.
177.5
.B83 Machinery alignment
1984 tables

DATE

© THE BAKER & TAYLOR CO.

Gulf Publishing Company
Book Division
Houston, London, Paris, Tokyo

TJ
177.5
.B83
1984

TSTI-Waco
3 6705 00026256 0

Machinery Alignment Tables

Face-OD and Reverse Indicator Methods

Gordon Buck

Library
Texas State Technical Institute

Dedication

To my family—Gregory, Laura, and especially Jean for her loving support on this book and in all my activities.

MACHINERY ALIGNMENT TABLES

Copyright © 1984 by Gulf Publishing Company, Houston, Texas. All rights reserved. Printed in the United States of America. This book, or parts thereof, may not be reproduced in any form without permission of the publisher.

Library of Congress Cataloging in Publication Data

Buck, Gordon S.

Machinery alignment tables.

Includes bibliographical references.

1. Machinery—Alignment—Tables. 2. Shafting—Tables.

I. Title.

TJ177.5.B83 1984 621.8′23 **83-22692**

ISBN 0-87201-015-5

Acknowledgments

I would particularly like to acknowledge the support given me by my former co-workers at Exxon Baton Rouge Refinery. Mr. O.A. Durand and E.J. Labat especially contributed to my education in machinery and alignment; Mr. C.R. Miannay challenged me to improve on existing techniques for machinery reliability in general; and Mr. J.M. Eeds continues to support and encourage my endeavors.

The enthusiastic response of Mr. J.W. Thornton of United Centrifugal Pump Company encouraged me to complete the tables.

Jean Buck encouraged, critiqued, edited, and in general provided all the support anyone could ask of a loving wife.

179419

Contents

	Hub Diameter (in.)	Leg Distance (in.)
Table 3-4	1.5	20
Table 3-5	1.5	25
Table 3-6	1.5	30
Table 3-7	1.5	35
Table 3-8	1.5	40
Table 3-9	1.75	5
Table 3-10	1.75	10
Table 3-11	1.75	15
Table 3-12	1.75	20
Table 3-13	1.75	25
Table 3-14	1.75	30
Table 3-15	1.75	35
Table 3-16	1.75	40
Table 3-17	2	5
Table 3-18	2	10
Table 3-19	2	15
Table 3-20	2	20
Table 3-21	2	25
Table 3-22	2	30
Table 3-23	2	35
Table 3-24	2	40
Table 3-25	2.25	15
Table 3-26	2.25	20
Table 3-27	2.25	25
Table 3-28	2.25	30
Table 3-29	2.25	35
Table 3-30	2.25	40
Table 3-31	2.25	50
Table 3-32	2.25	60
Table 3-33	2.25	70
Table 3-34	2.25	80
Table 3-35	2.5	15
Table 3-36	2.5	20
Table 3-37	2.5	25
Table 3-38	2.5	30
Table 3-39	2.5	35
Table 3-40	2.5	40

Chapter 4 Reverse Indicator Method

Chapter 5 Reverse Indicator Alignment Tables

	Separation (in.)	Leg Distance (in.)
Table 5-3	2	15
Table 5-4	2	20
Table 5-5	2	25
Table 5-6	2	30
Table 5-7	2.5	5
Table 5-8	2.5	10
Table 5-9	2.5	15
Table 5-10	2.5	20
Table 5-11	2.5	25
Table 5-12	2.5	30
Table 5-13	3	10
Table 5-14	3	15
Table 5-15	3	20
Table 5-16	3	25
Table 5-17	3	30
Table 5-18	3	35
Table 5-19	3	40
Table 5-20	3	50
Table 5-21	3	60
Table 5-22	3.5	10
Table 5-23	3.5	15
Table 5-24	3.5	20
Table 5-25	3.5	25
Table 5-26	3.5	30
Table 5-27	3.5	35
Table 5-28	3.5	40
Table 5-29	3.5	50
Table 5-30	3.5	60
Table 5-31	4	10
Table 5-32	4	15
Table 5-33	4	20
Table 5-34	4	25
Table 5-35	4	30
Table 5-36	4	35
Table 5-37	4	40
Table 5-38	4	50
Table 5-39	4	60

	Separation (in.)	Leg Distance (in.)
Table 5-40	5	15
Table 5-41	5	20
Table 5-42	5	25
Table 5-43	5	30
Table 5-44	5	35
Table 5-45	5	40
Table 5-46	5	50
Table 5-47	5	60
Table 5-48	5	70
Table 5-49	6	15
Table 5-50	6	20
Table 5-51	6	25
Table 5-52	6	30
Table 5-53	6	35
Table 5-54	6	40
Table 5-55	6	50
Table 5-56	6	60
Table 5-57	6	70
Table 5-58	7	15
Table 5-59	7	20
Table 5-60	7	25
Table 5-61	7	30
Table 5-62	7	35
Table 5-63	7	40
Table 5-64	7	50
Table 5-65	7	60
Table 5-66	7	70
Table 5-67	8	25
Table 5-68	8	30
Table 5-69	8	35
Table 5-70	8	40
Table 5-71	8	50
Table 5-72	8	60
Table 5-73	8	70
Table 5-74	8	80
Table 5-75	8	90
Table 5-76	8	100

	Separation (in.)	Leg Distance (in.)
Table 5-77	9	25
Table 5-78	9	30
Table 5-79	9	35
Table 5-80	9	40
Table 5-81	9	50
Table 5-82	9	60
Table 5-83	9	70
Table 5-84	9	80
Table 5-85	9	90
Table 5-86	9	100
Table 5-87	10	25
Table 5-88	10	30
Table 5-89	10	35
Table 5-90	10	40
Table 5-91	10	50
Table 5-92	10	60
Table 5-93	10	70
Table 5-94	10	80
Table 5-95	10	90
Table 5-96	10	100
Table 5-97	12	25
Table 5-98	12	30
Table 5-99	12	35
Table 5-100	12	40
Table 5-101	12	50
Table 5-102	12	60
Table 5-103	12	70
Table 5-104	12	80
Table 5-105	12	90
Table 5-106	12	100
Table 5-107	14	25
Table 5-108	14	30
Table 5-109	14	35
Table 5-110	14	40
Table 5-111	14	50
Table 5-112	14	60
Table 5-113	14	70

Preface

Machinery alignment has always been an interesting subject to me. One of the unique aspects of alignment is that calculations are done in the field. As a result, even though the calculations are not that difficult, trial and error methods are most commonly used. This is especially true for general purpose equipment.

Over the years, I have tried memorizing the equations, carrying a card with the equations, developing a nomogram, graphing, and using a programmable calculator. This compilation of machinery alignment tables actually is a result of this frustration. I believe the tables are the best way of converting alignment data into the required movement of the equipment.

The tables are not intended to cover all possible combinations of misalignment and machines. In particular, very large or unusual machines may require special attention. For these instances, please contact me directly. Special tables can be designed for your particular equipment.

Gordon S. Buck
5801 Parkhaven Drive
Baton Rouge, Louisiana 70816

Chapter 1

Introduction

Good alignment between DriveR and DriveN (this style is used to better distinguish the two terms) equipment is an essential part of any rotating equipment reliability program. Vertical alignment is normally obtained by adding shims beneath the equipment. Horizontal alignment is obtained by adjusting positioning screws. The amount of shim or movement can be determined from precise measurements. This handbook replaces the formulas, calculations, and graphs used to convert these measurements to the required amount of shim.

Many techniques are available for the precise alignment of rotating shafts. In general, a measurement is made from one shaft to the other from which the necessary correction is determined. The measurements may be taken by scale, caliper, dial indicator, micrometer, or electronic transducer. They may even be taken optically. The corrections may be determined by guessing, graphing, calculating, or by computer program; and they may be made by adding or removing shims, adjusting jackscrews or machining. In this book, the two most commonly used techniques, the Face-OD Method and the Reverse Indicator Method are described and simplified through the use of tables to interpret the measurements.

Chapter 2

Face—OD Method

The Face-OD Method is a relatively simple technique for aligning two shafts. It has the advantages of being easy to set up and requiring few special tools. In addition, it is easily understood by machinists and millwrights familiar with dial indicator measurements.

In the Face-OD Method, a dial indicator is mounted on one shaft and rotated so that it indicates on the other shaft. Two sets of readings are taken: one on the face of the stationary shaft and the other on the outside diameter of the stationary shaft. Hence the name "Face-OD" (sometimes called "Face-Rim").

In actual practice, the coupling hubs are usually on the shafts and the readings are taken on the coupling hubs. Normally, a single dial indicator is used. The indicator is mounted to read the face and then moved to read the OD. Of course, two indicators could be used simultaneously with the proper setup.

Setting Up

The equations, forms, and tables described for the Face-OD Method are based on the following assumptions:

1. The indicator is mounted on the shaft of the machine that will be shimmed.
2. The indicator rotates with the shaft and reads a stationary shaft (or coupling).
3. The indicator is zeroed at the top of the stationary shaft (or coupling).
4. There is no indicator sag, or the readings have been corrected for sag before using the tables. (See "Indicator sag" later in this chapter.)

A common problem is to align a centrifugal pump with a motor driver. The pump is connected to piping and baseplate properly and will not be moved. The coupling spacer is removed so that the motor shaft may be rotated independently. The dial indicator is mounted on the motor coupling hub and extended to read the pump coupling hub. The indicator is adjusted to read zero at the top of the pump coupling hub OD. Readings are taken at horizontal and vertical centerlines. The indicator is then changed to read the pump coupling hub face. The indicator is zeroed at the top of the face and readings are taken at horizontal and vertical centerlines. Shims are added (or removed) from the motor driver feet.

Many people prefer to use an alignment jig, which allows both shafts to rotate together. This can speed up gathering data and improve accuracy. The tables are also valid for this procedure. However, for purposes of explanation, the shafts are considered to be separated with one rotating and one stationary.

A typical setup is shown in Figure 2-1.

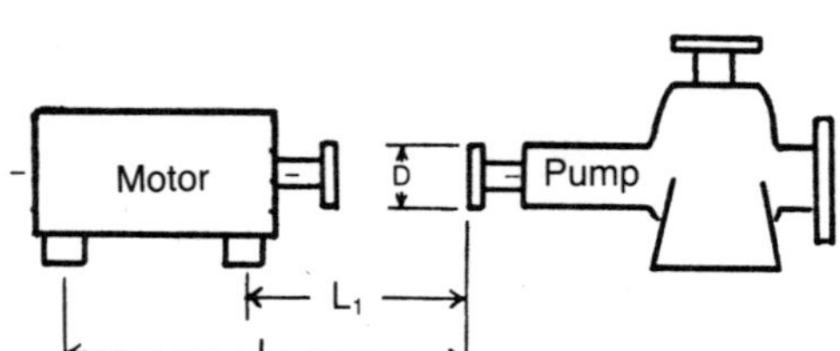

Figure 2-1. A typical pump—motor arrangement showing hub diameter (D) and leg distances (L_1 and L_2) used in the Face—OD Method.

Data Forms

Figure 2-2 is a useful form for recording alignment data. It is good practice to record both preliminary and final readings as well as operating (hot) readings. This can help speed up future re-alignments.

FACE — OD ALIGNMENT RECORD

Area ______________________ Date ____________

DriveN Equipment ____________ Type ____________

DriveR ______________________ Type ____________

Coupling ______________________________________

Size __________ Distance between shaft ends __________

Ambient temperature, °F ____ Operating temperature, °F ____

Expected growth of DriveR leg, mils: IB ______ OB ______

DriveN coupling face to DriveR leg, in.: IB ______ OB ______

Indicator sag, mils: Face TIR ________ OD TIR ________

Mount indicators on the DriveR and read the DriveN hub:

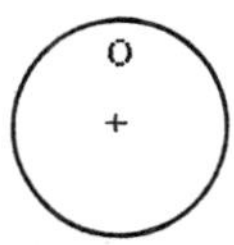

Face Readings, TIR
(Diameter: ______ inch)

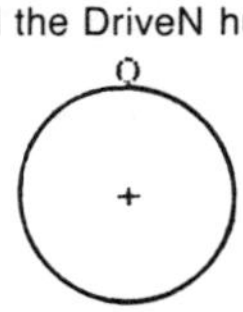

OD Readings, TIR

The vertical and horizontal sums of the readings should be equal.

Sum of Face readings: Vertical ______ Horizontal ______

Sum of OD readings: Vertical ______ Horizontal ______

Data Used in Tables

TIR (corrected for sag), mils: Face ________ OD ________

Shim required for IB leg, mils ______ (Actual used ______)

Shim required for OB leg, mils ______ (Actual used ______)

Remarks

____

____

____

Data taken by ________

Figure 2-2. This form is helpful in recording Face—OD alignment data.

Equations and Calculations

Once the Face-OD readings have been taken, the calculation of the required shim is actually straightforward:

$$\text{Shim} = \left(\frac{\text{Leg}}{\text{Hub Diameter}}\right)\left(\text{Face TIR}\right) - \left(\frac{\text{OD TIR}}{2}\right)$$

Face TIR = Face Total Indicator Reading, mils

OD TIR = OD Total Indicator Reading, mils

Hub diameter = Diameter for face readings, in.

Leg = Distance to leg which is to be shimmed, in.

Shim = Required shim, mils (Add shim for +, remove for −)

Be sure to use the + or − sign of the reading in the calculation.

The alignment tables in this book were constructed using this equation. Tables are much handier to use, and reduce the chances of error in calculation.

To use the tables, first find the section for your coupling hub diameter. Then go to the page for the leg you need to shim. Find your face and OD readings and read the shim required in the proper column.

Remember that you can set up the face reading to be on any diameter you choose by adjusting your alignment jig. Also, the OD TIR will be the same no matter what the hub diameter is (try it). Therefore, the diameter used in the tables should be the diameter from which you get the face readings. In general, try to use as large a diameter as possible to improve the accuracy of your data.

If the table does not have your exact readings, you can determine the required shim by looking in the nearby rows and columns and estimating. In some cases, you may find that apparently large changes in readings require very little change in shim.

If you wish to be more accurate than an estimate, you can break up your readings into parts, get the shim for each partial reading, and combine the partial shim to get the total. For example, suppose your face reading is +17 mils TIR and the OD reading is 0. This is not in the tables. However, you can look up the shim for +15 mils and +2 mils. The shim required for 17 mils is the sum of these two partial readings.

If the OD reading is not zero, you can still break up your data into two or more equivalent partial sets. To prove this to yourself, turn to Table 3-8 for a 1.5-in. diameter hub and 40-in. leg. Suppose your data showed −7 on the face and +20 on the OD. The shim would be −196 mils. Now break up the data into two sets. Let the first set be −4 on the face and +30 on the OD. This would require −121 mils of shim. The second set is −3 on the face and −10 on the OD. This would require −75 mils of shim. The total shim is still −196 mils, just as before.

For large misalignments, you can divide the face and OD readings by 10, enter the proper table, find the shim, and multiply it by 10 to get the correct shim.

Examples

Example 1: The Face-OD Method is used to align a motor to a pump. The indicators are zeroed at the top. The face reading is −2 mils TIR on a 3-in. circle. The OD reading is 5 mils on the 3.5-in.-diameter coupling hub. The distance from the pump coupling face to the near motor leg is 25 in.; to the far leg is 50 in. Indicator sag is zero. What shim is required?

Since the face reading was taken on a 3-in. diameter instead of the maximum of 3.5 in., use the tables for a 3-in. diameter.

Reading directly from Table 3-47 for a 25-in. leg shows to remove 19 mils from this leg. Reading directly from Table 3-51 for a 50-in. leg shows to remove 35 mils from this leg.

Example 2: The Face-OD Method is used to align a steam turbine to a pump. The indicators are zeroed at the top. The face reading is taken on a 3.5-in. circle; it is +3.5 mils TIR. The OD reading is 17 mils TIR. The turbine legs are 31 in. and 67 in. from the pump coupling face. The indicator jig has a sag of 5 mils TIR for OD readings and 1 mil TIR for face readings. From past experience, you want to set the front legs 5 mils low and the back legs 15 mils low. What shim is needed?

First, correct the readings for sag. The OD reading should be $17 + 5 = 22$ mils TIR. The face readings should be $3.5 + 1 = 4.5$ mils TIR.

Now, turn to Table 3-58 for a 3.5-in. diameter and 30-in. leg. The table does not have 4.5 mils for the face and 22 mils for the OD. You will have to estimate for these values. For 4 mils face and 20 mils OD the shim is +24. For 5 mils face and 20 mils OD the shim is +32. This means that for 4.5 mils face and 20 mils OD the shim would be 28 mils. Now notice that for either 4 or 5 mils on the face that going from 20 to 30 mils on the OD subtracts 5 mils from the shim. Therefore, going from 20 to 22 mils on the OD would subtract 1 mil. The shim required for the 30-in. leg would be 27 mils; however, the actual leg distance is 31 in.

Next, turn to Table 3-59 for a 3.5-in. diameter and 35-in. leg. You should estimate about 34 mils of shim for 4.5 mils on the face and 22 mils on the OD. Since you would add 27 mils for a 30-in. leg and 34 mils for a 35-in. leg, you should add about 28 or 29 mils for the 31-in. leg. (The calculated exact value is 28.9 mils.)

Remember that you want to set this leg to be 5 mils low. Therefore, you actually add only 24 mils of shim.

In a similar way, you use the 60-in. leg and 70-in. leg tables to determine a "cold" shim of 75 mils for the 67-in. leg. To set this leg 15 mils low, you actually add only 60 mils of shim.

Alignment Tolerances

You will not be able to align the shafts perfectly. If you ever think you have achieved that perfect alignment, re-check it the next day or at night, or during a rain. Alignment is a dynamic condition; it is constantly changing. Your objective is to obtain an alignment that is within the operating range of the coupling and drive train system. Since you do not know the direction or amount of change, your goal should be to align to within the tolerances of your setup and your ability to shim.

A common reason for not obtaining close alignment is that the required shim is not properly calculated. This means that the alignment procedure is actually a trial and error method. The alignment crew members become discouraged after a few tries and rationalize that the alignment is acceptable.

By using the alignment tables in this book, it should be easier to get "perfect" cold alignment rather than previous "acceptable" alignment with trial and error routines. Since the alignment tables should make complete correction possible the first time (and certainly the second!), there should be no reason to settle for less. However, things being the way they are, here are some criteria currently in use:

1. Reference 8 recommends 2 mils TIR per inch of face diameter and 2 mils TIR on the OD.
2. Reference 9 recommends 2 mils TIR on the face and 3 mils TIR on the OD.
3. Reference 6 recommends 3 mils TIR on the face and 3 mils TIR on the OD.
4. Reference 5 gives a graph of acceptable face-OD combinations for various shaft speeds. Some of these combinations are listed as follows:

Shaft RPM	Face TIR (mils)	OD TIR (mils)
2000	2	12
	10	2
4000	2	6
	5	2
6000	2	4
	6	1

A number of flexible disc coupling manufacturers give the maximum misalignment capability of their couplings as 1/2 degree of angularity. This is based on misalignment before fatiguing the discs and should not be used as an operating alignment criteria. The 1/2 degree is about 40 mils TIR on a 5-in. face. In practice, metal disc failures have been observed with disc packs at less than 1/4 degree.

With the tables, it should not be difficult to obtain alignment well within the "2 and 2" criterion. After all, this would still mean about 25 mils of shim are needed for a 3-in. diameter and 40-in. leg distance. However, remember that for a zero face reading, the shim is half the OD TIR. If you get to a zero face and have a 6-mil OD TIR, are you going to remove 3 mils of shim?

Errors in Taking Data

There are ample opportunities to err while taking alignment data. Some of the more common ones include:

1. Poor clamps or clamps not tight
2. Broken dial indicators
3. Indicator bar sag
4. Bent shaft
5. Coupling hub not fitted properly to shaft
6. Read dial indicator wrong (especially + and − readings)
7. Shaft moved in axial direction

Fortunately, the machines will serve as a final check. That is, if the calculated correction does not make a significant improvement in alignment, you or your equipment have made a mistake.

There is a simple way of checking your readings before adding or removing shims. The horizontal readings and the vertical readings should have the same sum. This applies to both face and OD values. For example, consider the following readings:

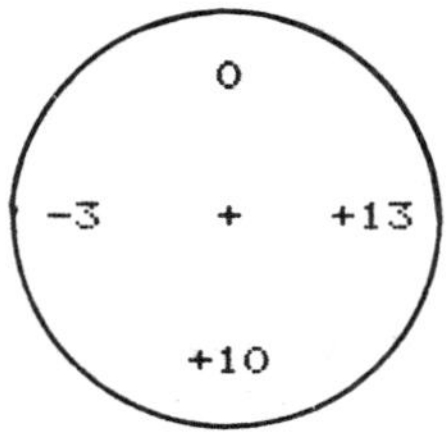

Face

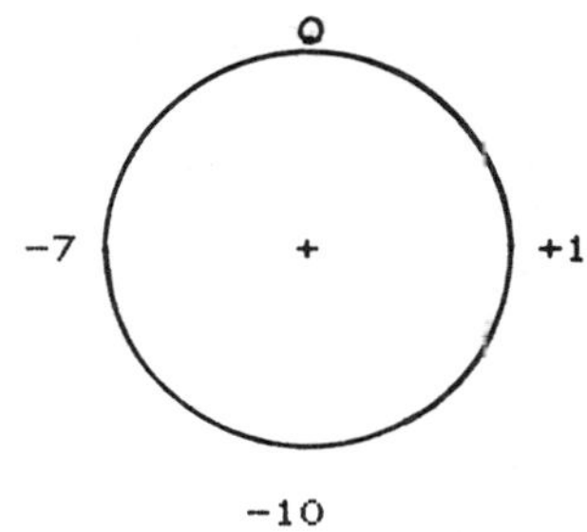

OD

For the face, the vertical readings total $0 + 10 = +10$. The horizontal readings total $+13 - 3 = +10$. Therefore, this is probably good data. However, the OD readings total -10 vertically and -6 horizontally. Since the vertical and horizontal sums are different, the OD readings are suspect and should be checked. The cause of the error may also have affected the face reading, so the face readings should also be checked.

If the accuracy of the data cannot be improved, check the significance of the error by determining the shim required for either case.

Indicator Sag

Indicator sag is caused by gravity and will always be present to some degree. You can minimize the amount of sag by using light indicators with stiff clamps and extensions. Table 2-1 gives some idea of the amount of sag to expect and the effect of weight and extension length. The table assumes a rigid clamping arrangement. To know the actual sag, you should test your alignment jig.

Table 2-1
Typical Indicator Sag
(OD TIR, mils, for 1/2-in. steel rod)

Indicator Weight (oz)	Length of Extension Rod (in.) 5	10	15	20	25
0	−.1	−2	−8	−24	−59
3	−.3	−3	−12	−35	−80
6	−.4	−4	−17	−46	−101
9	−.6	−6	−21	−57	−122

A slight correction for face readings may also be required, depending on your alignment jig arrangement.

To test your alignment jig, set up on a shaft that is between centers in a lathe. Arrange the indicator to read the OD of the same shaft it is mounted on. Zero the indicator at the top and rotate the shaft 180 degrees. The indicator now shows TIR sag for the OD readings. Rearrange the jig to read the face of a shoulder on the shaft (or a clamp) and repeat to get the face "sag" correction.

If you test the jig as described, you will find that the indicator shows a negative reading when the shaft is rotated. In other words, a situation that you expected to read zero TIR actually reads as a minus TIR. For example, the reading might be −11 mils on the sag test shaft. This is why typical sag values are also shown as negative values.

Now, suppose you take that same jig and use it to align equipment. When the OD indicator shows −11 mils TIR, you know that the true reading is 0. If the indicator shows −20 mils, the true reading would be −9 mils (−20 + 11). If the indicator shows a +7 mils, the true reading would be 18 mils. Mathematically, this is:

$$\text{Corrected Reading} = \text{Reading} - \text{Sag}$$

Remember to take the negative sign of the sag into account.

Hot Alignment

So far, the techniques presented have been for alignment of "cold" (that is, not operating) equipment. Good cold alignment is the basis for establishing the proper hot or operating alignment. If you know the true effects of piping and temperature, the cold alignment can be the final alignment.

A word of caution: Take manufacturers' suggested cold and hot alignments with a lump of salt. They do not know the effect of your piping, foundation, and operations as well as you do.

With the alignment tables, offsetting for thermal growth is simple. For example, suppose you have read +20 mils TIR on the OD of a 2.5-in.-diameter coupling and +4 mils TIR on the face. The leg distance is 50 in. to a leg that you think will "grow" by 30 mils when at operating temperature. Referring to the alignment tables, you see that the required cold correction is to add 70 mils of shim to that leg. Since you think the leg will grow 30 mils, you actually only add 40 mils of shim. If you check the face and OD after adding the 40 mils of shim, you probably will get readings of about +2 mils TIR on the face and +17 mils TIR on the OD. For these readings, the alignment table will indicate an additional 30 mils is required as necessary to allow for thermal growth.

In the same example, suppose you take hot alignment readings after adding the 40 mils of shim and the OD is now −5; the face is −1. According to the alignment table, you should take off 17 mils of shim. This means that the leg actually grew by 47 mils instead of 30 mils. You should record that the leg grows by 47 mils and use this correction in the future.

Chapter 3

Face—OD Alignment Tables

Table 3-1

Hub Diameter, inch: 1.5

Leg Distance, inch: 5

SHIM REQUIRED FOR ALIGNMENT

Face - OD Method

FACE TIR MILS	OD TIR, MILS -40	-30	-20	-15	-10	-5	0	5	10	15	20	30	40
-40	-113	-118	-123	-125	-128	-130	-133	-135	-138	-140	-143	-148	-153
-35	-96	-101	-106	-109	-111	-114	-116	-119	-121	-124	-126	-131	-136
-30	-80	-85	-90	-92	-95	-97	-100	-102	-105	-107	-110	-115	-120
-25	-63	-68	-73	-75	-78	-80	-83	-85	-88	-90	-93	-98	-103
-20	-46	-51	-56	-59	-61	-64	-66	-69	-71	-74	-76	-81	-86
-15	-30	-35	-40	-42	-45	-47	-50	-52	-55	-57	-60	-65	-70
-12	-20	-25	-30	-32	-35	-37	-40	-42	-45	-47	-50	-55	-60
-10	-13	-18	-23	-25	-28	-30	-33	-35	-38	-40	-43	-48	-53
-9	-10	-15	-20	-22	-25	-27	-30	-32	-35	-37	-40	-45	-50
-8	-6	-11	-16	-19	-21	-24	-26	-29	-31	-34	-36	-41	-46
-7	-3	-8	-13	-15	-18	-20	-23	-25	-28	-30	-33	-38	-43
-6	0	-5	-10	-12	-15	-17	-20	-22	-25	-27	-30	-35	-40
-5	3	-1	-6	-9	-11	-14	-16	-19	-21	-24	-26	-31	-36
-4	6	1	-3	-5	-8	-10	-13	-15	-18	-20	-23	-28	-33
-3	10	5	0	-2	-5	-7	-10	-12	-15	-17	-20	-25	-30
-2	13	8	3	0	-1	-4	-6	-9	-11	-14	-16	-21	-26
-1	16	11	6	4	1	0	-3	-5	-8	-10	-13	-18	-23
0	20	15	10	7	5	2	0	-2	-5	-7	-10	-15	-20
1	23	18	13	10	8	5	3	0	-1	-4	-6	-11	-16
2	26	21	16	14	11	9	6	4	1	0	-3	-8	-13
3	30	25	20	17	15	12	10	7	5	2	0	-5	-10
4	33	28	23	20	18	15	13	10	8	5	3	-1	-6
5	36	31	26	24	21	19	16	14	11	9	6	1	-3
6	40	35	30	27	25	22	20	17	15	12	10	5	0
7	43	38	33	30	28	25	23	20	18	15	13	8	3
8	46	41	36	34	31	29	26	24	21	19	16	11	6
9	50	45	40	37	35	32	30	27	25	22	20	15	10
10	53	48	43	40	38	35	33	30	28	25	23	18	13
12	60	55	50	47	45	42	40	37	35	32	30	25	20
15	70	65	60	57	55	52	50	47	45	42	40	35	30
20	86	81	76	74	71	69	66	64	61	59	56	51	46
25	103	98	93	90	88	85	83	80	78	75	73	68	63
30	120	115	110	107	105	102	100	97	95	92	90	85	80
35	136	131	126	124	121	119	116	114	111	109	106	101	96
40	153	148	143	140	138	135	133	130	128	125	123	118	113

Table 3-2

Hub Diameter, inch: 1.5

Leg Distance, inch: 10

SHIM REQUIRED FOR ALIGNMENT

Face - OD Method

FACE TIR MILS	OD TIR, MILS -40	-30	-20	-15	-10	-5	0	5	10	15	20	30	40
-40	-246	-251	-256	-259	-261	-264	-266	-269	-271	-274	-276	-281	-286
-35	-213	-218	-223	-225	-228	-230	-233	-235	-238	-240	-243	-248	-253
-30	-180	-185	-190	-192	-195	-197	-200	-202	-205	-207	-210	-215	-220
-25	-146	-151	-156	-159	-161	-164	-166	-169	-171	-174	-176	-181	-186
-20	-113	-118	-123	-125	-128	-130	-133	-135	-138	-140	-143	-148	-153
-15	-80	-85	-90	-92	-95	-97	-100	-102	-105	-107	-110	-115	-120
-12	-60	-65	-70	-72	-75	-77	-80	-82	-85	-87	-90	-95	-100
-10	-46	-51	-56	-59	-61	-64	-66	-69	-71	-74	-76	-81	-86
-9	-40	-45	-50	-52	-55	-57	-60	-62	-65	-67	-70	-75	-80
-8	-33	-38	-43	-45	-48	-50	-53	-55	-58	-60	-63	-68	-73
-7	-26	-31	-36	-39	-41	-44	-46	-49	-51	-54	-56	-61	-66
-6	-20	-25	-30	-32	-35	-37	-40	-42	-45	-47	-50	-55	-60
-5	-13	-18	-23	-25	-28	-30	-33	-35	-38	-40	-43	-48	-53
-4	-6	-11	-16	-19	-21	-24	-26	-29	-31	-34	-36	-41	-46
-3	0	-5	-10	-12	-15	-17	-20	-22	-25	-27	-30	-35	-40
-2	6	1	-3	-5	-8	-10	-13	-15	-18	-20	-23	-28	-33
-1	13	8	3	0	-1	-4	-6	-9	-11	-14	-16	-21	-26
0	20	15	10	7	5	2	0	-2	-5	-7	-10	-15	-20
1	26	21	16	14	11	9	6	4	1	0	-3	-8	-13
2	33	28	23	20	18	15	13	10	8	5	3	-1	-6
3	40	35	30	27	25	22	20	17	15	12	10	5	0
4	46	41	36	34	31	29	26	24	21	19	16	11	6
5	53	48	43	40	38	35	33	30	28	25	23	18	13
6	60	55	50	47	45	42	40	37	35	32	30	25	20
7	66	61	56	54	51	49	46	44	41	39	36	31	26
8	73	68	63	60	58	55	53	50	48	45	43	38	33
9	80	75	70	67	65	62	60	57	55	52	50	45	40
10	86	81	76	74	71	69	66	64	61	59	56	51	46
12	100	95	90	87	85	82	80	77	75	72	70	65	60
15	120	115	110	107	105	102	100	97	95	92	90	85	80
20	153	148	143	140	138	135	133	130	128	125	123	118	113
25	186	181	176	174	171	169	166	164	161	159	156	151	146
30	220	215	210	207	205	202	200	197	195	192	190	185	180
35	253	248	243	240	238	235	233	230	228	225	223	218	213
40	286	281	276	274	271	269	266	264	261	259	256	251	246

Table 3-3

Hub Diameter, inch: 1.5

Leg Distance, inch: 15

SHIM REQUIRED FOR ALIGNMENT

Face - OD Method

FACE TIR MILS	OD TIR, MILS -40	-30	-20	-15	-10	-5	0	5	10	15	20	30	40
-40	-380	-385	-390	-392	-395	-397	-400	-402	-405	-407	-410	-415	-420
-35	-330	-335	-340	-342	-345	-347	-350	-352	-355	-357	-360	-365	-370
-30	-280	-285	-290	-292	-295	-297	-300	-302	-305	-307	-310	-315	-320
-25	-230	-235	-240	-242	-245	-247	-250	-252	-255	-257	-260	-265	-270
-20	-180	-185	-190	-192	-195	-197	-200	-202	-205	-207	-210	-215	-220
-15	-130	-135	-140	-142	-145	-147	-150	-152	-155	-157	-160	-165	-170
-12	-100	-105	-110	-112	-115	-117	-120	-122	-125	-127	-130	-135	-140
-10	-80	-85	-90	-92	-95	-97	-100	-102	-105	-107	-110	-115	-120
-9	-70	-75	-80	-82	-85	-87	-90	-92	-95	-97	-100	-105	-110
-8	-60	-65	-70	-72	-75	-77	-80	-82	-85	-87	-90	-95	-100
-7	-50	-55	-60	-62	-65	-67	-70	-72	-75	-77	-80	-85	-90
-6	-40	-45	-50	-52	-55	-57	-60	-62	-65	-67	-70	-75	-80
-5	-30	-35	-40	-42	-45	-47	-50	-52	-55	-57	-60	-65	-70
-4	-20	-25	-30	-32	-35	-37	-40	-42	-45	-47	-50	-55	-60
-3	-10	-15	-20	-22	-25	-27	-30	-32	-35	-37	-40	-45	-50
-2	0	-5	-10	-12	-15	-17	-20	-22	-25	-27	-30	-35	-40
-1	10	5	0	-2	-5	-7	-10	-12	-15	-17	-20	-25	-30
0	20	15	10	7	5	2	0	-2	-5	-7	-10	-15	-20
1	30	25	20	17	15	12	10	7	5	2	0	-5	-10
2	40	35	30	27	25	22	20	17	15	12	10	5	0
3	50	45	40	37	35	32	30	27	25	22	20	15	10
4	60	55	50	47	45	42	40	37	35	32	30	25	20
5	70	65	60	57	55	52	50	47	45	42	40	35	30
6	80	75	70	67	65	62	60	57	55	52	50	45	40
7	90	85	80	77	75	72	70	67	65	62	60	55	50
8	100	95	90	87	85	82	80	77	75	72	70	65	60
9	110	105	100	97	95	92	90	87	85	82	80	75	70
10	120	115	110	107	105	102	100	97	95	92	90	85	80
12	140	135	130	127	125	122	120	117	115	112	110	105	100
15	170	165	160	157	155	152	150	147	145	142	140	135	130
20	220	215	210	207	205	202	200	197	195	192	190	185	180
25	270	265	260	257	255	252	250	247	245	242	240	235	230
30	320	315	310	307	305	302	300	297	295	292	290	285	280
35	370	365	360	357	355	352	350	347	345	342	340	335	330
40	420	415	410	407	405	402	400	397	395	392	390	385	380

Table 3-4

Hub Diameter, inch: 1.5

Leg Distance, inch: 20

SHIM REQUIRED FOR ALIGNMENT

Face - OD Method

FACE TIR MILS	OD TIR, MILS -40	-30	-20	-15	-10	-5	0	5	10	15	20	30	40
-40	-513	-518	-523	-525	-528	-530	-533	-535	-538	-540	-543	-548	-553
-35	-446	-451	-456	-459	-461	-464	-466	-469	-471	-474	-476	-481	-486
-30	-380	-385	-390	-392	-395	-397	-400	-402	-405	-407	-410	-415	-420
-25	-313	-318	-323	-325	-328	-330	-333	-335	-338	-340	-343	-348	-353
-20	-246	-251	-256	-259	-261	-264	-266	-269	-271	-274	-276	-281	-286
-15	-180	-185	-190	-192	-195	-197	-200	-202	-205	-207	-210	-215	-220
-12	-140	-145	-150	-152	-155	-157	-160	-162	-165	-167	-170	-175	-180
-10	-113	-118	-123	-125	-128	-130	-133	-135	-138	-140	-143	-148	-153
-9	-100	-105	-110	-112	-115	-117	-120	-122	-125	-127	-130	-135	-140
-8	-86	-91	-96	-99	-101	-104	-106	-109	-111	-114	-116	-121	-126
-7	-73	-78	-83	-85	-88	-90	-93	-95	-98	-100	-103	-108	-113
-6	-60	-65	-70	-72	-75	-77	-80	-82	-85	-87	-90	-95	-100
-5	-46	-51	-56	-59	-61	-64	-66	-69	-71	-74	-76	-81	-86
-4	-33	-38	-43	-45	-48	-50	-53	-55	-58	-60	-63	-68	-73
-3	-20	-25	-30	-32	-35	-37	-40	-42	-45	-47	-50	-55	-60
-2	-6	-11	-16	-19	-21	-24	-26	-29	-31	-34	-36	-41	-46
-1	6	1	-3	-5	-8	-10	-13	-15	-18	-20	-23	-28	-33
0	20	15	10	7	5	2	0	-2	-5	-7	-10	-15	-20
1	33	28	23	20	18	15	13	10	8	5	3	-1	-6
2	46	41	36	34	31	29	26	24	21	19	16	11	6
3	60	55	50	47	45	42	40	37	35	32	30	25	20
4	73	68	63	60	58	55	53	50	48	45	43	38	33
5	86	81	76	74	71	69	66	64	61	59	56	51	46
6	100	95	90	87	85	82	80	77	75	72	70	65	60
7	113	108	103	100	98	95	93	90	88	85	83	78	73
8	126	121	116	114	111	109	106	104	101	99	96	91	86
9	140	135	130	127	125	122	120	117	115	112	110	105	100
10	153	148	143	140	138	135	133	130	128	125	123	118	113
12	180	175	170	167	165	162	160	157	155	152	150	145	140
15	220	215	210	207	205	202	200	197	195	192	190	185	180
20	286	281	276	274	271	269	266	264	261	259	256	251	246
25	353	348	343	340	338	335	333	330	328	325	323	318	313
30	420	415	410	407	405	402	400	397	395	392	390	385	380
35	486	481	476	474	471	469	466	464	461	459	456	451	446
40	553	548	543	540	538	535	533	530	528	525	523	518	513

Table 3-5

Hub Diameter, inch: 1.5

Leg Distance, inch: 25

SHIM REQUIRED FOR ALIGNMENT

Face - OD Method

FACE TIR MILS	OD TIR, MILS -40	-30	-20	-15	-10	-5	0	5	10	15	20	30	40
-40	-646	-651	-656	-659	-661	-664	-666	-669	-671	-674	-676	-681	-686
-35	-563	-568	-573	-575	-578	-580	-583	-585	-588	-590	-593	-598	-603
-30	-480	-485	-490	-492	-495	-497	-500	-502	-505	-507	-510	-515	-520
-25	-396	-401	-406	-409	-411	-414	-416	-419	-421	-424	-426	-431	-436
-20	-313	-318	-323	-325	-328	-330	-333	-335	-338	-340	-343	-348	-353
-15	-230	-235	-240	-242	-245	-247	-250	-252	-255	-257	-260	-265	-270
-12	-180	-185	-190	-192	-195	-197	-200	-202	-205	-207	-210	-215	-220
-10	-146	-151	-156	-159	-161	-164	-166	-169	-171	-174	-176	-181	-186
-9	-130	-135	-140	-142	-145	-147	-150	-152	-155	-157	-160	-165	-170
-8	-113	-118	-123	-125	-128	-130	-133	-135	-138	-140	-143	-148	-153
-7	-96	-101	-106	-109	-111	-114	-116	-119	-121	-124	-126	-131	-136
-6	-80	-85	-90	-92	-95	-97	-100	-102	-105	-107	-110	-115	-120
-5	-63	-68	-73	-75	-78	-80	-83	-85	-88	-90	-93	-98	-103
-4	-46	-51	-56	-59	-61	-64	-66	-69	-71	-74	-76	-81	-86
-3	-30	-35	-40	-42	-45	-47	-50	-52	-55	-57	-60	-65	-70
-2	-13	-18	-23	-25	-28	-30	-33	-35	-38	-40	-43	-48	-53
-1	3	-1	-6	-9	-11	-14	-16	-19	-21	-24	-26	-31	-36
0	20	15	10	7	5	2	0	-2	-5	-7	-10	-15	-20
1	36	31	26	24	21	19	16	14	11	9	6	1	-3
2	53	48	43	40	38	35	33	30	28	25	23	18	13
3	70	65	60	57	55	52	50	47	45	42	40	35	30
4	86	81	76	74	71	69	66	64	61	59	56	51	46
5	103	98	93	90	88	85	83	80	78	75	73	68	63
6	120	115	110	107	105	102	100	97	95	92	90	85	80
7	136	131	126	124	121	119	116	114	111	109	106	101	96
8	153	148	143	140	138	135	133	130	128	125	123	118	113
9	170	165	160	157	155	152	150	147	145	142	140	135	130
10	186	181	176	174	171	169	166	164	161	159	156	151	146
12	220	215	210	207	205	202	200	197	195	192	190	185	180
15	270	265	260	257	255	252	250	247	245	242	240	235	230
20	353	348	343	340	338	335	333	330	328	325	323	318	313
25	436	431	426	424	421	419	416	414	411	409	406	401	396
30	520	515	510	507	505	502	500	497	495	492	490	485	480
35	603	598	593	590	588	585	583	580	578	575	573	568	563
40	686	681	676	674	671	669	666	664	661	659	656	651	646

Table 3-6

Hub Diameter, inch: 1.5

Leg Distance, inch: 30

SHIM REQUIRED FOR ALIGNMENT

Face - OD Method

FACE TIR MILS	OD TIR, MILS -40	-30	-20	-15	-10	-5	0	5	10	15	20	30	40
-40	-780	-785	-790	-792	-795	-797	-800	-802	-805	-807	-810	-815	-820
-35	-680	-685	-690	-692	-695	-697	-700	-702	-705	-707	-710	-715	-720
-30	-580	-585	-590	-592	-595	-597	-600	-602	-605	-607	-610	-615	-620
-25	-480	-485	-490	-492	-495	-497	-500	-502	-505	-507	-510	-515	-520
-20	-380	-385	-390	-392	-395	-397	-400	-402	-405	-407	-410	-415	-420
-15	-280	-285	-290	-292	-295	-297	-300	-302	-305	-307	-310	-315	-320
-12	-220	-225	-230	-232	-235	-237	-240	-242	-245	-247	-250	-255	-260
-10	-180	-185	-190	-192	-195	-197	-200	-202	-205	-207	-210	-215	-220
-9	-160	-165	-170	-172	-175	-177	-180	-182	-185	-187	-190	-195	-200
-8	-140	-145	-150	-152	-155	-157	-160	-162	-165	-167	-170	-175	-180
-7	-120	-125	-130	-132	-135	-137	-140	-142	-145	-147	-150	-155	-160
-6	-100	-105	-110	-112	-115	-117	-120	-122	-125	-127	-130	-135	-140
-5	-80	-85	-90	-92	-95	-97	-100	-102	-105	-107	-110	-115	-120
-4	-60	-65	-70	-72	-75	-77	-80	-82	-85	-87	-90	-95	-100
-3	-40	-45	-50	-52	-55	-57	-60	-62	-65	-67	-70	-75	-80
-2	-20	-25	-30	-32	-35	-37	-40	-42	-45	-47	-50	-55	-60
-1	0	-5	-10	-12	-15	-17	-20	-22	-25	-27	-30	-35	-40
0	20	15	10	7	5	2	0	-2	-5	-7	-10	-15	-20
1	40	35	30	27	25	22	20	17	15	12	10	5	0
2	60	55	50	47	45	42	40	37	35	32	30	25	20
3	80	75	70	67	65	62	60	57	55	52	50	45	40
4	100	95	90	87	85	82	80	77	75	72	70	65	60
5	120	115	110	107	105	102	100	97	95	92	90	85	80
6	140	135	130	127	125	122	120	117	115	112	110	105	100
7	160	155	150	147	145	142	140	137	135	132	130	125	120
8	180	175	170	167	165	162	160	157	155	152	150	145	140
9	200	195	190	187	185	182	180	177	175	172	170	165	160
10	220	215	210	207	205	202	200	197	195	192	190	185	180
12	260	255	250	247	245	242	240	237	235	232	230	225	220
15	320	315	310	307	305	302	300	297	295	292	290	285	280
20	420	415	410	407	405	402	400	397	395	392	390	385	380
25	520	515	510	507	505	502	500	497	495	492	490	485	480
30	620	615	610	607	605	602	600	597	595	592	590	585	580
35	720	715	710	707	705	702	700	697	695	692	690	685	680
40	820	815	810	807	805	802	800	797	795	792	790	785	780

Table 3-7

Hub Diameter, inch: 1.5

Leg Distance, inch: 35

SHIM REQUIRED FOR ALIGNMENT

Face - OD Method

FACE TIR MILS	OD TIR, MILS -40	-30	-20	-15	-10	-5	0	5	10	15	20	30	40
-40	-913	-918	-923	-925	-928	-930	-933	-935	-938	-940	-943	-948	-953
-35	-796	-801	-806	-809	-811	-814	-816	-819	-821	-824	-826	-831	-836
-30	-680	-685	-690	-692	-695	-697	-700	-702	-705	-707	-710	-715	-720
-25	-563	-568	-573	-575	-578	-580	-583	-585	-588	-590	-593	-598	-603
-20	-446	-451	-456	-459	-461	-464	-466	-469	-471	-474	-476	-481	-486
-15	-330	-335	-340	-342	-345	-347	-350	-352	-355	-357	-360	-365	-370
-12	-260	-265	-270	-272	-275	-277	-280	-282	-285	-287	-290	-295	-300
-10	-213	-218	-223	-225	-228	-230	-233	-235	-238	-240	-243	-248	-253
-9	-190	-195	-200	-202	-205	-207	-210	-212	-215	-217	-220	-225	-230
-8	-166	-171	-176	-179	-181	-184	-186	-189	-191	-194	-196	-201	-206
-7	-143	-148	-153	-155	-158	-160	-163	-165	-168	-170	-173	-178	-183
-6	-120	-125	-130	-132	-135	-137	-140	-142	-145	-147	-150	-155	-160
-5	-96	-101	-106	-109	-111	-114	-116	-119	-121	-124	-126	-131	-136
-4	-73	-78	-83	-85	-88	-90	-93	-95	-98	-100	-103	-108	-113
-3	-50	-55	-60	-62	-65	-67	-70	-72	-75	-77	-80	-85	-90
-2	-26	-31	-36	-39	-41	-44	-46	-49	-51	-54	-56	-61	-66
-1	-3	-8	-13	-15	-18	-20	-23	-25	-28	-30	-33	-38	-43
0	20	15	10	7	5	2	0	-2	-5	-7	-10	-15	-20
1	43	38	33	30	28	25	23	20	18	15	13	8	3
2	66	61	56	54	51	49	46	44	41	39	36	31	26
3	90	85	80	77	75	72	70	67	65	62	60	55	50
4	113	108	103	100	98	95	93	90	88	85	83	78	73
5	136	131	126	124	121	119	116	114	111	109	106	101	96
6	160	155	150	147	145	142	140	137	135	132	130	125	120
7	183	178	173	170	168	165	163	160	158	155	153	148	143
8	206	201	196	194	191	189	186	184	181	179	176	171	166
9	230	225	220	217	215	212	210	207	205	202	200	195	190
10	253	248	243	240	238	235	233	230	228	225	223	218	213
12	300	295	290	287	285	282	280	277	275	272	270	265	260
15	370	365	360	357	355	352	350	347	345	342	340	335	330
20	486	481	476	474	471	469	466	464	461	459	456	451	446
25	603	598	593	590	588	585	583	580	578	575	573	568	563
30	720	715	710	707	705	702	700	697	695	692	690	685	680
35	836	831	826	824	821	819	816	814	811	809	806	801	796
40	953	948	943	940	938	935	933	930	928	925	923	918	913

Table 3-8

Hub Diameter, inch: 1.5

Leg Distance, inch: 40

SHIM REQUIRED FOR ALIGNMENT

Face - OD Method

FACE TIR MILS	OD TIR, MILS -40	-30	-20	-15	-10	-5	0	5	10	15	20	30	40
-40	-1046	-1051	-1056	-1059	-1061	-1064	-1066	-1069	-1071	-1074	-1076	-1081	-1086
-35	-913	-918	-923	-925	-928	-930	-933	-935	-938	-940	-943	-948	-953
-30	-780	-785	-790	-792	-795	-797	-800	-802	-805	-807	-810	-815	-820
-25	-646	-651	-656	-659	-661	-664	-666	-669	-671	-674	-676	-681	-686
-20	-513	-518	-523	-525	-528	-530	-533	-535	-538	-540	-543	-548	-553
-15	-380	-385	-390	-392	-395	-397	-400	-402	-405	-407	-410	-415	-420
-12	-300	-305	-310	-312	-315	-317	-320	-322	-325	-327	-330	-335	-340
-10	-246	-251	-256	-259	-261	-264	-266	-269	-271	-274	-276	-281	-286
-9	-220	-225	-230	-232	-235	-237	-240	-242	-245	-247	-250	-255	-260
-8	-193	-198	-203	-205	-208	-210	-213	-215	-218	-220	-223	-228	-233
-7	-166	-171	-176	-179	-181	-184	-186	-189	-191	-194	-196	-201	-206
-6	-140	-145	-150	-152	-155	-157	-160	-162	-165	-167	-170	-175	-180
-5	-113	-118	-123	-125	-128	-130	-133	-135	-138	-140	-143	-148	-153
-4	-86	-91	-96	-99	-101	-104	-106	-109	-111	-114	-116	-121	-126
-3	-60	-65	-70	-72	-75	-77	-80	-82	-85	-87	-90	-95	-100
-2	-33	-38	-43	-45	-48	-50	-53	-55	-58	-60	-63	-68	-73
-1	-6	-11	-16	-19	-21	-24	-26	-29	-31	-34	-36	-41	-46
0	20	15	10	7	5	2	0	-2	-5	-7	-10	-15	-20
1	46	41	36	34	31	29	26	24	21	19	16	11	6
2	73	68	63	60	58	55	53	50	48	45	43	38	33
3	100	95	90	87	85	82	80	77	75	72	70	65	60
4	126	121	116	114	111	109	106	104	101	99	96	91	86
5	153	148	143	140	138	135	133	130	128	125	123	118	113
6	180	175	170	167	165	162	160	157	155	152	150	145	140
7	206	201	196	194	191	189	186	184	181	179	176	171	166
8	233	228	223	220	218	215	213	210	208	205	203	198	193
9	260	255	250	247	245	242	240	237	235	232	230	225	220
10	286	281	276	274	271	269	266	264	261	259	256	251	246
12	340	335	330	327	325	322	320	317	315	312	310	305	300
15	420	415	410	407	405	402	400	397	395	392	390	385	380
20	553	548	543	540	538	535	533	530	528	525	523	518	513
25	686	681	676	674	671	669	666	664	661	659	656	651	646
30	820	815	810	807	805	802	800	797	795	792	790	785	780
35	953	948	943	940	938	935	933	930	928	925	923	918	913
40	1086	1081	1076	1074	1071	1069	1066	1064	1061	1059	1056	1051	1046

Table 3-9

Hub Diameter, inch: 1.75

Leg Distance, inch: 5

SHIM REQUIRED FOR ALIGNMENT

Face - OD Method

FACE TIR MILS	OD TIR, MILS -40	-30	-20	-15	-10	-5	0	5	10	15	20	30	40
-40	-94	-99	-104	-106	-109	-111	-114	-116	-119	-121	-124	-129	-134
-35	-80	-85	-90	-92	-95	-97	-100	-102	-105	-107	-110	-115	-120
-30	-65	-70	-75	-78	-80	-83	-85	-88	-90	-93	-95	-100	-105
-25	-51	-56	-61	-63	-66	-68	-71	-73	-76	-78	-81	-86	-91
-20	-37	-42	-47	-49	-52	-54	-57	-59	-62	-64	-67	-72	-77
-15	-22	-27	-32	-35	-37	-40	-42	-45	-47	-50	-52	-57	-62
-12	-14	-19	-24	-26	-29	-31	-34	-36	-39	-41	-44	-49	-54
-10	-8	-13	-18	-21	-23	-26	-28	-31	-33	-36	-38	-43	-48
-9	-5	-10	-15	-18	-20	-23	-25	-28	-30	-33	-35	-40	-45
-8	-2	-7	-12	-15	-17	-20	-22	-25	-27	-30	-32	-37	-42
-7	0	-5	-10	-12	-15	-17	-20	-22	-25	-27	-30	-35	-40
-6	2	-2	-7	-9	-12	-14	-17	-19	-22	-24	-27	-32	-37
-5	5	0	-4	-6	-9	-11	-14	-16	-19	-21	-24	-29	-34
-4	8	3	-1	-3	-6	-8	-11	-13	-16	-18	-21	-26	-31
-3	11	6	1	-1	-3	-6	-8	-11	-13	-16	-18	-23	-28
-2	14	9	4	1	0	-3	-5	-8	-10	-13	-15	-20	-25
-1	17	12	7	4	2	0	-2	-5	-7	-10	-12	-17	-22
0	20	15	10	7	5	2	0	-2	-5	-7	-10	-15	-20
1	22	17	12	10	7	5	2	0	-2	-4	-7	-12	-17
2	25	20	15	13	10	8	5	3	0	-1	-4	-9	-14
3	28	23	18	16	13	11	8	6	3	1	-1	-6	-11
4	31	26	21	18	16	13	11	8	6	3	1	-3	-8
5	34	29	24	21	19	16	14	11	9	6	4	0	-5
6	37	32	27	24	22	19	17	14	12	9	7	2	-2
7	40	35	30	27	25	22	20	17	15	12	10	5	0
8	42	37	32	30	27	25	22	20	17	15	12	7	2
9	45	40	35	33	30	28	25	23	20	18	15	10	5
10	48	43	38	36	33	31	28	26	23	21	18	13	8
12	54	49	44	41	39	36	34	31	29	26	24	19	14
15	62	57	52	50	47	45	42	40	37	35	32	27	22
20	77	72	67	64	62	59	57	54	52	49	47	42	37
25	91	86	81	78	76	73	71	68	66	63	61	56	51
30	105	100	95	93	90	88	85	83	80	78	75	70	65
35	120	115	110	107	105	102	100	97	95	92	90	85	80
40	134	129	124	121	119	116	114	111	109	106	104	99	94

Table 3-10

Hub Diameter, inch: 1.75

Leg Distance, inch: 10

SHIM REQUIRED FOR ALIGNMENT

Face - OD Method

FACE TIR MILS	OD TIR, MILS -40	-30	-20	-15	-10	-5	0	5	10	15	20	30	40
-40	-208	-213	-218	-221	-223	-226	-228	-231	-233	-236	-238	-243	-248
-35	-180	-185	-190	-192	-195	-197	-200	-202	-205	-207	-210	-215	-220
-30	-151	-156	-161	-163	-166	-168	-171	-173	-176	-178	-181	-186	-191
-25	-122	-127	-132	-135	-137	-140	-142	-145	-147	-150	-152	-157	-162
-20	-94	-99	-104	-106	-109	-111	-114	-116	-119	-121	-124	-129	-134
-15	-65	-70	-75	-78	-80	-83	-85	-88	-90	-93	-95	-100	-105
-12	-48	-53	-58	-61	-63	-66	-68	-71	-73	-76	-78	-83	-88
-10	-37	-42	-47	-49	-52	-54	-57	-59	-62	-64	-67	-72	-77
-9	-31	-36	-41	-43	-46	-48	-51	-53	-56	-58	-61	-66	-71
-8	-25	-30	-35	-38	-40	-43	-45	-48	-50	-53	-55	-60	-65
-7	-20	-25	-30	-32	-35	-37	-40	-42	-45	-47	-50	-55	-60
-6	-14	-19	-24	-26	-29	-31	-34	-36	-39	-41	-44	-49	-54
-5	-8	-13	-18	-21	-23	-26	-28	-31	-33	-36	-38	-43	-48
-4	-2	-7	-12	-15	-17	-20	-22	-25	-27	-30	-32	-37	-42
-3	2	-2	-7	-9	-12	-14	-17	-19	-22	-24	-27	-32	-37
-2	8	3	-1	-3	-6	-8	-11	-13	-16	-18	-21	-26	-31
-1	14	9	4	1	0	-3	-5	-8	-10	-13	-15	-20	-25
0	20	15	10	7	5	2	0	-2	-5	-7	-10	-15	-20
1	25	20	15	13	10	8	5	3	0	-1	-4	-9	-14
2	31	26	21	18	16	13	11	8	6	3	1	-3	-8
3	37	32	27	24	22	19	17	14	12	9	7	2	-2
4	42	37	32	30	27	25	22	20	17	15	12	7	2
5	48	43	38	36	33	31	28	26	23	21	18	13	8
6	54	49	44	41	39	36	34	31	29	26	24	19	14
7	60	55	50	47	45	42	40	37	35	32	30	25	20
8	65	60	55	53	50	48	45	43	40	38	35	30	25
9	71	66	61	58	56	53	51	48	46	43	41	36	31
10	77	72	67	64	62	59	57	54	52	49	47	42	37
12	88	83	78	76	73	71	68	66	63	61	58	53	48
15	105	100	95	93	90	88	85	83	80	78	75	70	65
20	134	129	124	121	119	116	114	111	109	106	104	99	94
25	162	157	152	150	147	145	142	140	137	135	132	127	122
30	191	186	181	178	176	173	171	168	166	163	161	156	151
35	220	215	210	207	205	202	200	197	195	192	190	185	180
40	248	243	238	236	233	231	228	226	223	221	218	213	208

Table 3-11

Hub Diameter, inch: 1.75

Leg Distance, inch: 15

SHIM REQUIRED FOR ALIGNMENT

Face - OD Method

FACE TIR MILS	OD TIR, MILS -40	-30	-20	-15	-10	-5	0	5	10	15	20	30	40
-40	-322	-327	-332	-335	-337	-340	-342	-345	-347	-350	-352	-357	-362
-35	-280	-285	-290	-292	-295	-297	-300	-302	-305	-307	-310	-315	-320
-30	-237	-242	-247	-249	-252	-254	-257	-259	-262	-264	-267	-272	-277
-25	-194	-199	-204	-206	-209	-211	-214	-216	-219	-221	-224	-229	-234
-20	-151	-156	-161	-163	-166	-168	-171	-173	-176	-178	-181	-186	-191
-15	-108	-113	-118	-121	-123	-126	-128	-131	-133	-136	-138	-143	-148
-12	-82	-87	-92	-95	-97	-100	-102	-105	-107	-110	-112	-117	-122
-10	-65	-70	-75	-78	-80	-83	-85	-88	-90	-93	-95	-100	-105
-9	-57	-62	-67	-69	-72	-74	-77	-79	-82	-84	-87	-92	-97
-8	-48	-53	-58	-61	-63	-66	-68	-71	-73	-76	-78	-83	-88
-7	-40	-45	-50	-52	-55	-57	-60	-62	-65	-67	-70	-75	-80
-6	-31	-36	-41	-43	-46	-48	-51	-53	-56	-58	-61	-66	-71
-5	-22	-27	-32	-35	-37	-40	-42	-45	-47	-50	-52	-57	-62
-4	-14	-19	-24	-26	-29	-31	-34	-36	-39	-41	-44	-49	-54
-3	-5	-10	-15	-18	-20	-23	-25	-28	-30	-33	-35	-40	-45
-2	2	-2	-7	-9	-12	-14	-17	-19	-22	-24	-27	-32	-37
-1	11	6	1	-1	-3	-6	-8	-11	-13	-16	-18	-23	-28
0	20	15	10	7	5	2	0	-2	-5	-7	-10	-15	-20
1	28	23	18	16	13	11	8	6	3	1	-1	-6	-11
2	37	32	27	24	22	19	17	14	12	9	7	2	-2
3	45	40	35	33	30	28	25	23	20	18	15	10	5
4	54	49	44	41	39	36	34	31	29	26	24	19	14
5	62	57	52	50	47	45	42	40	37	35	32	27	22
6	71	66	61	58	56	53	51	48	46	43	41	36	31
7	80	75	70	67	65	62	60	57	55	52	50	45	40
8	88	83	78	76	73	71	68	66	63	61	58	53	48
9	97	92	87	84	82	79	77	74	72	69	67	62	57
10	105	100	95	93	90	88	85	83	80	78	75	70	65
12	122	117	112	110	107	105	102	100	97	95	92	87	82
15	148	143	138	136	133	131	128	126	123	121	118	113	108
20	191	186	181	178	176	173	171	168	166	163	161	156	151
25	234	229	224	221	219	216	214	211	209	206	204	199	194
30	277	272	267	264	262	259	257	254	252	249	247	242	237
35	320	315	310	307	305	302	300	297	295	292	290	285	280
40	362	357	352	350	347	345	342	340	337	335	332	327	322

Table 3-12

Hub Diameter, inch: 1.75

Leg Distance, inch: 20

SHIM REQUIRED FOR ALIGNMENT

Face - OD Method

FACE TIR MILS	OD TIR, MILS -40	-30	-20	-15	-10	-5	0	5	10	15	20	30	40
-40	-437	-442	-447	-449	-452	-454	-457	-459	-462	-464	-467	-472	-477
-35	-380	-385	-390	-392	-395	-397	-400	-402	-405	-407	-410	-415	-420
-30	-322	-327	-332	-335	-337	-340	-342	-345	-347	-350	-352	-357	-362
-25	-265	-270	-275	-278	-280	-283	-285	-288	-290	-293	-295	-300	-305
-20	-208	-213	-218	-221	-223	-226	-228	-231	-233	-236	-238	-243	-248
-15	-151	-156	-161	-163	-166	-168	-171	-173	-176	-178	-181	-186	-191
-12	-117	-122	-127	-129	-132	-134	-137	-139	-142	-144	-147	-152	-157
-10	-94	-99	-104	-106	-109	-111	-114	-116	-119	-121	-124	-129	-134
-9	-82	-87	-92	-95	-97	-100	-102	-105	-107	-110	-112	-117	-122
-8	-71	-76	-81	-83	-86	-88	-91	-93	-96	-98	-101	-106	-111
-7	-60	-65	-70	-72	-75	-77	-80	-82	-85	-87	-90	-95	-100
-6	-48	-53	-58	-61	-63	-66	-68	-71	-73	-76	-78	-83	-88
-5	-37	-42	-47	-49	-52	-54	-57	-59	-62	-64	-67	-72	-77
-4	-25	-30	-35	-38	-40	-43	-45	-48	-50	-53	-55	-60	-65
-3	-14	-19	-24	-26	-29	-31	-34	-36	-39	-41	-44	-49	-54
-2	-2	-7	-12	-15	-17	-20	-22	-25	-27	-30	-32	-37	-42
-1	8	3	-1	-3	-6	-8	-11	-13	-16	-18	-21	-26	-31
0	20	15	10	7	5	2	0	-2	-5	-7	-10	-15	-20
1	31	26	21	18	16	13	11	8	6	3	1	-3	-8
2	42	37	32	30	27	25	22	20	17	15	12	7	2
3	54	49	44	41	39	36	34	31	29	26	24	19	14
4	65	60	55	53	50	48	45	43	40	38	35	30	25
5	77	72	67	64	62	59	57	54	52	49	47	42	37
6	88	83	78	76	73	71	68	66	63	61	58	53	48
7	100	95	90	87	85	82	80	77	75	72	70	65	60
8	111	106	101	98	96	93	91	88	86	83	81	76	71
9	122	117	112	110	107	105	102	100	97	95	92	87	82
10	134	129	124	121	119	116	114	111	109	106	104	99	94
12	157	152	147	144	142	139	137	134	132	129	127	122	117
15	191	186	181	178	176	173	171	168	166	163	161	156	151
20	248	243	238	236	233	231	228	226	223	221	218	213	208
25	305	300	295	293	290	288	285	283	280	278	275	270	265
30	362	357	352	350	347	345	342	340	337	335	332	327	322
35	420	415	410	407	405	402	400	397	395	392	390	385	380
40	477	472	467	464	462	459	457	454	452	449	447	442	437

Table 3-13

Hub Diameter, inch: 1.75

Leg Distance, inch: 25

SHIM REQUIRED FOR ALIGNMENT

Face - OD Method

FACE TIR MILS	OD TIR, MILS -40	-30	-20	-15	-10	-5	0	5	10	15	20	30	40
-40	-551	-556	-561	-563	-566	-568	-571	-573	-576	-578	-581	-586	-591
-35	-480	-485	-490	-492	-495	-497	-500	-502	-505	-507	-510	-515	-520
-30	-408	-413	-418	-421	-423	-426	-428	-431	-433	-436	-438	-443	-448
-25	-337	-342	-347	-349	-352	-354	-357	-359	-362	-364	-367	-372	-377
-20	-265	-270	-275	-278	-280	-283	-285	-288	-290	-293	-295	-300	-305
-15	-194	-199	-204	-206	-209	-211	-214	-216	-219	-221	-224	-229	-234
-12	-151	-156	-161	-163	-166	-168	-171	-173	-176	-178	-181	-186	-191
-10	-122	-127	-132	-135	-137	-140	-142	-145	-147	-150	-152	-157	-162
-9	-108	-113	-118	-121	-123	-126	-128	-131	-133	-136	-138	-143	-148
-8	-94	-99	-104	-106	-109	-111	-114	-116	-119	-121	-124	-129	-134
-7	-80	-85	-90	-92	-95	-97	-100	-102	-105	-107	-110	-115	-120
-6	-65	-70	-75	-78	-80	-83	-85	-88	-90	-93	-95	-100	-105
-5	-51	-56	-61	-63	-66	-68	-71	-73	-76	-78	-81	-86	-91
-4	-37	-42	-47	-49	-52	-54	-57	-59	-62	-64	-67	-72	-77
-3	-22	-27	-32	-35	-37	-40	-42	-45	-47	-50	-52	-57	-62
-2	-8	-13	-18	-21	-23	-26	-28	-31	-33	-36	-38	-43	-48
-1	5	0	-4	-6	-9	-11	-14	-16	-19	-21	-24	-29	-34
0	20	15	10	7	5	2	0	-2	-5	-7	-10	-15	-20
1	34	29	24	21	19	16	14	11	9	6	4	0	-5
2	48	43	38	36	33	31	28	26	23	21	18	13	8
3	62	57	52	50	47	45	42	40	37	35	32	27	22
4	77	72	67	64	62	59	57	54	52	49	47	42	37
5	91	86	81	78	76	73	71	68	66	63	61	56	51
6	105	100	95	93	90	88	85	83	80	78	75	70	65
7	120	115	110	107	105	102	100	97	95	92	90	85	80
8	134	129	124	121	119	116	114	111	109	106	104	99	94
9	148	143	138	136	133	131	128	126	123	121	118	113	108
10	162	157	152	150	147	145	142	140	137	135	132	127	122
12	191	186	181	178	176	173	171	168	166	163	161	156	151
15	234	229	224	221	219	216	214	211	209	206	204	199	194
20	305	300	295	293	290	288	285	283	280	278	275	270	265
25	377	372	367	364	362	359	357	354	352	349	347	342	337
30	448	443	438	436	433	431	428	426	423	421	418	413	408
35	520	515	510	507	505	502	500	497	495	492	490	485	480
40	591	586	581	578	576	573	571	568	566	563	561	556	551

Table 3-14

Hub Diameter, inch: 1.75

Leg Distance, inch: 30

SHIM REQUIRED FOR ALIGNMENT

Face - OD Method

FACE TIR MILS	OD TIR, MILS -40	-30	-20	-15	-10	-5	0	5	10	15	20	30	40
-40	-665	-670	-675	-678	-680	-683	-685	-688	-690	-693	-695	-700	-705
-35	-580	-585	-590	-592	-595	-597	-600	-602	-605	-607	-610	-615	-620
-30	-494	-499	-504	-506	-509	-511	-514	-516	-519	-521	-524	-529	-534
-25	-408	-413	-418	-421	-423	-426	-428	-431	-433	-436	-438	-443	-448
-20	-322	-327	-332	-335	-337	-340	-342	-345	-347	-350	-352	-357	-362
-15	-237	-242	-247	-249	-252	-254	-257	-259	-262	-264	-267	-272	-277
-12	-185	-190	-195	-198	-200	-203	-205	-208	-210	-213	-215	-220	-225
-10	-151	-156	-161	-163	-166	-168	-171	-173	-176	-178	-181	-186	-191
-9	-134	-139	-144	-146	-149	-151	-154	-156	-159	-161	-164	-169	-174
-8	-117	-122	-127	-129	-132	-134	-137	-139	-142	-144	-147	-152	-157
-7	-100	-105	-110	-112	-115	-117	-120	-122	-125	-127	-130	-135	-140
-6	-82	-87	-92	-95	-97	-100	-102	-105	-107	-110	-112	-117	-122
-5	-65	-70	-75	-78	-80	-83	-85	-88	-90	-93	-95	-100	-105
-4	-48	-53	-58	-61	-63	-66	-68	-71	-73	-76	-78	-83	-88
-3	-31	-36	-41	-43	-46	-48	-51	-53	-56	-58	-61	-66	-71
-2	-14	-19	-24	-26	-29	-31	-34	-36	-39	-41	-44	-49	-54
-1	2	-2	-7	-9	-12	-14	-17	-19	-22	-24	-27	-32	-37
0	20	15	10	7	5	2	0	-2	-5	-7	-10	-15	-20
1	37	32	27	24	22	19	17	14	12	9	7	2	-2
2	54	49	44	41	39	36	34	31	29	26	24	19	14
3	71	66	61	58	56	53	51	48	46	43	41	36	31
4	88	83	78	76	73	71	68	66	63	61	58	53	48
5	105	100	95	93	90	88	85	83	80	78	75	70	65
6	122	117	112	110	107	105	102	100	97	95	92	87	82
7	140	135	130	127	125	122	120	117	115	112	110	105	100
8	157	152	147	144	142	139	137	134	132	129	127	122	117
9	174	169	164	161	159	156	154	151	149	146	144	139	134
10	191	186	181	178	176	173	171	168	166	163	161	156	151
12	225	220	215	213	210	208	205	203	200	198	195	190	185
15	277	272	267	264	262	259	257	254	252	249	247	242	237
20	362	357	352	350	347	345	342	340	337	335	332	327	322
25	448	443	438	436	433	431	428	426	423	421	418	413	408
30	534	529	524	521	519	516	514	511	509	506	504	499	494
35	620	615	610	607	605	602	600	597	595	592	590	585	580
40	705	700	695	693	690	688	685	683	680	678	675	670	665

Table 3-15

Hub Diameter, inch: 1.75

Leg Distance, inch: 35

SHIM REQUIRED FOR ALIGNMENT

Face - OD Method

FACE TIR MILS	OD TIR, MILS -40	-30	-20	-15	-10	-5	0	5	10	15	20	30	40
-40	-780	-785	-790	-792	-795	-797	-800	-802	-805	-807	-810	-815	-820
-35	-680	-685	-690	-692	-695	-697	-700	-702	-705	-707	-710	-715	-720
-30	-580	-585	-590	-592	-595	-597	-600	-602	-605	-607	-610	-615	-620
-25	-480	-485	-490	-492	-495	-497	-500	-502	-505	-507	-510	-515	-520
-20	-380	-385	-390	-392	-395	-397	-400	-402	-405	-407	-410	-415	-420
-15	-280	-285	-290	-292	-295	-297	-300	-302	-305	-307	-310	-315	-320
-12	-220	-225	-230	-232	-235	-237	-240	-242	-245	-247	-250	-255	-260
-10	-180	-185	-190	-192	-195	-197	-200	-202	-205	-207	-210	-215	-220
-9	-160	-165	-170	-172	-175	-177	-180	-182	-185	-187	-190	-195	-200
-8	-140	-145	-150	-152	-155	-157	-160	-162	-165	-167	-170	-175	-180
-7	-120	-125	-130	-132	-135	-137	-140	-142	-145	-147	-150	-155	-160
-6	-100	-105	-110	-112	-115	-117	-120	-122	-125	-127	-130	-135	-140
-5	-80	-85	-90	-92	-95	-97	-100	-102	-105	-107	-110	-115	-120
-4	-60	-65	-70	-72	-75	-77	-80	-82	-85	-87	-90	-95	-100
-3	-40	-45	-50	-52	-55	-57	-60	-62	-65	-67	-70	-75	-80
-2	-20	-25	-30	-32	-35	-37	-40	-42	-45	-47	-50	-55	-60
-1	0	-5	-10	-12	-15	-17	-20	-22	-25	-27	-30	-35	-40
0	20	15	10	7	5	2	0	-2	-5	-7	-10	-15	-20
1	40	35	30	27	25	22	20	17	15	12	10	5	0
2	60	55	50	47	45	42	40	37	35	32	30	25	20
3	80	75	70	67	65	62	60	57	55	52	50	45	40
4	100	95	90	87	85	82	80	77	75	72	70	65	60
5	120	115	110	107	105	102	100	97	95	92	90	85	80
6	140	135	130	127	125	122	120	117	115	112	110	105	100
7	160	155	150	147	145	142	140	137	135	132	130	125	120
8	180	175	170	167	165	162	160	157	155	152	150	145	140
9	200	195	190	187	185	182	180	177	175	172	170	165	160
10	220	215	210	207	205	202	200	197	195	192	190	185	180
12	260	255	250	247	245	242	240	237	235	232	230	225	220
15	320	315	310	307	305	302	300	297	295	292	290	285	280
20	420	415	410	407	405	402	400	397	395	392	390	385	380
25	520	515	510	507	505	502	500	497	495	492	490	485	480
30	620	615	610	607	605	602	600	597	595	592	590	585	580
35	720	715	710	707	705	702	700	697	695	692	690	685	680
40	820	815	810	807	805	802	800	797	795	792	790	785	780

Table 3-16

Hub Diameter, inch: 1.75

Leg Distance, inch: 40

SHIM REQUIRED FOR ALIGNMENT

Face - OD Method

FACE TIR MILS	OD TIR, MILS -40	-30	-20	-15	-10	-5	0	5	10	15	20	30	40
-40	-894	-899	-904	-906	-909	-911	-914	-916	-919	-921	-924	-929	-934
-35	-780	-785	-790	-792	-795	-797	-800	-802	-805	-807	-810	-815	-820
-30	-665	-670	-675	-678	-680	-683	-685	-688	-690	-693	-695	-700	-705
-25	-551	-556	-561	-563	-566	-568	-571	-573	-576	-578	-581	-586	-591
-20	-437	-442	-447	-449	-452	-454	-457	-459	-462	-464	-467	-472	-477
-15	-322	-327	-332	-335	-337	-340	-342	-345	-347	-350	-352	-357	-362
-12	-254	-259	-264	-266	-269	-271	-274	-276	-279	-281	-284	-289	-294
-10	-208	-213	-218	-221	-223	-226	-228	-231	-233	-236	-238	-243	-248
-9	-185	-190	-195	-198	-200	-203	-205	-208	-210	-213	-215	-220	-225
-8	-162	-167	-172	-175	-177	-180	-182	-185	-187	-190	-192	-197	-202
-7	-140	-145	-150	-152	-155	-157	-160	-162	-165	-167	-170	-175	-180
-6	-117	-122	-127	-129	-132	-134	-137	-139	-142	-144	-147	-152	-157
-5	-94	-99	-104	-106	-109	-111	-114	-116	-119	-121	-124	-129	-134
-4	-71	-76	-81	-83	-86	-88	-91	-93	-96	-98	-101	-106	-111
-3	-48	-53	-58	-61	-63	-66	-68	-71	-73	-76	-78	-83	-88
-2	-25	-30	-35	-38	-40	-43	-45	-48	-50	-53	-55	-60	-65
-1	-2	-7	-12	-15	-17	-20	-22	-25	-27	-30	-32	-37	-42
0	20	15	10	7	5	2	0	-2	-5	-7	-10	-15	-20
1	42	37	32	30	27	25	22	20	17	15	12	7	2
2	65	60	55	53	50	48	45	43	40	38	35	30	25
3	88	83	78	76	73	71	68	66	63	61	58	53	48
4	111	106	101	98	96	93	91	88	86	83	81	76	71
5	134	129	124	121	119	116	114	111	109	106	104	99	94
6	157	152	147	144	142	139	137	134	132	129	127	122	117
7	180	175	170	167	165	162	160	157	155	152	150	145	140
8	202	197	192	190	187	185	182	180	177	175	172	167	162
9	225	220	215	213	210	208	205	203	200	198	195	190	185
10	248	243	238	236	233	231	228	226	223	221	218	213	208
12	294	289	284	281	279	276	274	271	269	266	264	259	254
15	362	357	352	350	347	345	342	340	337	335	332	327	322
20	477	472	467	464	462	459	457	454	452	449	447	442	437
25	591	586	581	578	576	573	571	568	566	563	561	556	551
30	705	700	695	693	690	688	685	683	680	678	675	670	665
35	820	815	810	807	805	802	800	797	795	792	790	785	780
40	934	929	924	921	919	916	914	911	909	906	904	899	894

Table 3-17

Hub Diameter, inch: 2
Leg Distance, inch: 5

SHIM REQUIRED FOR ALIGNMENT

Face - OD Method

FACE TIR MILS	OD TIR, MILS -40	-30	-20	-15	-10	-5	0	5	10	15	20	30	40
-40	-80	-85	-90	-92	-95	-97	-100	-102	-105	-107	-110	-115	-120
-35	-67	-72	-77	-80	-82	-85	-87	-90	-92	-95	-97	-102	-107
-30	-55	-60	-65	-67	-70	-72	-75	-77	-80	-82	-85	-90	-95
-25	-42	-47	-52	-55	-57	-60	-62	-65	-67	-70	-72	-77	-82
-20	-30	-35	-40	-42	-45	-47	-50	-52	-55	-57	-60	-65	-70
-15	-17	-22	-27	-30	-32	-35	-37	-40	-42	-45	-47	-52	-57
-12	-10	-15	-20	-22	-25	-27	-30	-32	-35	-37	-40	-45	-50
-10	-5	-10	-15	-17	-20	-22	-25	-27	-30	-32	-35	-40	-45
-9	-2	-7	-12	-15	-17	-20	-22	-25	-27	-30	-32	-37	-42
-8	0	-5	-10	-12	-15	-17	-20	-22	-25	-27	-30	-35	-40
-7	2	-2	-7	-10	-12	-15	-17	-20	-22	-25	-27	-32	-37
-6	5	0	-5	-7	-10	-12	-15	-17	-20	-22	-25	-30	-35
-5	7	2	-2	-5	-7	-10	-12	-15	-17	-20	-22	-27	-32
-4	10	5	0	-2	-5	-7	-10	-12	-15	-17	-20	-25	-30
-3	12	7	2	0	-2	-5	-7	-10	-12	-15	-17	-22	-27
-2	15	10	5	2	0	-2	-5	-7	-10	-12	-15	-20	-25
-1	17	12	7	5	2	0	-2	-5	-7	-10	-12	-17	-22
0	20	15	10	7	5	2	0	-2	-5	-7	-10	-15	-20
1	22	17	12	10	7	5	2	0	-2	-5	-7	-12	-17
2	25	20	15	12	10	7	5	2	0	-2	-5	-10	-15
3	27	22	17	15	12	10	7	5	2	0	-2	-7	-12
4	30	25	20	17	15	12	10	7	5	2	0	-5	-10
5	32	27	22	20	17	15	12	10	7	5	2	-2	-7
6	35	30	25	22	20	17	15	12	10	7	5	0	-5
7	37	32	27	25	22	20	17	15	12	10	7	2	-2
8	40	35	30	27	25	22	20	17	15	12	10	5	0
9	42	37	32	30	27	25	22	20	17	15	12	7	2
10	45	40	35	32	30	27	25	22	20	17	15	10	5
12	50	45	40	37	35	32	30	27	25	22	20	15	10
15	57	52	47	45	42	40	37	35	32	30	27	22	17
20	70	65	60	57	55	52	50	47	45	42	40	35	30
25	82	77	72	70	67	65	62	60	57	55	52	47	42
30	95	90	85	82	80	77	75	72	70	67	65	60	55
35	107	102	97	95	92	90	87	85	82	80	77	72	67
40	120	115	110	107	105	102	100	97	95	92	90	85	80

Table 3-18

Hub Diameter, inch: 2

Leg Distance, inch: 10

SHIM REQUIRED FOR ALIGNMENT

Face - OD Method

FACE TIR MILS	OD TIR, MILS -40	-30	-20	-15	-10	-5	0	5	10	15	20	30	40
-40	-180	-185	-190	-192	-195	-197	-200	-202	-205	-207	-210	-215	-220
-35	-155	-160	-165	-167	-170	-172	-175	-177	-180	-182	-185	-190	-195
-30	-130	-135	-140	-142	-145	-147	-150	-152	-155	-157	-160	-165	-170
-25	-105	-110	-115	-117	-120	-122	-125	-127	-130	-132	-135	-140	-145
-20	-80	-85	-90	-92	-95	-97	-100	-102	-105	-107	-110	-115	-120
-15	-55	-60	-65	-67	-70	-72	-75	-77	-80	-82	-85	-90	-95
-12	-40	-45	-50	-52	-55	-57	-60	-62	-65	-67	-70	-75	-80
-10	-30	-35	-40	-42	-45	-47	-50	-52	-55	-57	-60	-65	-70
-9	-25	-30	-35	-37	-40	-42	-45	-47	-50	-52	-55	-60	-65
-8	-20	-25	-30	-32	-35	-37	-40	-42	-45	-47	-50	-55	-60
-7	-15	-20	-25	-27	-30	-32	-35	-37	-40	-42	-45	-50	-55
-6	-10	-15	-20	-22	-25	-27	-30	-32	-35	-37	-40	-45	-50
-5	-5	-10	-15	-17	-20	-22	-25	-27	-30	-32	-35	-40	-45
-4	0	-5	-10	-12	-15	-17	-20	-22	-25	-27	-30	-35	-40
-3	5	0	-5	-7	-10	-12	-15	-17	-20	-22	-25	-30	-35
-2	10	5	0	-2	-5	-7	-10	-12	-15	-17	-20	-25	-30
-1	15	10	5	2	0	-2	-5	-7	-10	-12	-15	-20	-25
0	20	15	10	7	5	2	0	-2	-5	-7	-10	-15	-20
1	25	20	15	12	10	7	5	2	0	-2	-5	-10	-15
2	30	25	20	17	15	12	10	7	5	2	0	-5	-10
3	35	30	25	22	20	17	15	12	10	7	5	0	-5
4	40	35	30	27	25	22	20	17	15	12	10	5	0
5	45	40	35	32	30	27	25	22	20	17	15	10	5
6	50	45	40	37	35	32	30	27	25	22	20	15	10
7	55	50	45	42	40	37	35	32	30	27	25	20	15
8	60	55	50	47	45	42	40	37	35	32	30	25	20
9	65	60	55	52	50	47	45	42	40	37	35	30	25
10	70	65	60	57	55	52	50	47	45	42	40	35	30
12	80	75	70	67	65	62	60	57	55	52	50	45	40
15	95	90	85	82	80	77	75	72	70	67	65	60	55
20	120	115	110	107	105	102	100	97	95	92	90	85	80
25	145	140	135	132	130	127	125	122	120	117	115	110	105
30	170	165	160	157	155	152	150	147	145	142	140	135	130
35	195	190	185	182	180	177	175	172	170	167	165	160	155
40	220	215	210	207	205	202	200	197	195	192	190	185	180

Table 3-19

Hub Diameter, inch: 2
Leg Distance, inch: 15

SHIM REQUIRED FOR ALIGNMENT

Face - OD Method

FACE TIR MILS	OD TIR, MILS -40	-30	-20	-15	-10	-5	0	5	10	15	20	30	40
-40	-280	-285	-290	-292	-295	-297	-300	-302	-305	-307	-310	-315	-320
-35	-242	-247	-252	-255	-257	-260	-262	-265	-267	-270	-272	-277	-282
-30	-205	-210	-215	-217	-220	-222	-225	-227	-230	-232	-235	-240	-245
-25	-167	-172	-177	-180	-182	-185	-187	-190	-192	-195	-197	-202	-207
-20	-130	-135	-140	-142	-145	-147	-150	-152	-155	-157	-160	-165	-170
-15	-92	-97	-102	-105	-107	-110	-112	-115	-117	-120	-122	-127	-132
-12	-70	-75	-80	-82	-85	-87	-90	-92	-95	-97	-100	-105	-110
-10	-55	-60	-65	-67	-70	-72	-75	-77	-80	-82	-85	-90	-95
-9	-47	-52	-57	-60	-62	-65	-67	-70	-72	-75	-77	-82	-87
-8	-40	-45	-50	-52	-55	-57	-60	-62	-65	-67	-70	-75	-80
-7	-32	-37	-42	-45	-47	-50	-52	-55	-57	-60	-62	-67	-72
-6	-25	-30	-35	-37	-40	-42	-45	-47	-50	-52	-55	-60	-65
-5	-17	-22	-27	-30	-32	-35	-37	-40	-42	-45	-47	-52	-57
-4	-10	-15	-20	-22	-25	-27	-30	-32	-35	-37	-40	-45	-50
-3	-2	-7	-12	-15	-17	-20	-22	-25	-27	-30	-32	-37	-42
-2	5	0	-5	-7	-10	-12	-15	-17	-20	-22	-25	-30	-35
-1	12	7	2	0	-2	-5	-7	-10	-12	-15	-17	-22	-27
0	20	15	10	7	5	2	0	-2	-5	-7	-10	-15	-20
1	27	22	17	15	12	10	7	5	2	0	-2	-7	-12
2	35	30	25	22	20	17	15	12	10	7	5	0	-5
3	42	37	32	30	27	25	22	20	17	15	12	7	2
4	50	45	40	37	35	32	30	27	25	22	20	15	10
5	57	52	47	45	42	40	37	35	32	30	27	22	17
6	65	60	55	52	50	47	45	42	40	37	35	30	25
7	72	67	62	60	57	55	52	50	47	45	42	37	32
8	80	75	70	67	65	62	60	57	55	52	50	45	40
9	87	82	77	75	72	70	67	65	62	60	57	52	47
10	95	90	85	82	80	77	75	72	70	67	65	60	55
12	110	105	100	97	95	92	90	87	85	82	80	75	70
15	132	127	122	120	117	115	112	110	107	105	102	97	92
20	170	165	160	157	155	152	150	147	145	142	140	135	130
25	207	202	197	195	192	190	187	185	182	180	177	172	167
30	245	240	235	232	230	227	225	222	220	217	215	210	205
35	282	277	272	270	267	265	262	260	257	255	252	247	242
40	320	315	310	307	305	302	300	297	295	292	290	285	280

Table 3-20

Hub Diameter, inch: 2

Leg Distance, inch: 20

SHIM REQUIRED FOR ALIGNMENT

Face - OD Method

FACE TIR MILS	OD TIR, MILS -40	-30	-20	-15	-10	-5	0	5	10	15	20	30	40
-40	-380	-385	-390	-392	-395	-397	-400	-402	-405	-407	-410	-415	-420
-35	-330	-335	-340	-342	-345	-347	-350	-352	-355	-357	-360	-365	-370
-30	-280	-285	-290	-292	-295	-297	-300	-302	-305	-307	-310	-315	-320
-25	-230	-235	-240	-242	-245	-247	-250	-252	-255	-257	-260	-265	-270
-20	-180	-185	-190	-192	-195	-197	-200	-202	-205	-207	-210	-215	-220
-15	-130	-135	-140	-142	-145	-147	-150	-152	-155	-157	-160	-165	-170
-12	-100	-105	-110	-112	-115	-117	-120	-122	-125	-127	-130	-135	-140
-10	-80	-85	-90	-92	-95	-97	-100	-102	-105	-107	-110	-115	-120
-9	-70	-75	-80	-82	-85	-87	-90	-92	-95	-97	-100	-105	-110
-8	-60	-65	-70	-72	-75	-77	-80	-82	-85	-87	-90	-95	-100
-7	-50	-55	-60	-62	-65	-67	-70	-72	-75	-77	-80	-85	-90
-6	-40	-45	-50	-52	-55	-57	-60	-62	-65	-67	-70	-75	-80
-5	-30	-35	-40	-42	-45	-47	-50	-52	-55	-57	-60	-65	-70
-4	-20	-25	-30	-32	-35	-37	-40	-42	-45	-47	-50	-55	-60
-3	-10	-15	-20	-22	-25	-27	-30	-32	-35	-37	-40	-45	-50
-2	0	-5	-10	-12	-15	-17	-20	-22	-25	-27	-30	-35	-40
-1	10	5	0	-2	-5	-7	-10	-12	-15	-17	-20	-25	-30
0	20	15	10	7	5	2	0	-2	-5	-7	-10	-15	-20
1	30	25	20	17	15	12	10	7	5	2	0	-5	-10
2	40	35	30	27	25	22	20	17	15	12	10	5	0
3	50	45	40	37	35	32	30	27	25	22	20	15	10
4	60	55	50	47	45	42	40	37	35	32	30	25	20
5	70	65	60	57	55	52	50	47	45	42	40	35	30
6	80	75	70	67	65	62	60	57	55	52	50	45	40
7	90	85	80	77	75	72	70	67	65	62	60	55	50
8	100	95	90	87	85	82	80	77	75	72	70	65	60
9	110	105	100	97	95	92	90	87	85	82	80	75	70
10	120	115	110	107	105	102	100	97	95	92	90	85	80
12	140	135	130	127	125	122	120	117	115	112	110	105	100
15	170	165	160	157	155	152	150	147	145	142	140	135	130
20	220	215	210	207	205	202	200	197	195	192	190	185	180
25	270	265	260	257	255	252	250	247	245	242	240	235	230
30	320	315	310	307	305	302	300	297	295	292	290	285	280
35	370	365	360	357	355	352	350	347	345	342	340	335	330
40	420	415	410	407	405	402	400	397	395	392	390	385	380

Table 3-21

Hub Diameter, inch: 2

Leg Distance, inch: 25

SHIM REQUIRED FOR ALIGNMENT

Face – OD Method

FACE TIR MILS	OD TIR, MILS -40	-30	-20	-15	-10	-5	0	5	10	15	20	30	40
-40	-480	-485	-490	-492	-495	-497	-500	-502	-505	-507	-510	-515	-520
-35	-417	-422	-427	-430	-432	-435	-437	-440	-442	-445	-447	-452	-457
-30	-355	-360	-365	-367	-370	-372	-375	-377	-380	-382	-385	-390	-395
-25	-292	-297	-302	-305	-307	-310	-312	-315	-317	-320	-322	-327	-332
-20	-230	-235	-240	-242	-245	-247	-250	-252	-255	-257	-260	-265	-270
-15	-167	-172	-177	-180	-182	-185	-187	-190	-192	-195	-197	-202	-207
-12	-130	-135	-140	-142	-145	-147	-150	-152	-155	-157	-160	-165	-170
-10	-105	-110	-115	-117	-120	-122	-125	-127	-130	-132	-135	-140	-145
-9	-92	-97	-102	-105	-107	-110	-112	-115	-117	-120	-122	-127	-132
-8	-80	-85	-90	-92	-95	-97	-100	-102	-105	-107	-110	-115	-120
-7	-67	-72	-77	-80	-82	-85	-87	-90	-92	-95	-97	-102	-107
-6	-55	-60	-65	-67	-70	-72	-75	-77	-80	-82	-85	-90	-95
-5	-42	-47	-52	-55	-57	-60	-62	-65	-67	-70	-72	-77	-82
-4	-30	-35	-40	-42	-45	-47	-50	-52	-55	-57	-60	-65	-70
-3	-17	-22	-27	-30	-32	-35	-37	-40	-42	-45	-47	-52	-57
-2	-5	-10	-15	-17	-20	-22	-25	-27	-30	-32	-35	-40	-45
-1	7	2	-2	-5	-7	-10	-12	-15	-17	-20	-22	-27	-32
0	20	15	10	7	5	2	0	-2	-5	-7	-10	-15	-20
1	32	27	22	20	17	15	12	10	7	5	2	-2	-7
2	45	40	35	32	30	27	25	22	20	17	15	10	5
3	57	52	47	45	42	40	37	35	32	30	27	22	17
4	70	65	60	57	55	52	50	47	45	42	40	35	30
5	82	77	72	70	67	65	62	60	57	55	52	47	42
6	95	90	85	82	80	77	75	72	70	67	65	60	55
7	107	102	97	95	92	90	87	85	82	80	77	72	67
8	120	115	110	107	105	102	100	97	95	92	90	85	80
9	132	127	122	120	117	115	112	110	107	105	102	97	92
10	145	140	135	132	130	127	125	122	120	117	115	110	105
12	170	165	160	157	155	152	150	147	145	142	140	135	130
15	207	202	197	195	192	190	187	185	182	180	177	172	167
20	270	265	260	257	255	252	250	247	245	242	240	235	230
25	332	327	322	320	317	315	312	310	307	305	302	297	292
30	395	390	385	382	380	377	375	372	370	367	365	360	355
35	457	452	447	445	442	440	437	435	432	430	427	422	417
40	520	515	510	507	505	502	500	497	495	492	490	485	480

Table 3-22

Hub Diameter, inch: 2

Leg Distance, inch: 30

SHIM REQUIRED FOR ALIGNMENT

Face - OD Method

FACE TIR MILS	OD TIR, MILS -40	-30	-20	-15	-10	-5	0	5	10	15	20	30	40
-40	-580	-585	-590	-592	-595	-597	-600	-602	-605	-607	-610	-615	-620
-35	-505	-510	-515	-517	-520	-522	-525	-527	-530	-532	-535	-540	-545
-30	-430	-435	-440	-442	-445	-447	-450	-452	-455	-457	-460	-465	-470
-25	-355	-360	-365	-367	-370	-372	-375	-377	-380	-382	-385	-390	-395
-20	-280	-285	-290	-292	-295	-297	-300	-302	-305	-307	-310	-315	-320
-15	-205	-210	-215	-217	-220	-222	-225	-227	-230	-232	-235	-240	-245
-12	-160	-165	-170	-172	-175	-177	-180	-182	-185	-187	-190	-195	-200
-10	-130	-135	-140	-142	-145	-147	-150	-152	-155	-157	-160	-165	-170
-9	-115	-120	-125	-127	-130	-132	-135	-137	-140	-142	-145	-150	-155
-8	-100	-105	-110	-112	-115	-117	-120	-122	-125	-127	-130	-135	-140
-7	-85	-90	-95	-97	-100	-102	-105	-107	-110	-112	-115	-120	-125
-6	-70	-75	-80	-82	-85	-87	-90	-92	-95	-97	-100	-105	-110
-5	-55	-60	-65	-67	-70	-72	-75	-77	-80	-82	-85	-90	-95
-4	-40	-45	-50	-52	-55	-57	-60	-62	-65	-67	-70	-75	-80
-3	-25	-30	-35	-37	-40	-42	-45	-47	-50	-52	-55	-60	-65
-2	-10	-15	-20	-22	-25	-27	-30	-32	-35	-37	-40	-45	-50
-1	5	0	-5	-7	-10	-12	-15	-17	-20	-22	-25	-30	-35
0	20	15	10	7	5	2	0	-2	-5	-7	-10	-15	-20
1	35	30	25	22	20	17	15	12	10	7	5	0	-5
2	50	45	40	37	35	32	30	27	25	22	20	15	10
3	65	60	55	52	50	47	45	42	40	37	35	30	25
4	80	75	70	67	65	62	60	57	55	52	50	45	40
5	95	90	85	82	80	77	75	72	70	67	65	60	55
6	110	105	100	97	95	92	90	87	85	82	80	75	70
7	125	120	115	112	110	107	105	102	100	97	95	90	85
8	140	135	130	127	125	122	120	117	115	112	110	105	100
9	155	150	145	142	140	137	135	132	130	127	125	120	115
10	170	165	160	157	155	152	150	147	145	142	140	135	130
12	200	195	190	187	185	182	180	177	175	172	170	165	160
15	245	240	235	232	230	227	225	222	220	217	215	210	205
20	320	315	310	307	305	302	300	297	295	292	290	285	280
25	395	390	385	382	380	377	375	372	370	367	365	360	355
30	470	465	460	457	455	452	450	447	445	442	440	435	430
35	545	540	535	532	530	527	525	522	520	517	515	510	505
40	620	615	610	607	605	602	600	597	595	592	590	585	580

Table 3-23

Hub Diameter, inch: 2

Leg Distance, inch: 35

SHIM REQUIRED FOR ALIGNMENT

Face - OD Method

FACE TIR MILS	OD TIR, MILS -40	-30	-20	-15	-10	-5	0	5	10	15	20	30	40
-40	-680	-685	-690	-692	-695	-697	-700	-702	-705	-707	-710	-715	-720
-35	-592	-597	-602	-605	-607	-610	-612	-615	-617	-620	-622	-627	-632
-30	-505	-510	-515	-517	-520	-522	-525	-527	-530	-532	-535	-540	-545
-25	-417	-422	-427	-430	-432	-435	-437	-440	-442	-445	-447	-452	-457
-20	-330	-335	-340	-342	-345	-347	-350	-352	-355	-357	-360	-365	-370
-15	-242	-247	-252	-255	-257	-260	-262	-265	-267	-270	-272	-277	-282
-12	-190	-195	-200	-202	-205	-207	-210	-212	-215	-217	-220	-225	-230
-10	-155	-160	-165	-167	-170	-172	-175	-177	-180	-182	-185	-190	-195
-9	-137	-142	-147	-150	-152	-155	-157	-160	-162	-165	-167	-172	-177
-8	-120	-125	-130	-132	-135	-137	-140	-142	-145	-147	-150	-155	-160
-7	-102	-107	-112	-115	-117	-120	-122	-125	-127	-130	-132	-137	-142
-6	-85	-90	-95	-97	-100	-102	-105	-107	-110	-112	-115	-120	-125
-5	-67	-72	-77	-80	-82	-85	-87	-90	-92	-95	-97	-102	-107
-4	-50	-55	-60	-62	-65	-67	-70	-72	-75	-77	-80	-85	-90
-3	-32	-37	-42	-45	-47	-50	-52	-55	-57	-60	-62	-67	-72
-2	-15	-20	-25	-27	-30	-32	-35	-37	-40	-42	-45	-50	-55
-1	2	-2	-7	-10	-12	-15	-17	-20	-22	-25	-27	-32	-37
0	20	15	10	7	5	2	0	-2	-5	-7	-10	-15	-20
1	37	32	27	25	22	20	17	15	12	10	7	2	-2
2	55	50	45	42	40	37	35	32	30	27	25	20	15
3	72	67	62	60	57	55	52	50	47	45	42	37	32
4	90	85	80	77	75	72	70	67	65	62	60	55	50
5	107	102	97	95	92	90	87	85	82	80	77	72	67
6	125	120	115	112	110	107	105	102	100	97	95	90	85
7	142	137	132	130	127	125	122	120	117	115	112	107	102
8	160	155	150	147	145	142	140	137	135	132	130	125	120
9	177	172	167	165	162	160	157	155	152	150	147	142	137
10	195	190	185	182	180	177	175	172	170	167	165	160	155
12	230	225	220	217	215	212	210	207	205	202	200	195	190
15	282	277	272	270	267	265	262	260	257	255	252	247	242
20	370	365	360	357	355	352	350	347	345	342	340	335	330
25	457	452	447	445	442	440	437	435	432	430	427	422	417
30	545	540	535	532	530	527	525	522	520	517	515	510	505
35	632	627	622	620	617	615	612	610	607	605	602	597	592
40	720	715	710	707	705	702	700	697	695	692	690	685	680

Table 3-24

Hub Diameter, inch: 2

Leg Distance, inch: 40

SHIM REQUIRED FOR ALIGNMENT

Face - OD Method

FACE TIR MILS	OD TIR, MILS -40	-30	-20	-15	-10	-5	0	5	10	15	20	30	40
-40	-780	-785	-790	-792	-795	-797	-800	-802	-805	-807	-810	-815	-820
-35	-680	-685	-690	-692	-695	-697	-700	-702	-705	-707	-710	-715	-720
-30	-580	-585	-590	-592	-595	-597	-600	-602	-605	-607	-610	-615	-620
-25	-480	-485	-490	-492	-495	-497	-500	-502	-505	-507	-510	-515	-520
-20	-380	-385	-390	-392	-395	-397	-400	-402	-405	-407	-410	-415	-420
-15	-280	-285	-290	-292	-295	-297	-300	-302	-305	-307	-310	-315	-320
-12	-220	-225	-230	-232	-235	-237	-240	-242	-245	-247	-250	-255	-260
-10	-180	-185	-190	-192	-195	-197	-200	-202	-205	-207	-210	-215	-220
-9	-160	-165	-170	-172	-175	-177	-180	-182	-185	-187	-190	-195	-200
-8	-140	-145	-150	-152	-155	-157	-160	-162	-165	-167	-170	-175	-180
-7	-120	-125	-130	-132	-135	-137	-140	-142	-145	-147	-150	-155	-160
-6	-100	-105	-110	-112	-115	-117	-120	-122	-125	-127	-130	-135	-140
-5	-80	-85	-90	-92	-95	-97	-100	-102	-105	-107	-110	-115	-120
-4	-60	-65	-70	-72	-75	-77	-80	-82	-85	-87	-90	-95	-100
-3	-40	-45	-50	-52	-55	-57	-60	-62	-65	-67	-70	-75	-80
-2	-20	-25	-30	-32	-35	-37	-40	-42	-45	-47	-50	-55	-60
-1	0	-5	-10	-12	-15	-17	-20	-22	-25	-27	-30	-35	-40
0	20	15	10	7	5	2	0	-2	-5	-7	-10	-15	-20
1	40	35	30	27	25	22	20	17	15	12	10	5	0
2	60	55	50	47	45	42	40	37	35	32	30	25	20
3	80	75	70	67	65	62	60	57	55	52	50	45	40
4	100	95	90	87	85	82	80	77	75	72	70	65	60
5	120	115	110	107	105	102	100	97	95	92	90	85	80
6	140	135	130	127	125	122	120	117	115	112	110	105	100
7	160	155	150	147	145	142	140	137	135	132	130	125	120
8	180	175	170	167	165	162	160	157	155	152	150	145	140
9	200	195	190	187	185	182	180	177	175	172	170	165	160
10	220	215	210	207	205	202	200	197	195	192	190	185	180
12	260	255	250	247	245	242	240	237	235	232	230	225	220
15	320	315	310	307	305	302	300	297	295	292	290	285	280
20	420	415	410	407	405	402	400	397	395	392	390	385	380
25	520	515	510	507	505	502	500	497	495	492	490	485	480
30	620	615	610	607	605	602	600	597	595	592	590	585	580
35	720	715	710	707	705	702	700	697	695	692	690	685	680
40	820	815	810	807	805	802	800	797	795	792	790	785	780

Table 3-25

Hub Diameter, inch: 2.25

Leg Distance, inch: 15

SHIM REQUIRED FOR ALIGNMENT

Face - OD Method

FACE TIR MILS	OD TIR, MILS -40	-30	-20	-15	-10	-5	0	5	10	15	20	30	40
-40	-246	-251	-256	-259	-261	-264	-266	-269	-271	-274	-276	-281	-286
-35	-213	-218	-223	-225	-228	-230	-233	-235	-238	-240	-243	-248	-253
-30	-180	-185	-190	-192	-195	-197	-200	-202	-205	-207	-210	-215	-220
-25	-146	-151	-156	-159	-161	-164	-166	-169	-171	-174	-176	-181	-186
-20	-113	-118	-123	-125	-128	-130	-133	-135	-138	-140	-143	-148	-153
-15	-80	-85	-90	-92	-95	-97	-100	-102	-105	-107	-110	-115	-120
-12	-60	-65	-70	-72	-75	-77	-80	-82	-85	-87	-90	-95	-100
-10	-46	-51	-56	-59	-61	-64	-66	-69	-71	-74	-76	-81	-86
-9	-40	-45	-50	-52	-55	-57	-60	-62	-65	-67	-70	-75	-80
-8	-33	-38	-43	-45	-48	-50	-53	-55	-58	-60	-63	-68	-73
-7	-26	-31	-36	-39	-41	-44	-46	-49	-51	-54	-56	-61	-66
-6	-20	-25	-30	-32	-35	-37	-40	-42	-45	-47	-50	-55	-60
-5	-13	-18	-23	-25	-28	-30	-33	-35	-38	-40	-43	-48	-53
-4	-6	-11	-16	-19	-21	-24	-26	-29	-31	-34	-36	-41	-46
-3	0	-5	-10	-12	-15	-17	-20	-22	-25	-27	-30	-35	-40
-2	6	1	-3	-5	-8	-10	-13	-15	-18	-20	-23	-28	-33
-1	13	8	3	0	-1	-4	-6	-9	-11	-14	-16	-21	-26
0	20	15	10	7	5	2	0	-2	-5	-7	-10	-15	-20
1	26	21	16	14	11	9	6	4	1	0	-3	-8	-13
2	33	28	23	20	18	15	13	10	8	5	3	-1	-6
3	40	35	30	27	25	22	20	17	15	12	10	5	0
4	46	41	36	34	31	29	26	24	21	19	16	11	6
5	53	48	43	40	38	35	33	30	28	25	23	18	13
6	60	55	50	47	45	42	40	37	35	32	30	25	20
7	66	61	56	54	51	49	46	44	41	39	36	31	26
8	73	68	63	60	58	55	53	50	48	45	43	38	33
9	80	75	70	67	65	62	60	57	55	52	50	45	40
10	86	81	76	74	71	69	66	64	61	59	56	51	46
12	100	95	90	87	85	82	80	77	75	72	70	65	60
15	120	115	110	107	105	102	100	97	95	92	90	85	80
20	153	148	143	140	138	135	133	130	128	125	123	118	113
25	186	181	176	174	171	169	166	164	161	159	156	151	146
30	220	215	210	207	205	202	200	197	195	192	190	185	180
35	253	248	243	240	238	235	233	230	228	225	223	218	213
40	286	281	276	274	271	269	266	264	261	259	256	251	246

Table 3-26

Hub Diameter, inch: 2.25

Leg Distance, inch: 20

SHIM REQUIRED FOR ALIGNMENT

Face - OD Method

FACE TIR MILS	OD TIR, MILS -40	-30	-20	-15	-10	-5	0	5	10	15	20	30	40
-40	-335	-340	-345	-348	-350	-353	-355	-358	-360	-363	-365	-370	-375
-35	-291	-296	-301	-303	-306	-308	-311	-313	-316	-318	-321	-326	-331
-30	-246	-251	-256	-259	-261	-264	-266	-269	-271	-274	-276	-281	-286
-25	-202	-207	-212	-214	-217	-219	-222	-224	-227	-229	-232	-237	-242
-20	-157	-162	-167	-170	-172	-175	-177	-180	-182	-185	-187	-192	-197
-15	-113	-118	-123	-125	-128	-130	-133	-135	-138	-140	-143	-148	-153
-12	-86	-91	-96	-99	-101	-104	-106	-109	-111	-114	-116	-121	-126
-10	-68	-73	-78	-81	-83	-86	-88	-91	-93	-96	-98	-103	-108
-9	-60	-65	-70	-72	-75	-77	-80	-82	-85	-87	-90	-95	-100
-8	-51	-56	-61	-63	-66	-68	-71	-73	-76	-78	-81	-86	-91
-7	-42	-47	-52	-54	-57	-59	-62	-64	-67	-69	-72	-77	-82
-6	-33	-38	-43	-45	-48	-50	-53	-55	-58	-60	-63	-68	-73
-5	-24	-29	-34	-36	-39	-41	-44	-46	-49	-51	-54	-59	-64
-4	-15	-20	-25	-28	-30	-33	-35	-38	-40	-43	-45	-50	-55
-3	-6	-11	-16	-19	-21	-24	-26	-29	-31	-34	-36	-41	-46
-2	2	-2	-7	-10	-12	-15	-17	-20	-22	-25	-27	-32	-37
-1	11	6	1	-1	-3	-6	-8	-11	-13	-16	-18	-23	-28
0	20	15	10	7	5	2	0	-2	-5	-7	-10	-15	-20
1	28	23	18	16	13	11	8	6	3	1	-1	-6	-11
2	37	32	27	25	22	20	17	15	12	10	7	2	-2
3	46	41	36	34	31	29	26	24	21	19	16	11	6
4	55	50	45	43	40	38	35	33	30	28	25	20	15
5	64	59	54	51	49	46	44	41	39	36	34	29	24
6	73	68	63	60	58	55	53	50	48	45	43	38	33
7	82	77	72	69	67	64	62	59	57	54	52	47	42
8	91	86	81	78	76	73	71	68	66	63	61	56	51
9	100	95	90	87	85	82	80	77	75	72	70	65	60
10	108	103	98	96	93	91	88	86	83	81	78	73	68
12	126	121	116	114	111	109	106	104	101	99	96	91	86
15	153	148	143	140	138	135	133	130	128	125	123	118	113
20	197	192	187	185	182	180	177	175	172	170	167	162	157
25	242	237	232	229	227	224	222	219	217	214	212	207	202
30	286	281	276	274	271	269	266	264	261	259	256	251	246
35	331	326	321	318	316	313	311	308	306	303	301	296	291
40	375	370	365	363	360	358	355	353	350	348	345	340	335

Table 3-27

Hub Diameter, inch: 2.25

Leg Distance, inch: 25

SHIM REQUIRED FOR ALIGNMENT

Face - OD Method

FACE TIR MILS	OD TIR, MILS -40	-30	-20	-15	-10	-5	0	5	10	15	20	30	40
-40	-424	-429	-434	-436	-439	-441	-444	-446	-449	-451	-454	-459	-464
-35	-368	-373	-378	-381	-383	-386	-388	-391	-393	-396	-398	-403	-408
-30	-313	-318	-323	-325	-328	-330	-333	-335	-338	-340	-343	-348	-353
-25	-257	-262	-267	-270	-272	-275	-277	-280	-282	-285	-287	-292	-297
-20	-202	-207	-212	-214	-217	-219	-222	-224	-227	-229	-232	-237	-242
-15	-146	-151	-156	-159	-161	-164	-166	-169	-171	-174	-176	-181	-186
-12	-113	-118	-123	-125	-128	-130	-133	-135	-138	-140	-143	-148	-153
-10	-91	-96	-101	-103	-106	-108	-111	-113	-116	-118	-121	-126	-131
-9	-80	-85	-90	-92	-95	-97	-100	-102	-105	-107	-110	-115	-120
-8	-68	-73	-78	-81	-83	-86	-88	-91	-93	-96	-98	-103	-108
-7	-57	-62	-67	-70	-72	-75	-77	-80	-82	-85	-87	-92	-97
-6	-46	-51	-56	-59	-61	-64	-66	-69	-71	-74	-76	-81	-86
-5	-35	-40	-45	-48	-50	-53	-55	-58	-60	-63	-65	-70	-75
-4	-24	-29	-34	-36	-39	-41	-44	-46	-49	-51	-54	-59	-64
-3	-13	-18	-23	-25	-28	-30	-33	-35	-38	-40	-43	-48	-53
-2	-2	-7	-12	-14	-17	-19	-22	-24	-27	-29	-32	-37	-42
-1	8	3	-1	-3	-6	-8	-11	-13	-16	-18	-21	-26	-31
0	20	15	10	7	5	2	0	-2	-5	-7	-10	-15	-20
1	31	26	21	18	16	13	11	8	6	3	1	-3	-8
2	42	37	32	29	27	24	22	19	17	14	12	7	2
3	53	48	43	40	38	35	33	30	28	25	23	18	13
4	64	59	54	51	49	46	44	41	39	36	34	29	24
5	75	70	65	63	60	58	55	53	50	48	45	40	35
6	86	81	76	74	71	69	66	64	61	59	56	51	46
7	97	92	87	85	82	80	77	75	72	70	67	62	57
8	108	103	98	96	93	91	88	86	83	81	78	73	68
9	120	115	110	107	105	102	100	97	95	92	90	85	80
10	131	126	121	118	116	113	111	108	106	103	101	96	91
12	153	148	143	140	138	135	133	130	128	125	123	118	113
15	186	181	176	174	171	169	166	164	161	159	156	151	146
20	242	237	232	229	227	224	222	219	217	214	212	207	202
25	297	292	287	285	282	280	277	275	272	270	267	262	257
30	353	348	343	340	338	335	333	330	328	325	323	318	313
35	408	403	398	396	393	391	388	386	383	381	378	373	368
40	464	459	454	451	449	446	444	441	439	436	434	429	424

Table 3-28

Hub Diameter, inch: 2.25

Leg Distance, inch: 30

SHIM REQUIRED FOR ALIGNMENT

Face - OD Method

FACE TIR MILS	OD TIR, MILS -40	-30	-20	-15	-10	-5	0	5	10	15	20	30	40
-40	-513	-518	-523	-525	-528	-530	-533	-535	-538	-540	-543	-548	-553
-35	-446	-451	-456	-459	-461	-464	-466	-469	-471	-474	-476	-481	-486
-30	-380	-385	-390	-392	-395	-397	-400	-402	-405	-407	-410	-415	-420
-25	-313	-318	-323	-325	-328	-330	-333	-335	-338	-340	-343	-348	-353
-20	-246	-251	-256	-259	-261	-264	-266	-269	-271	-274	-276	-281	-286
-15	-180	-185	-190	-192	-195	-197	-200	-202	-205	-207	-210	-215	-220
-12	-140	-145	-150	-152	-155	-157	-160	-162	-165	-167	-170	-175	-180
-10	-113	-118	-123	-125	-128	-130	-133	-135	-138	-140	-143	-148	-153
-9	-100	-105	-110	-112	-115	-117	-120	-122	-125	-127	-130	-135	-140
-8	-86	-91	-96	-99	-101	-104	-106	-109	-111	-114	-116	-121	-126
-7	-73	-78	-83	-85	-88	-90	-93	-95	-98	-100	-103	-108	-113
-6	-60	-65	-70	-72	-75	-77	-80	-82	-85	-87	-90	-95	-100
-5	-46	-51	-56	-59	-61	-64	-66	-69	-71	-74	-76	-81	-86
-4	-33	-38	-43	-45	-48	-50	-53	-55	-58	-60	-63	-68	-73
-3	-20	-25	-30	-32	-35	-37	-40	-42	-45	-47	-50	-55	-60
-2	-6	-11	-16	-19	-21	-24	-26	-29	-31	-34	-36	-41	-46
-1	6	1	-3	-5	-8	-10	-13	-15	-18	-20	-23	-28	-33
0	20	15	10	7	5	2	0	-2	-5	-7	-10	-15	-20
1	33	28	23	20	18	15	13	10	8	5	3	-1	-6
2	46	41	36	34	31	29	26	24	21	19	16	11	6
3	60	55	50	47	45	42	40	37	35	32	30	25	20
4	73	68	63	60	58	55	53	50	48	45	43	38	33
5	86	81	76	74	71	69	66	64	61	59	56	51	46
6	100	95	90	87	85	82	80	77	75	72	70	65	60
7	113	108	103	100	98	95	93	90	88	85	83	78	73
8	126	121	116	114	111	109	106	104	101	99	96	91	86
9	140	135	130	127	125	122	120	117	115	112	110	105	100
10	153	148	143	140	138	135	133	130	128	125	123	118	113
12	180	175	170	167	165	162	160	157	155	152	150	145	140
15	220	215	210	207	205	202	200	197	195	192	190	185	180
20	286	281	276	274	271	269	266	264	261	259	256	251	246
25	353	348	343	340	338	335	333	330	328	325	323	318	313
30	420	415	410	407	405	402	400	397	395	392	390	385	380
35	486	481	476	474	471	469	466	464	461	459	456	451	446
40	553	548	543	540	538	535	533	530	528	525	523	518	513

Table 3-29

Hub Diameter, inch: 2.25
Leg Distance, inch: 35

SHIM REQUIRED FOR ALIGNMENT

Face - OD Method

FACE TIR MILS	OD TIR, MILS -40	-30	-20	-15	-10	-5	0	5	10	15	20	30	40
-40	-602	-607	-612	-614	-617	-619	-622	-624	-627	-629	-632	-637	-642
-35	-524	-529	-534	-536	-539	-541	-544	-546	-549	-551	-554	-559	-564
-30	-446	-451	-456	-459	-461	-464	-466	-469	-471	-474	-476	-481	-486
-25	-368	-373	-378	-381	-383	-386	-388	-391	-393	-396	-398	-403	-408
-20	-291	-296	-301	-303	-306	-308	-311	-313	-316	-318	-321	-326	-331
-15	-213	-218	-223	-225	-228	-230	-233	-235	-238	-240	-243	-248	-253
-12	-166	-171	-176	-179	-181	-184	-186	-189	-191	-194	-196	-201	-206
-10	-135	-140	-145	-148	-150	-153	-155	-158	-160	-163	-165	-170	-175
-9	-120	-125	-130	-132	-135	-137	-140	-142	-145	-147	-150	-155	-160
-8	-104	-109	-114	-116	-119	-121	-124	-126	-129	-131	-134	-139	-144
-7	-88	-93	-98	-101	-103	-106	-108	-111	-113	-116	-118	-123	-128
-6	-73	-78	-83	-85	-88	-90	-93	-95	-98	-100	-103	-108	-113
-5	-57	-62	-67	-70	-72	-75	-77	-80	-82	-85	-87	-92	-97
-4	-42	-47	-52	-54	-57	-59	-62	-64	-67	-69	-72	-77	-82
-3	-26	-31	-36	-39	-41	-44	-46	-49	-51	-54	-56	-61	-66
-2	-11	-16	-21	-23	-26	-28	-31	-33	-36	-38	-41	-46	-51
-1	4	0	-5	-8	-10	-13	-15	-18	-20	-23	-25	-30	-35
0	20	15	10	7	5	2	0	-2	-5	-7	-10	-15	-20
1	35	30	25	23	20	18	15	13	10	8	5	0	-4
2	51	46	41	38	36	33	31	28	26	23	21	16	11
3	66	61	56	54	51	49	46	44	41	39	36	31	26
4	82	77	72	69	67	64	62	59	57	54	52	47	42
5	97	92	87	85	82	80	77	75	72	70	67	62	57
6	113	108	103	100	98	95	93	90	88	85	83	78	73
7	128	123	118	116	113	111	108	106	103	101	98	93	88
8	144	139	134	131	129	126	124	121	119	116	114	109	104
9	160	155	150	147	145	142	140	137	135	132	130	125	120
10	175	170	165	163	160	158	155	153	150	148	145	140	135
12	206	201	196	194	191	189	186	184	181	179	176	171	166
15	253	248	243	240	238	235	233	230	228	225	223	218	213
20	331	326	321	318	316	313	311	308	306	303	301	296	291
25	408	403	398	396	393	391	388	386	383	381	378	373	368
30	486	481	476	474	471	469	466	464	461	459	456	451	446
35	564	559	554	551	549	546	544	541	539	536	534	529	524
40	642	637	632	629	627	624	622	619	617	614	612	607	602

Table 3-30

Hub Diameter, inch: 2.25

Leg Distance, inch: 40

SHIM REQUIRED FOR ALIGNMENT

Face - OD Method

FACE TIR MILS	OD TIR, MILS -40	-30	-20	-15	-10	-5	0	5	10	15	20	30	40
-40	-691	-696	-701	-703	-706	-708	-711	-713	-716	-718	-721	-726	-731
-35	-602	-607	-612	-614	-617	-619	-622	-624	-627	-629	-632	-637	-642
-30	-513	-518	-523	-525	-528	-530	-533	-535	-538	-540	-543	-548	-553
-25	-424	-429	-434	-436	-439	-441	-444	-446	-449	-451	-454	-459	-464
-20	-335	-340	-345	-348	-350	-353	-355	-358	-360	-363	-365	-370	-375
-15	-246	-251	-256	-259	-261	-264	-266	-269	-271	-274	-276	-281	-286
-12	-193	-198	-203	-205	-208	-210	-213	-215	-218	-220	-223	-228	-233
-10	-157	-162	-167	-170	-172	-175	-177	-180	-182	-185	-187	-192	-197
-9	-140	-145	-150	-152	-155	-157	-160	-162	-165	-167	-170	-175	-180
-8	-122	-127	-132	-134	-137	-139	-142	-144	-147	-149	-152	-157	-162
-7	-104	-109	-114	-116	-119	-121	-124	-126	-129	-131	-134	-139	-144
-6	-86	-91	-96	-99	-101	-104	-106	-109	-111	-114	-116	-121	-126
-5	-68	-73	-78	-81	-83	-86	-88	-91	-93	-96	-98	-103	-108
-4	-51	-56	-61	-63	-66	-68	-71	-73	-76	-78	-81	-86	-91
-3	-33	-38	-43	-45	-48	-50	-53	-55	-58	-60	-63	-68	-73
-2	-15	-20	-25	-28	-30	-33	-35	-38	-40	-43	-45	-50	-55
-1	2	-2	-7	-10	-12	-15	-17	-20	-22	-25	-27	-32	-37
0	20	15	10	7	5	2	0	-2	-5	-7	-10	-15	-20
1	37	32	27	25	22	20	17	15	12	10	7	2	-2
2	55	50	45	43	40	38	35	33	30	28	25	20	15
3	73	68	63	60	58	55	53	50	48	45	43	38	33
4	91	86	81	78	76	73	71	68	66	63	61	56	51
5	108	103	98	96	93	91	88	86	83	81	78	73	68
6	126	121	116	114	111	109	106	104	101	99	96	91	86
7	144	139	134	131	129	126	124	121	119	116	114	109	104
8	162	157	152	149	147	144	142	139	137	134	132	127	122
9	180	175	170	167	165	162	160	157	155	152	150	145	140
10	197	192	187	185	182	180	177	175	172	170	167	162	157
12	233	228	223	220	218	215	213	210	208	205	203	198	193
15	286	281	276	274	271	269	266	264	261	259	256	251	246
20	375	370	365	363	360	358	355	353	350	348	345	340	335
25	464	459	454	451	449	446	444	441	439	436	434	429	424
30	553	548	543	540	538	535	533	530	528	525	523	518	513
35	642	637	632	629	627	624	622	619	617	614	612	607	602
40	731	726	721	718	716	713	711	708	706	703	701	696	691

Table 3-31

Hub Diameter, inch: 2.25
Leg Distance, inch: 50

SHIM REQUIRED FOR ALIGNMENT

Face - OD Method

FACE TIR MILS	OD TIR, MILS -40	-30	-20	-15	-10	-5	0	5	10	15	20	30	40
-40	-868	-873	-878	-881	-883	-886	-888	-891	-893	-896	-898	-903	-908
-35	-757	-762	-767	-770	-772	-775	-777	-780	-782	-785	-787	-792	-797
-30	-646	-651	-656	-659	-661	-664	-666	-669	-671	-674	-676	-681	-686
-25	-535	-540	-545	-548	-550	-553	-555	-558	-560	-563	-565	-570	-575
-20	-424	-429	-434	-436	-439	-441	-444	-446	-449	-451	-454	-459	-464
-15	-313	-318	-323	-325	-328	-330	-333	-335	-338	-340	-343	-348	-353
-12	-246	-251	-256	-259	-261	-264	-266	-269	-271	-274	-276	-281	-286
-10	-202	-207	-212	-214	-217	-219	-222	-224	-227	-229	-232	-237	-242
-9	-180	-185	-190	-192	-195	-197	-200	-202	-205	-207	-210	-215	-220
-8	-157	-162	-167	-170	-172	-175	-177	-180	-182	-185	-187	-192	-197
-7	-135	-140	-145	-148	-150	-153	-155	-158	-160	-163	-165	-170	-175
-6	-113	-118	-123	-125	-128	-130	-133	-135	-138	-140	-143	-148	-153
-5	-91	-96	-101	-103	-106	-108	-111	-113	-116	-118	-121	-126	-131
-4	-68	-73	-78	-81	-83	-86	-88	-91	-93	-96	-98	-103	-108
-3	-46	-51	-56	-59	-61	-64	-66	-69	-71	-74	-76	-81	-86
-2	-24	-29	-34	-36	-39	-41	-44	-46	-49	-51	-54	-59	-64
-1	-2	-7	-12	-14	-17	-19	-22	-24	-27	-29	-32	-37	-42
0	20	15	10	7	5	2	0	-2	-5	-7	-10	-15	-20
1	42	37	32	29	27	24	22	19	17	14	12	7	2
2	64	59	54	51	49	46	44	41	39	36	34	29	24
3	86	81	76	74	71	69	66	64	61	59	56	51	46
4	108	103	98	96	93	91	88	86	83	81	78	73	68
5	131	126	121	118	116	113	111	108	106	103	101	96	91
6	153	148	143	140	138	135	133	130	128	125	123	118	113
7	175	170	165	163	160	158	155	153	150	148	145	140	135
8	197	192	187	185	182	180	177	175	172	170	167	162	157
9	220	215	210	207	205	202	200	197	195	192	190	185	180
10	242	237	232	229	227	224	222	219	217	214	212	207	202
12	286	281	276	274	271	269	266	264	261	259	256	251	246
15	353	348	343	340	338	335	333	330	328	325	323	318	313
20	464	459	454	451	449	446	444	441	439	436	434	429	424
25	575	570	565	563	560	558	555	553	550	548	545	540	535
30	686	681	676	674	671	669	666	664	661	659	656	651	646
35	797	792	787	785	782	780	777	775	772	770	767	762	757
40	908	903	898	896	893	891	888	886	883	881	878	873	868

Table 3-32

Hub Diameter, inch: 2.25

Leg Distance, inch: 60

SHIM REQUIRED FOR ALIGNMENT

Face - OD Method

FACE TIR MILS	OD TIR, MILS -40	-30	-20	-15	-10	-5	0	5	10	15	20	30	40
-40	-1046	-1051	-1056	-1059	-1061	-1064	-1066	-1069	-1071	-1074	-1076	-1081	-1086
-35	-913	-918	-923	-925	-928	-930	-933	-935	-938	-940	-943	-948	-953
-30	-780	-785	-790	-792	-795	-797	-800	-802	-805	-807	-810	-815	-820
-25	-646	-651	-656	-659	-661	-664	-666	-669	-671	-674	-676	-681	-686
-20	-513	-518	-523	-525	-528	-530	-533	-535	-538	-540	-543	-548	-553
-15	-380	-385	-390	-392	-395	-397	-400	-402	-405	-407	-410	-415	-420
-12	-300	-305	-310	-312	-315	-317	-320	-322	-325	-327	-330	-335	-340
-10	-246	-251	-256	-259	-261	-264	-266	-269	-271	-274	-276	-281	-286
-9	-220	-225	-230	-232	-235	-237	-240	-242	-245	-247	-250	-255	-260
-8	-193	-198	-203	-205	-208	-210	-213	-215	-218	-220	-223	-228	-233
-7	-166	-171	-176	-179	-181	-184	-186	-189	-191	-194	-196	-201	-206
-6	-140	-145	-150	-152	-155	-157	-160	-162	-165	-167	-170	-175	-180
-5	-113	-118	-123	-125	-128	-130	-133	-135	-138	-140	-143	-148	-153
-4	-86	-91	-96	-99	-101	-104	-106	-109	-111	-114	-116	-121	-126
-3	-60	-65	-70	-72	-75	-77	-80	-82	-85	-87	-90	-95	-100
-2	-33	-38	-43	-45	-48	-50	-53	-55	-58	-60	-63	-68	-73
-1	-6	-11	-16	-19	-21	-24	-26	-29	-31	-34	-36	-41	-46
0	20	15	10	7	5	2	0	-2	-5	-7	-10	-15	-20
1	46	41	36	34	31	29	26	24	21	19	16	11	6
2	73	68	63	60	58	55	53	50	48	45	43	38	33
3	100	95	90	87	85	82	80	77	75	72	70	65	60
4	126	121	116	114	111	109	106	104	101	99	96	91	86
5	153	148	143	140	138	135	133	130	128	125	123	118	113
6	180	175	170	167	165	162	160	157	155	152	150	145	140
7	206	201	196	194	191	189	186	184	181	179	176	171	166
8	233	228	223	220	218	215	213	210	208	205	203	198	193
9	260	255	250	247	245	242	240	237	235	232	230	225	220
10	286	281	276	274	271	269	266	264	261	259	256	251	246
12	340	335	330	327	325	322	320	317	315	312	310	305	300
15	420	415	410	407	405	402	400	397	395	392	390	385	380
20	553	548	543	540	538	535	533	530	528	525	523	518	513
25	686	681	676	674	671	669	666	664	661	659	656	651	646
30	820	815	810	807	805	802	800	797	795	792	790	785	780
35	953	948	943	940	938	935	933	930	928	925	923	918	913
40	1086	1081	1076	1074	1071	1069	1066	1064	1061	1059	1056	1051	1046

Table 3-33

Hub Diameter, inch: 2.25

Leg Distance, inch: 70

SHIM REQUIRED FOR ALIGNMENT

Face - OD Method

FACE TIR MILS	OD TIR, MILS -40	-30	-20	-15	-10	-5	0	5	10	15	20	30	40
-40	-1224	-1229	-1234	-1236	-1239	-1241	-1244	-1246	-1249	-1251	-1254	-1259	-1264
-35	-1068	-1073	-1078	-1081	-1083	-1086	-1088	-1091	-1093	-1096	-1098	-1103	-1108
-30	-913	-918	-923	-925	-928	-930	-933	-935	-938	-940	-943	-948	-953
-25	-757	-762	-767	-770	-772	-775	-777	-780	-782	-785	-787	-792	-797
-20	-602	-607	-612	-614	-617	-619	-622	-624	-627	-629	-632	-637	-642
-15	-446	-451	-456	-459	-461	-464	-466	-469	-471	-474	-476	-481	-486
-12	-353	-358	-363	-365	-368	-370	-373	-375	-378	-380	-383	-388	-393
-10	-291	-296	-301	-303	-306	-308	-311	-313	-316	-318	-321	-326	-331
-9	-260	-265	-270	-272	-275	-277	-280	-282	-285	-287	-290	-295	-300
-8	-228	-233	-238	-241	-243	-246	-248	-251	-253	-256	-258	-263	-268
-7	-197	-202	-207	-210	-212	-215	-217	-220	-222	-225	-227	-232	-237
-6	-166	-171	-176	-179	-181	-184	-186	-189	-191	-194	-196	-201	-206
-5	-135	-140	-145	-148	-150	-153	-155	-158	-160	-163	-165	-170	-175
-4	-104	-109	-114	-116	-119	-121	-124	-126	-129	-131	-134	-139	-144
-3	-73	-78	-83	-85	-88	-90	-93	-95	-98	-100	-103	-108	-113
-2	-42	-47	-52	-54	-57	-59	-62	-64	-67	-69	-72	-77	-82
-1	-11	-16	-21	-23	-26	-28	-31	-33	-36	-38	-41	-46	-51
0	20	15	10	7	5	2	0	-2	-5	-7	-10	-15	-20
1	51	46	41	38	36	33	31	28	26	23	21	16	11
2	82	77	72	69	67	64	62	59	57	54	52	47	42
3	113	108	103	100	98	95	93	90	88	85	83	78	73
4	144	139	134	131	129	126	124	121	119	116	114	109	104
5	175	170	165	163	160	158	155	153	150	148	145	140	135
6	206	201	196	194	191	189	186	184	181	179	176	171	166
7	237	232	227	225	222	220	217	215	212	210	207	202	197
8	268	263	258	256	253	251	248	246	243	241	238	233	228
9	300	295	290	287	285	282	280	277	275	272	270	265	260
10	331	326	321	318	316	313	311	308	306	303	301	296	291
12	393	388	383	380	378	375	373	370	368	365	363	358	353
15	486	481	476	474	471	469	466	464	461	459	456	451	446
20	642	637	632	629	627	624	622	619	617	614	612	607	602
25	797	792	787	785	782	780	777	775	772	770	767	762	757
30	953	948	943	940	938	935	933	930	928	925	923	918	913
35	1108	1103	1098	1096	1093	1091	1088	1086	1083	1081	1078	1073	1068
40	1264	1259	1254	1251	1249	1246	1244	1241	1239	1236	1234	1229	1224

Table 3-34

Hub Diameter, inch: 2.25

Leg Distance, inch: 80

SHIM REQUIRED FOR ALIGNMENT

Face - OD Method

FACE TIR MILS	OD TIR, MILS -40	-30	-20	-15	-10	-5	0	5	10	15	20	30	40
-40	-1402	-1407	-1412	-1414	-1417	-1419	-1422	-1424	-1427	-1429	-1432	-1437	-1442
-35	-1224	-1229	-1234	-1236	-1239	-1241	-1244	-1246	-1249	-1251	-1254	-1259	-1264
-30	-1046	-1051	-1056	-1059	-1061	-1064	-1066	-1069	-1071	-1074	-1076	-1081	-1086
-25	-868	-873	-878	-881	-883	-886	-888	-891	-893	-896	-898	-903	-908
-20	-691	-696	-701	-703	-706	-708	-711	-713	-716	-718	-721	-726	-731
-15	-513	-518	-523	-525	-528	-530	-533	-535	-538	-540	-543	-548	-553
-12	-406	-411	-416	-419	-421	-424	-426	-429	-431	-434	-436	-441	-446
-10	-335	-340	-345	-348	-350	-353	-355	-358	-360	-363	-365	-370	-375
-9	-300	-305	-310	-312	-315	-317	-320	-322	-325	-327	-330	-335	-340
-8	-264	-269	-274	-276	-279	-281	-284	-286	-289	-291	-294	-299	-304
-7	-228	-233	-238	-241	-243	-246	-248	-251	-253	-256	-258	-263	-268
-6	-193	-198	-203	-205	-208	-210	-213	-215	-218	-220	-223	-228	-233
-5	-157	-162	-167	-170	-172	-175	-177	-180	-182	-185	-187	-192	-197
-4	-122	-127	-132	-134	-137	-139	-142	-144	-147	-149	-152	-157	-162
-3	-86	-91	-96	-99	-101	-104	-106	-109	-111	-114	-116	-121	-126
-2	-51	-56	-61	-63	-66	-68	-71	-73	-76	-78	-81	-86	-91
-1	-15	-20	-25	-28	-30	-33	-35	-38	-40	-43	-45	-50	-55
0	20	15	10	7	5	2	0	-2	-5	-7	-10	-15	-20
1	55	50	45	43	40	38	35	33	30	28	25	20	15
2	91	86	81	78	76	73	71	68	66	63	61	56	51
3	126	121	116	114	111	109	106	104	101	99	96	91	86
4	162	157	152	149	147	144	142	139	137	134	132	127	122
5	197	192	187	185	182	180	177	175	172	170	167	162	157
6	233	228	223	220	218	215	213	210	208	205	203	198	193
7	268	263	258	256	253	251	248	246	243	241	238	233	228
8	304	299	294	291	289	286	284	281	279	276	274	269	264
9	340	335	330	327	325	322	320	317	315	312	310	305	300
10	375	370	365	363	360	358	355	353	350	348	345	340	335
12	446	441	436	434	431	429	426	424	421	419	416	411	406
15	553	548	543	540	538	535	533	530	528	525	523	518	513
20	731	726	721	718	716	713	711	708	706	703	701	696	691
25	908	903	898	896	893	891	888	886	883	881	878	873	868
30	1086	1081	1076	1074	1071	1069	1066	1064	1061	1059	1056	1051	1046
35	1264	1259	1254	1251	1249	1246	1244	1241	1239	1236	1234	1229	1224
40	1442	1437	1432	1429	1427	1424	1422	1419	1417	1414	1412	1407	1402

Table 3-35

Hub Diameter, inch: 2.5

Leg Distance, inch: 15

SHIM REQUIRED FOR ALIGNMENT

Face - OD Method

FACE TIR MILS	OD TIR, MILS -40	-30	-20	-15	-10	-5	0	5	10	15	20	30	40
-40	-220	-225	-230	-232	-235	-237	-240	-242	-245	-247	-250	-255	-260
-35	-190	-195	-200	-202	-205	-207	-210	-212	-215	-217	-220	-225	-230
-30	-160	-165	-170	-172	-175	-177	-180	-182	-185	-187	-190	-195	-200
-25	-130	-135	-140	-142	-145	-147	-150	-152	-155	-157	-160	-165	-170
-20	-100	-105	-110	-112	-115	-117	-120	-122	-125	-127	-130	-135	-140
-15	-70	-75	-80	-82	-85	-87	-90	-92	-95	-97	-100	-105	-110
-12	-52	-57	-62	-64	-67	-69	-72	-74	-77	-79	-82	-87	-92
-10	-40	-45	-50	-52	-55	-57	-60	-62	-65	-67	-70	-75	-80
-9	-34	-39	-44	-46	-49	-51	-54	-56	-59	-61	-64	-69	-74
-8	-28	-33	-38	-40	-43	-45	-48	-50	-53	-55	-58	-63	-68
-7	-22	-27	-32	-34	-37	-39	-42	-44	-47	-49	-52	-57	-62
-6	-16	-21	-26	-28	-31	-33	-36	-38	-41	-43	-46	-51	-56
-5	-10	-15	-20	-22	-25	-27	-30	-32	-35	-37	-40	-45	-50
-4	-4	-9	-14	-16	-19	-21	-24	-26	-29	-31	-34	-39	-44
-3	2	-3	-8	-10	-13	-15	-18	-20	-23	-25	-28	-33	-38
-2	8	3	-2	-4	-7	-9	-12	-14	-17	-19	-22	-27	-32
-1	14	9	4	1	-1	-3	-6	-8	-11	-13	-16	-21	-26
0	20	15	10	7	5	2	0	-2	-5	-7	-10	-15	-20
1	26	21	16	13	11	8	6	3	1	-1	-4	-9	-14
2	32	27	22	19	17	14	12	9	7	4	2	-3	-8
3	38	33	28	25	23	20	18	15	13	10	8	3	-2
4	44	39	34	31	29	26	24	21	19	16	14	9	4
5	50	45	40	37	35	32	30	27	25	22	20	15	10
6	56	51	46	43	41	38	36	33	31	28	26	21	16
7	62	57	52	49	47	44	42	39	37	34	32	27	22
8	68	63	58	55	53	50	48	45	43	40	38	33	28
9	74	69	64	61	59	56	54	51	49	46	44	39	34
10	80	75	70	67	65	62	60	57	55	52	50	45	40
12	92	87	82	79	77	74	72	69	67	64	62	57	52
15	110	105	100	97	95	92	90	87	85	82	80	75	70
20	140	135	130	127	125	122	120	117	115	112	110	105	100
25	170	165	160	157	155	152	150	147	145	142	140	135	130
30	200	195	190	187	185	182	180	177	175	172	170	165	160
35	230	225	220	217	215	212	210	207	205	202	200	195	190
40	260	255	250	247	245	242	240	237	235	232	230	225	220

Table 3-36

Hub Diameter, inch: 2.5
Leg Distance, inch: 20

SHIM REQUIRED FOR ALIGNMENT

Face - OD Method

FACE TIR MILS	OD TIR, MILS -40	-30	-20	-15	-10	-5	0	5	10	15	20	30	40
-40	-300	-305	-310	-312	-315	-317	-320	-322	-325	-327	-330	-335	-340
-35	-260	-265	-270	-272	-275	-277	-280	-282	-285	-287	-290	-295	-300
-30	-220	-225	-230	-232	-235	-237	-240	-242	-245	-247	-250	-255	-260
-25	-180	-185	-190	-192	-195	-197	-200	-202	-205	-207	-210	-215	-220
-20	-140	-145	-150	-152	-155	-157	-160	-162	-165	-167	-170	-175	-180
-15	-100	-105	-110	-112	-115	-117	-120	-122	-125	-127	-130	-135	-140
-12	-76	-81	-86	-88	-91	-93	-96	-98	-101	-103	-106	-111	-116
-10	-60	-65	-70	-72	-75	-77	-80	-82	-85	-87	-90	-95	-100
-9	-52	-57	-62	-64	-67	-69	-72	-74	-77	-79	-82	-87	-92
-8	-44	-49	-54	-56	-59	-61	-64	-66	-69	-71	-74	-79	-84
-7	-36	-41	-46	-48	-51	-53	-56	-58	-61	-63	-66	-71	-76
-6	-28	-33	-38	-40	-43	-45	-48	-50	-53	-55	-58	-63	-68
-5	-20	-25	-30	-32	-35	-37	-40	-42	-45	-47	-50	-55	-60
-4	-12	-17	-22	-24	-27	-29	-32	-34	-37	-39	-42	-47	-52
-3	-4	-9	-14	-16	-19	-21	-24	-26	-29	-31	-34	-39	-44
-2	4	-1	-6	-8	-11	-13	-16	-18	-21	-23	-26	-31	-36
-1	12	7	2	0	-3	-5	-8	-10	-13	-15	-18	-23	-28
0	20	15	10	7	5	2	0	-2	-5	-7	-10	-15	-20
1	28	23	18	15	13	10	8	5	3	0	-2	-7	-12
2	36	31	26	23	21	18	16	13	11	8	6	1	-4
3	44	39	34	31	29	26	24	21	19	16	14	9	4
4	52	47	42	39	37	34	32	29	27	24	22	17	12
5	60	55	50	47	45	42	40	37	35	32	30	25	20
6	68	63	58	55	53	50	48	45	43	40	38	33	28
7	76	71	66	63	61	58	56	53	51	48	46	41	36
8	84	79	74	71	69	66	64	61	59	56	54	49	44
9	92	87	82	79	77	74	72	69	67	64	62	57	52
10	100	95	90	87	85	82	80	77	75	72	70	65	60
12	116	111	106	103	101	98	96	93	91	88	86	81	76
15	140	135	130	127	125	122	120	117	115	112	110	105	100
20	180	175	170	167	165	162	160	157	155	152	150	145	140
25	220	215	210	207	205	202	200	197	195	192	190	185	180
30	260	255	250	247	245	242	240	237	235	232	230	225	220
35	300	295	290	287	285	282	280	277	275	272	270	265	260
40	340	335	330	327	325	322	320	317	315	312	310	305	300

Table 3-37

Hub Diameter, inch: 2.5

Leg Distance, inch: 25

SHIM REQUIRED FOR ALIGNMENT

Face - OD Method

FACE TIR MILS	OD TIR, MILS -40	-30	-20	-15	-10	-5	0	5	10	15	20	30	40
-40	-380	-385	-390	-392	-395	-397	-400	-402	-405	-407	-410	-415	-420
-35	-330	-335	-340	-342	-345	-347	-350	-352	-355	-357	-360	-365	-370
-30	-280	-285	-290	-292	-295	-297	-300	-302	-305	-307	-310	-315	-320
-25	-230	-235	-240	-242	-245	-247	-250	-252	-255	-257	-260	-265	-270
-20	-180	-185	-190	-192	-195	-197	-200	-202	-205	-207	-210	-215	-220
-15	-130	-135	-140	-142	-145	-147	-150	-152	-155	-157	-160	-165	-170
-12	-100	-105	-110	-112	-115	-117	-120	-122	-125	-127	-130	-135	-140
-10	-80	-85	-90	-92	-95	-97	-100	-102	-105	-107	-110	-115	-120
-9	-70	-75	-80	-82	-85	-87	-90	-92	-95	-97	-100	-105	-110
-8	-60	-65	-70	-72	-75	-77	-80	-82	-85	-87	-90	-95	-100
-7	-50	-55	-60	-62	-65	-67	-70	-72	-75	-77	-80	-85	-90
-6	-40	-45	-50	-52	-55	-57	-60	-62	-65	-67	-70	-75	-80
-5	-30	-35	-40	-42	-45	-47	-50	-52	-55	-57	-60	-65	-70
-4	-20	-25	-30	-32	-35	-37	-40	-42	-45	-47	-50	-55	-60
-3	-10	-15	-20	-22	-25	-27	-30	-32	-35	-37	-40	-45	-50
-2	0	-5	-10	-12	-15	-17	-20	-22	-25	-27	-30	-35	-40
-1	10	5	0	-2	-5	-7	-10	-12	-15	-17	-20	-25	-30
0	20	15	10	7	5	2	0	-2	-5	-7	-10	-15	-20
1	30	25	20	17	15	12	10	7	5	2	0	-5	-10
2	40	35	30	27	25	22	20	17	15	12	10	5	0
3	50	45	40	37	35	32	30	27	25	22	20	15	10
4	60	55	50	47	45	42	40	37	35	32	30	25	20
5	70	65	60	57	55	52	50	47	45	42	40	35	30
6	80	75	70	67	65	62	60	57	55	52	50	45	40
7	90	85	80	77	75	72	70	67	65	62	60	55	50
8	100	95	90	87	85	82	80	77	75	72	70	65	60
9	110	105	100	97	95	92	90	87	85	82	80	75	70
10	120	115	110	107	105	102	100	97	95	92	90	85	80
12	140	135	130	127	125	122	120	117	115	112	110	105	100
15	170	165	160	157	155	152	150	147	145	142	140	135	130
20	220	215	210	207	205	202	200	197	195	192	190	185	180
25	270	265	260	257	255	252	250	247	245	242	240	235	230
30	320	315	310	307	305	302	300	297	295	292	290	285	280
35	370	365	360	357	355	352	350	347	345	342	340	335	330
40	420	415	410	407	405	402	400	397	395	392	390	385	380

Table 3-38

Hub Diameter, inch: 2.5

Leg Distance, inch: 30

SHIM REQUIRED FOR ALIGNMENT

Face - OD Method

FACE TIR MILS	OD TIR, MILS -40	-30	-20	-15	-10	-5	0	5	10	15	20	30	40
-40	-460	-465	-470	-472	-475	-477	-480	-482	-485	-487	-490	-495	-500
-35	-400	-405	-410	-412	-415	-417	-420	-422	-425	-427	-430	-435	-440
-30	-340	-345	-350	-352	-355	-357	-360	-362	-365	-367	-370	-375	-380
-25	-280	-285	-290	-292	-295	-297	-300	-302	-305	-307	-310	-315	-320
-20	-220	-225	-230	-232	-235	-237	-240	-242	-245	-247	-250	-255	-260
-15	-160	-165	-170	-172	-175	-177	-180	-182	-185	-187	-190	-195	-200
-12	-124	-129	-134	-136	-139	-141	-144	-146	-149	-151	-154	-159	-164
-10	-100	-105	-110	-112	-115	-117	-120	-122	-125	-127	-130	-135	-140
-9	-88	-93	-98	-100	-103	-105	-108	-110	-113	-115	-118	-123	-128
-8	-76	-81	-86	-88	-91	-93	-96	-98	-101	-103	-106	-111	-116
-7	-64	-69	-74	-76	-79	-81	-84	-86	-89	-91	-94	-99	-104
-6	-52	-57	-62	-64	-67	-69	-72	-74	-77	-79	-82	-87	-92
-5	-40	-45	-50	-52	-55	-57	-60	-62	-65	-67	-70	-75	-80
-4	-28	-33	-38	-40	-43	-45	-48	-50	-53	-55	-58	-63	-68
-3	-16	-21	-26	-28	-31	-33	-36	-38	-41	-43	-46	-51	-56
-2	-4	-9	-14	-16	-19	-21	-24	-26	-29	-31	-34	-39	-44
-1	8	3	-2	-4	-7	-9	-12	-14	-17	-19	-22	-27	-32
0	20	15	10	7	5	2	0	-2	-5	-7	-10	-15	-20
1	32	27	22	19	17	14	12	9	7	4	2	-3	-8
2	44	39	34	31	29	26	24	21	19	16	14	9	4
3	56	51	46	43	41	38	36	33	31	28	26	21	16
4	68	63	58	55	53	50	48	45	43	40	38	33	28
5	80	75	70	67	65	62	60	57	55	52	50	45	40
6	92	87	82	79	77	74	72	69	67	64	62	57	52
7	104	99	94	91	89	86	84	81	79	76	74	69	64
8	116	111	106	103	101	98	96	93	91	88	86	81	76
9	128	123	118	115	113	110	108	105	103	100	98	93	88
10	140	135	130	127	125	122	120	117	115	112	110	105	100
12	164	159	154	151	149	146	144	141	139	136	134	129	124
15	200	195	190	187	185	182	180	177	175	172	170	165	160
20	260	255	250	247	245	242	240	237	235	232	230	225	220
25	320	315	310	307	305	302	300	297	295	292	290	285	280
30	380	375	370	367	365	362	360	357	355	352	350	345	340
35	440	435	430	427	425	422	420	417	415	412	410	405	400
40	500	495	490	487	485	482	480	477	475	472	470	465	460

Table 3-39

Hub Diameter, inch: 2.5

Leg Distance, inch: 35

SHIM REQUIRED FOR ALIGNMENT

Face – OD Method

FACE TIR MILS	OD TIR, MILS -40	-30	-20	-15	-10	-5	0	5	10	15	20	30	40
-40	-540	-545	-550	-552	-555	-557	-560	-562	-565	-567	-570	-575	-580
-35	-470	-475	-480	-482	-485	-487	-490	-492	-495	-497	-500	-505	-510
-30	-400	-405	-410	-412	-415	-417	-420	-422	-425	-427	-430	-435	-440
-25	-330	-335	-340	-342	-345	-347	-350	-352	-355	-357	-360	-365	-370
-20	-260	-265	-270	-272	-275	-277	-280	-282	-285	-287	-290	-295	-300
-15	-190	-195	-200	-202	-205	-207	-210	-212	-215	-217	-220	-225	-230
-12	-148	-153	-158	-160	-163	-165	-168	-170	-173	-175	-178	-183	-188
-10	-120	-125	-130	-132	-135	-137	-140	-142	-145	-147	-150	-155	-160
-9	-106	-111	-116	-118	-121	-123	-126	-128	-131	-133	-136	-141	-146
-8	-92	-97	-102	-104	-107	-109	-112	-114	-117	-119	-122	-127	-132
-7	-78	-83	-88	-90	-93	-95	-98	-100	-103	-105	-108	-113	-118
-6	-64	-69	-74	-76	-79	-81	-84	-86	-89	-91	-94	-99	-104
-5	-50	-55	-60	-62	-65	-67	-70	-72	-75	-77	-80	-85	-90
-4	-36	-41	-46	-48	-51	-53	-56	-58	-61	-63	-66	-71	-76
-3	-22	-27	-32	-34	-37	-39	-42	-44	-47	-49	-52	-57	-62
-2	-8	-13	-18	-20	-23	-25	-28	-30	-33	-35	-38	-43	-48
-1	6	1	-4	-6	-9	-11	-14	-16	-19	-21	-24	-29	-34
0	20	15	10	7	5	2	0	-2	-5	-7	-10	-15	-20
1	34	29	24	21	19	16	14	11	9	6	4	-1	-6
2	48	43	38	35	33	30	28	25	23	20	18	13	8
3	62	57	52	49	47	44	42	39	37	34	32	27	22
4	76	71	66	63	61	58	56	53	51	48	46	41	36
5	90	85	80	77	75	72	70	67	65	62	60	55	50
6	104	99	94	91	89	86	84	81	79	76	74	69	64
7	118	113	108	105	103	100	98	95	93	90	88	83	78
8	132	127	122	119	117	114	112	109	107	104	102	97	92
9	146	141	136	133	131	128	126	123	121	118	116	111	106
10	160	155	150	147	145	142	140	137	135	132	130	125	120
12	188	183	178	175	173	170	168	165	163	160	158	153	148
15	230	225	220	217	215	212	210	207	205	202	200	195	190
20	300	295	290	287	285	282	280	277	275	272	270	265	260
25	370	365	360	357	355	352	350	347	345	342	340	335	330
30	440	435	430	427	425	422	420	417	415	412	410	405	400
35	510	505	500	497	495	492	490	487	485	482	480	475	470
40	580	575	570	567	565	562	560	557	555	552	550	545	540

Table 3-40

Hub Diameter, inch: 2.5

Leg Distance, inch: 40

SHIM REQUIRED FOR ALIGNMENT

Face - OD Method

FACE TIR MILS	OD TIR, MILS -40	-30	-20	-15	-10	-5	0	5	10	15	20	30	40
-40	-620	-625	-630	-632	-635	-637	-640	-642	-645	-647	-650	-655	-660
-35	-540	-545	-550	-552	-555	-557	-560	-562	-565	-567	-570	-575	-580
-30	-460	-465	-470	-472	-475	-477	-480	-482	-485	-487	-490	-495	-500
-25	-380	-385	-390	-392	-395	-397	-400	-402	-405	-407	-410	-415	-420
-20	-300	-305	-310	-312	-315	-317	-320	-322	-325	-327	-330	-335	-340
-15	-220	-225	-230	-232	-235	-237	-240	-242	-245	-247	-250	-255	-260
-12	-172	-177	-182	-184	-187	-189	-192	-194	-197	-199	-202	-207	-212
-10	-140	-145	-150	-152	-155	-157	-160	-162	-165	-167	-170	-175	-180
-9	-124	-129	-134	-136	-139	-141	-144	-146	-149	-151	-154	-159	-164
-8	-108	-113	-118	-120	-123	-125	-128	-130	-133	-135	-138	-143	-148
-7	-92	-97	-102	-104	-107	-109	-112	-114	-117	-119	-122	-127	-132
-6	-76	-81	-86	-88	-91	-93	-96	-98	-101	-103	-106	-111	-116
-5	-60	-65	-70	-72	-75	-77	-80	-82	-85	-87	-90	-95	-100
-4	-44	-49	-54	-56	-59	-61	-64	-66	-69	-71	-74	-79	-84
-3	-28	-33	-38	-40	-43	-45	-48	-50	-53	-55	-58	-63	-68
-2	-12	-17	-22	-24	-27	-29	-32	-34	-37	-39	-42	-47	-52
-1	4	-1	-6	-8	-11	-13	-16	-18	-21	-23	-26	-31	-36
0	20	15	10	7	5	2	0	-2	-5	-7	-10	-15	-20
1	36	31	26	23	21	18	16	13	11	8	6	1	-4
2	52	47	42	39	37	34	32	29	27	24	22	17	12
3	68	63	58	55	53	50	48	45	43	40	38	33	28
4	84	79	74	71	69	66	64	61	59	56	54	49	44
5	100	95	90	87	85	82	80	77	75	72	70	65	60
6	116	111	106	103	101	98	96	93	91	88	86	81	76
7	132	127	122	119	117	114	112	109	107	104	102	97	92
8	148	143	138	135	133	130	128	125	123	120	118	113	108
9	164	159	154	151	149	146	144	141	139	136	134	129	124
10	180	175	170	167	165	162	160	157	155	152	150	145	140
12	212	207	202	199	197	194	192	189	187	184	182	177	172
15	260	255	250	247	245	242	240	237	235	232	230	225	220
20	340	335	330	327	325	322	320	317	315	312	310	305	300
25	420	415	410	407	405	402	400	397	395	392	390	385	380
30	500	495	490	487	485	482	480	477	475	472	470	465	460
35	580	575	570	567	565	562	560	557	555	552	550	545	540
40	660	655	650	647	645	642	640	637	635	632	630	625	620

Table 3-41

Hub Diameter, inch: 2.5

Leg Distance, inch: 50

SHIM REQUIRED FOR ALIGNMENT

Face - OD Method

FACE TIR MILS	OD TIR, MILS -40	-30	-20	-15	-10	-5	0	5	10	15	20	30	40
-40	-780	-785	-790	-792	-795	-797	-800	-802	-805	-807	-810	-815	-820
-35	-680	-685	-690	-692	-695	-697	-700	-702	-705	-707	-710	-715	-720
-30	-580	-585	-590	-592	-595	-597	-600	-602	-605	-607	-610	-615	-620
-25	-480	-485	-490	-492	-495	-497	-500	-502	-505	-507	-510	-515	-520
-20	-380	-385	-390	-392	-395	-397	-400	-402	-405	-407	-410	-415	-420
-15	-280	-285	-290	-292	-295	-297	-300	-302	-305	-307	-310	-315	-320
-12	-220	-225	-230	-232	-235	-237	-240	-242	-245	-247	-250	-255	-260
-10	-180	-185	-190	-192	-195	-197	-200	-202	-205	-207	-210	-215	-220
-9	-160	-165	-170	-172	-175	-177	-180	-182	-185	-187	-190	-195	-200
-8	-140	-145	-150	-152	-155	-157	-160	-162	-165	-167	-170	-175	-180
-7	-120	-125	-130	-132	-135	-137	-140	-142	-145	-147	-150	-155	-160
-6	-100	-105	-110	-112	-115	-117	-120	-122	-125	-127	-130	-135	-140
-5	-80	-85	-90	-92	-95	-97	-100	-102	-105	-107	-110	-115	-120
-4	-60	-65	-70	-72	-75	-77	-80	-82	-85	-87	-90	-95	-100
-3	-40	-45	-50	-52	-55	-57	-60	-62	-65	-67	-70	-75	-80
-2	-20	-25	-30	-32	-35	-37	-40	-42	-45	-47	-50	-55	-60
-1	0	-5	-10	-12	-15	-17	-20	-22	-25	-27	-30	-35	-40
0	20	15	10	7	5	2	0	-2	-5	-7	-10	-15	-20
1	40	35	30	27	25	22	20	17	15	12	10	5	0
2	60	55	50	47	45	42	40	37	35	32	30	25	20
3	80	75	70	67	65	62	60	57	55	52	50	45	40
4	100	95	90	87	85	82	80	77	75	72	70	65	60
5	120	115	110	107	105	102	100	97	95	92	90	85	80
6	140	135	130	127	125	122	120	117	115	112	110	105	100
7	160	155	150	147	145	142	140	137	135	132	130	125	120
8	180	175	170	167	165	162	160	157	155	152	150	145	140
9	200	195	190	187	185	182	180	177	175	172	170	165	160
10	220	215	210	207	205	202	200	197	195	192	190	185	180
12	260	255	250	247	245	242	240	237	235	232	230	225	220
15	320	315	310	307	305	302	300	297	295	292	290	285	280
20	420	415	410	407	405	402	400	397	395	392	390	385	380
25	520	515	510	507	505	502	500	497	495	492	490	485	480
30	620	615	610	607	605	602	600	597	595	592	590	585	580
35	720	715	710	707	705	702	700	697	695	692	690	685	680
40	820	815	810	807	805	802	800	797	795	792	790	785	780

Table 3-42

Hub Diameter, inch: 2.5

Leg Distance, inch: 60

SHIM REQUIRED FOR ALIGNMENT

Face - OD Method

FACE TIR MILS	OD TIR, MILS -40	-30	-20	-15	-10	-5	0	5	10	15	20	30	40
-40	-940	-945	-950	-952	-955	-957	-960	-962	-965	-967	-970	-975	-980
-35	-820	-825	-830	-832	-835	-837	-840	-842	-845	-847	-850	-855	-860
-30	-700	-705	-710	-712	-715	-717	-720	-722	-725	-727	-730	-735	-740
-25	-580	-585	-590	-592	-595	-597	-600	-602	-605	-607	-610	-615	-620
-20	-460	-465	-470	-472	-475	-477	-480	-482	-485	-487	-490	-495	-500
-15	-340	-345	-350	-352	-355	-357	-360	-362	-365	-367	-370	-375	-380
-12	-268	-273	-278	-280	-283	-285	-288	-290	-293	-295	-298	-303	-308
-10	-220	-225	-230	-232	-235	-237	-240	-242	-245	-247	-250	-255	-260
-9	-196	-201	-206	-208	-211	-213	-216	-218	-221	-223	-226	-231	-236
-8	-172	-177	-182	-184	-187	-189	-192	-194	-197	-199	-202	-207	-212
-7	-148	-153	-158	-160	-163	-165	-168	-170	-173	-175	-178	-183	-188
-6	-124	-129	-134	-136	-139	-141	-144	-146	-149	-151	-154	-159	-164
-5	-100	-105	-110	-112	-115	-117	-120	-122	-125	-127	-130	-135	-140
-4	-76	-81	-86	-88	-91	-93	-96	-98	-101	-103	-106	-111	-116
-3	-52	-57	-62	-64	-67	-69	-72	-74	-77	-79	-82	-87	-92
-2	-28	-33	-38	-40	-43	-45	-48	-50	-53	-55	-58	-63	-68
-1	-4	-9	-14	-16	-19	-21	-24	-26	-29	-31	-34	-39	-44
0	20	15	10	7	5	2	0	-2	-5	-7	-10	-15	-20
1	44	39	34	31	29	26	24	21	19	16	14	9	4
2	68	63	58	55	53	50	48	45	43	40	38	33	28
3	92	87	82	79	77	74	72	69	67	64	62	57	52
4	116	111	106	103	101	98	96	93	91	88	86	81	76
5	140	135	130	127	125	122	120	117	115	112	110	105	100
6	164	159	154	151	149	146	144	141	139	136	134	129	124
7	188	183	178	175	173	170	168	165	163	160	158	153	148
8	212	207	202	199	197	194	192	189	187	184	182	177	172
9	236	231	226	223	221	218	216	213	211	208	206	201	196
10	260	255	250	247	245	242	240	237	235	232	230	225	220
12	308	303	298	295	293	290	288	285	283	280	278	273	268
15	380	375	370	367	365	362	360	357	355	352	350	345	340
20	500	495	490	487	485	482	480	477	475	472	470	465	460
25	620	615	610	607	605	602	600	597	595	592	590	585	580
30	740	735	730	727	725	722	720	717	715	712	710	705	700
35	860	855	850	847	845	842	840	837	835	832	830	825	820
40	980	975	970	967	965	962	960	957	955	952	950	945	940

Table 3-43

Hub Diameter, inch: 2.5

Leg Distance, inch: 70

SHIM REQUIRED FOR ALIGNMENT

Face - OD Method

FACE TIR MILS	OD TIR, MILS -40	-30	-20	-15	-10	-5	0	5	10	15	20	30	40
-40	-1100	-1105	-1110	-1112	-1115	-1117	-1120	-1122	-1125	-1127	-1130	-1135	-1140
-35	-960	-965	-970	-972	-975	-977	-980	-982	-985	-987	-990	-995	-1000
-30	-820	-825	-830	-832	-835	-837	-840	-842	-845	-847	-850	-355	-860
-25	-680	-685	-690	-692	-695	-697	-700	-702	-705	-707	-710	-715	-720
-20	-540	-545	-550	-552	-555	-557	-560	-562	-565	-567	-570	-575	-580
-15	-400	-405	-410	-412	-415	-417	-420	-422	-425	-427	-430	-435	-440
-12	-316	-321	-326	-328	-331	-333	-336	-338	-341	-343	-346	-351	-356
-10	-260	-265	-270	-272	-275	-277	-280	-282	-285	-287	-290	-295	-300
-9	-232	-237	-242	-244	-247	-249	-252	-254	-257	-259	-262	-267	-272
-8	-204	-209	-214	-216	-219	-221	-224	-226	-229	-231	-234	-239	-244
-7	-176	-181	-186	-188	-191	-193	-196	-198	-201	-203	-206	-211	-216
-6	-148	-153	-158	-160	-163	-165	-168	-170	-173	-175	-178	-183	-188
-5	-120	-125	-130	-132	-135	-137	-140	-142	-145	-147	-150	-155	-160
-4	-92	-97	-102	-104	-107	-109	-112	-114	-117	-119	-122	-127	-132
-3	-64	-69	-74	-76	-79	-81	-84	-86	-89	-91	-94	-99	-104
-2	-36	-41	-46	-48	-51	-53	-56	-58	-61	-63	-66	-71	-76
-1	-8	-13	-18	-20	-23	-25	-28	-30	-33	-35	-38	-43	-48
0	20	15	10	7	5	2	0	-2	-5	-7	-10	-15	-20
1	48	43	38	35	33	30	28	25	23	20	18	13	8
2	76	71	66	63	61	58	56	53	51	48	46	41	36
3	104	99	94	91	89	96	84	81	79	76	74	69	64
4	132	127	122	119	117	114	112	109	107	104	102	97	92
5	160	155	150	147	145	142	140	137	135	132	130	125	120
6	188	183	178	175	173	170	168	165	163	160	158	153	148
7	216	211	206	203	201	198	196	193	191	188	186	181	176
8	244	239	234	231	229	226	224	221	219	216	214	209	204
9	272	267	262	259	257	254	252	249	247	244	242	237	232
10	300	295	290	287	285	282	280	277	275	272	270	265	260
12	356	351	346	343	341	338	336	333	331	328	326	321	316
15	440	435	430	427	425	422	420	417	415	412	410	405	400
20	580	575	570	567	565	562	560	557	555	552	550	545	540
25	720	715	710	707	705	702	700	697	695	692	690	685	680
30	860	855	850	847	845	842	840	837	835	832	830	825	820
35	1000	995	990	987	985	982	980	977	975	972	970	965	960
40	1140	1135	1130	1127	1125	1122	1120	1117	1115	1112	1110	1105	1100

Table 3-44

Hub Diameter, inch: 2.5

Leg Distance, inch: 80

SHIM REQUIRED FOR ALIGNMENT

Face - OD Method

FACE TIR MILS	OD TIR, MILS -40	-30	-20	-15	-10	-5	0	5	10	15	20	30	40
-40	-1260	-1265	-1270	-1272	-1275	-1277	-1280	-1282	-1285	-1287	-1290	-1295	-1300
-35	-1100	-1105	-1110	-1112	-1115	-1117	-1120	-1122	-1125	-1127	-1130	-1135	-1140
-30	-940	-945	-950	-952	-955	-957	-960	-962	-965	-967	-970	-975	-980
-25	-780	-785	-790	-792	-795	-797	-800	-802	-805	-807	-810	-815	-820
-20	-620	-625	-630	-632	-635	-637	-640	-642	-645	-647	-650	-655	-660
-15	-460	-465	-470	-472	-475	-477	-480	-482	-485	-487	-490	-495	-500
-12	-364	-369	-374	-376	-379	-381	-384	-386	-389	-391	-394	-399	-404
-10	-300	-305	-310	-312	-315	-317	-320	-322	-325	-327	-330	-335	-340
-9	-268	-273	-278	-280	-283	-285	-288	-290	-293	-295	-298	-303	-308
-8	-236	-241	-246	-248	-251	-253	-256	-258	-261	-263	-266	-271	-276
-7	-204	-209	-214	-216	-219	-221	-224	-226	-229	-231	-234	-239	-244
-6	-172	-177	-182	-184	-187	-189	-192	-194	-197	-199	-202	-207	-212
-5	-140	-145	-150	-152	-155	-157	-160	-162	-165	-167	-170	-175	-180
-4	-108	-113	-118	-120	-123	-125	-128	-130	-133	-135	-138	-143	-148
-3	-76	-81	-86	-88	-91	-93	-96	-98	-101	-103	-106	-111	-116
-2	-44	-49	-54	-56	-59	-61	-64	-66	-69	-71	-74	-79	-84
-1	-12	-17	-22	-24	-27	-29	-32	-34	-37	-39	-42	-47	-52
0	20	15	10	7	5	2	0	-2	-5	-7	-10	-15	-20
1	52	47	42	39	37	34	32	29	27	24	22	17	12
2	84	79	74	71	69	66	64	61	59	56	54	49	44
3	116	111	106	103	101	98	96	93	91	88	86	81	76
4	148	143	138	135	133	130	128	125	123	120	118	113	108
5	180	175	170	167	165	162	160	157	155	152	150	145	140
6	212	207	202	199	197	194	192	189	187	184	182	177	172
7	244	239	234	231	229	226	224	221	219	216	214	209	204
8	276	271	266	263	261	258	256	253	251	248	246	241	236
9	308	303	298	295	293	290	288	285	283	280	278	273	268
10	340	335	330	327	325	322	320	317	315	312	310	305	300
12	404	399	394	391	389	386	384	381	379	376	374	369	364
15	500	495	490	487	485	482	480	477	475	472	470	465	460
20	660	655	650	647	645	642	640	637	635	632	630	625	620
25	820	815	810	807	805	802	800	797	795	792	790	785	780
30	980	975	970	967	965	962	960	957	955	952	950	945	940
35	1140	1135	1130	1127	1125	1122	1120	1117	1115	1112	1110	1105	1100
40	1300	1295	1290	1287	1285	1282	1280	1277	1275	1272	1270	1265	1260

Table 3-45

Hub Diameter, inch: 3

Leg Distance, inch: 15

SHIM REQUIRED FOR ALIGNMENT

Face - OD Method

FACE TIR MILS	OD TIR, MILS -40	-30	-20	-15	-10	-5	0	5	10	15	20	30	40
-40	-180	-185	-190	-192	-195	-197	-200	-202	-205	-207	-210	-215	-220
-35	-155	-160	-165	-167	-170	-172	-175	-177	-180	-182	-185	-190	-195
-30	-130	-135	-140	-142	-145	-147	-150	-152	-155	-157	-160	-165	-170
-25	-105	-110	-115	-117	-120	-122	-125	-127	-130	-132	-135	-140	-145
-20	-80	-85	-90	-92	-95	-97	-100	-102	-105	-107	-110	-115	-120
-15	-55	-60	-65	-67	-70	-72	-75	-77	-80	-82	-85	-90	-95
-12	-40	-45	-50	-52	-55	-57	-60	-62	-65	-67	-70	-75	-80
-10	-30	-35	-40	-42	-45	-47	-50	-52	-55	-57	-60	-65	-70
-9	-25	-30	-35	-37	-40	-42	-45	-47	-50	-52	-55	-60	-65
-8	-20	-25	-30	-32	-35	-37	-40	-42	-45	-47	-50	-55	-60
-7	-15	-20	-25	-27	-30	-32	-35	-37	-40	-42	-45	-50	-55
-6	-10	-15	-20	-22	-25	-27	-30	-32	-35	-37	-40	-45	-50
-5	-5	-10	-15	-17	-20	-22	-25	-27	-30	-32	-35	-40	-45
-4	0	-5	-10	-12	-15	-17	-20	-22	-25	-27	-30	-35	-40
-3	5	0	-5	-7	-10	-12	-15	-17	-20	-22	-25	-30	-35
-2	10	5	0	-2	-5	-7	-10	-12	-15	-17	-20	-25	-30
-1	15	10	5	2	0	-2	-5	-7	-10	-12	-15	-20	-25
0	20	15	10	7	5	2	0	-2	-5	-7	-10	-15	-20
1	25	20	15	12	10	7	5	2	0	-2	-5	-10	-15
2	30	25	20	17	15	12	10	7	5	2	0	-5	-10
3	35	30	25	22	20	17	15	12	10	7	5	0	-5
4	40	35	30	27	25	22	20	17	15	12	10	5	0
5	45	40	35	32	30	27	25	22	20	17	15	10	5
6	50	45	40	37	35	32	30	27	25	22	20	15	10
7	55	50	45	42	40	37	35	32	30	27	25	20	15
8	60	55	50	47	45	42	40	37	35	32	30	25	20
9	65	60	55	52	50	47	45	42	40	37	35	30	25
10	70	65	60	57	55	52	50	47	45	42	40	35	30
12	80	75	70	67	65	62	60	57	55	52	50	45	40
15	95	90	85	82	80	77	75	72	70	67	65	60	55
20	120	115	110	107	105	102	100	97	95	92	90	85	80
25	145	140	135	132	130	127	125	122	120	117	115	110	105
30	170	165	160	157	155	152	150	147	145	142	140	135	130
35	195	190	185	182	180	177	175	172	170	167	165	160	155
40	220	215	210	207	205	202	200	197	195	192	190	185	180

Table 3-46

Hub Diameter, inch: 3

Leg Distance, inch: 20

SHIM REQUIRED FOR ALIGNMENT

Face - OD Method

FACE TIR MILS	OD TIR, MILS -40	-30	-20	-15	-10	-5	0	5	10	15	20	30	40
-40	-246	-251	-256	-259	-261	-264	-266	-269	-271	-274	-276	-281	-286
-35	-213	-218	-223	-225	-228	-230	-233	-235	-238	-240	-243	-248	-253
-30	-180	-185	-190	-192	-195	-197	-200	-202	-205	-207	-210	-215	-220
-25	-146	-151	-156	-159	-161	-164	-166	-169	-171	-174	-176	-181	-186
-20	-113	-118	-123	-125	-128	-130	-133	-135	-138	-140	-143	-148	-153
-15	-80	-85	-90	-92	-95	-97	-100	-102	-105	-107	-110	-115	-120
-12	-60	-65	-70	-72	-75	-77	-80	-82	-85	-87	-90	-95	-100
-10	-46	-51	-56	-59	-61	-64	-66	-69	-71	-74	-76	-81	-86
-9	-40	-45	-50	-52	-55	-57	-60	-62	-65	-67	-70	-75	-80
-8	-33	-38	-43	-45	-48	-50	-53	-55	-58	-60	-63	-68	-73
-7	-26	-31	-36	-39	-41	-44	-46	-49	-51	-54	-56	-61	-66
-6	-20	-25	-30	-32	-35	-37	-40	-42	-45	-47	-50	-55	-60
-5	-13	-18	-23	-25	-28	-30	-33	-35	-38	-40	-43	-48	-53
-4	-6	-11	-16	-19	-21	-24	-26	-29	-31	-34	-36	-41	-46
-3	0	-5	-10	-12	-15	-17	-20	-22	-25	-27	-30	-35	-40
-2	6	1	-3	-5	-8	-10	-13	-15	-18	-20	-23	-28	-33
-1	13	8	3	0	-1	-4	-6	-9	-11	-14	-16	-21	-26
0	20	15	10	7	5	2	0	-2	-5	-7	-10	-15	-20
1	26	21	16	14	11	9	6	4	1	0	-3	-8	-13
2	33	28	23	20	18	15	13	10	8	5	3	-1	-6
3	40	35	30	27	25	22	20	17	15	12	10	5	0
4	46	41	36	34	31	29	26	24	21	19	16	11	6
5	53	48	43	40	38	35	33	30	28	25	23	18	13
6	60	55	50	47	45	42	40	37	35	32	30	25	20
7	66	61	56	54	51	49	46	44	41	39	36	31	26
8	73	68	63	60	58	55	53	50	48	45	43	38	33
9	80	75	70	67	65	62	60	57	55	52	50	45	40
10	86	81	76	74	71	69	66	64	61	59	56	51	46
12	100	95	90	87	85	82	80	77	75	72	70	65	60
15	120	115	110	107	105	102	100	97	95	92	90	85	80
20	153	148	143	140	138	135	133	130	128	125	123	118	113
25	186	181	176	174	171	169	166	164	161	159	156	151	146
30	220	215	210	207	205	202	200	197	195	192	190	185	180
35	253	248	243	240	238	235	233	230	228	225	223	218	213
40	286	281	276	274	271	269	266	264	261	259	256	251	246

Table 3-47

Hub Diameter, inch: 3

Leg Distance, inch: 25

SHIM REQUIRED FOR ALIGNMENT

Face - OD Method

FACE TIR MILS	OD TIR, MILS -40	-30	-20	-15	-10	-5	0	5	10	15	20	30	40
-40	-313	-318	-323	-325	-328	-330	-333	-335	-338	-340	-343	-348	-353
-35	-271	-276	-281	-284	-286	-289	-291	-294	-296	-299	-301	-306	-311
-30	-230	-235	-240	-242	-245	-247	-250	-252	-255	-257	-260	-265	-270
-25	-188	-193	-198	-200	-203	-205	-208	-210	-213	-215	-218	-223	-228
-20	-146	-151	-156	-159	-161	-164	-166	-169	-171	-174	-176	-181	-186
-15	-105	-110	-115	-117	-120	-122	-125	-127	-130	-132	-135	-140	-145
-12	-80	-85	-90	-92	-95	-97	-100	-102	-105	-107	-110	-115	-120
-10	-63	-68	-73	-75	-78	-80	-83	-85	-88	-90	-93	-98	-103
-9	-55	-60	-65	-67	-70	-72	-75	-77	-80	-82	-85	-90	-95
-8	-46	-51	-56	-59	-61	-64	-66	-69	-71	-74	-76	-81	-86
-7	-38	-43	-48	-50	-53	-55	-58	-60	-63	-65	-68	-73	-78
-6	-30	-35	-40	-42	-45	-47	-50	-52	-55	-57	-60	-65	-70
-5	-21	-26	-31	-34	-36	-39	-41	-44	-46	-49	-51	-56	-61
-4	-13	-18	-23	-25	-28	-30	-33	-35	-38	-40	-43	-48	-53
-3	-5	-10	-15	-17	-20	-22	-25	-27	-30	-32	-35	-40	-45
-2	3	-1	-6	-9	-11	-14	-16	-19	-21	-24	-26	-31	-36
-1	11	6	1	0	-3	-5	-8	-10	-13	-15	-18	-23	-28
0	20	15	10	7	5	2	0	-2	-5	-7	-10	-15	-20
1	28	23	18	15	13	10	8	5	3	0	-1	-6	-11
2	36	31	26	24	21	19	16	14	11	9	6	1	-3
3	45	40	35	32	30	27	25	22	20	17	15	10	5
4	53	48	43	40	38	35	33	30	28	25	23	18	13
5	61	56	51	49	46	44	41	39	36	34	31	26	21
6	70	65	60	57	55	52	50	47	45	42	40	35	30
7	78	73	68	65	63	60	58	55	53	50	48	43	38
8	86	81	76	74	71	69	66	64	61	59	56	51	46
9	95	90	85	82	80	77	75	72	70	67	65	60	55
10	103	98	93	90	88	85	83	80	78	75	73	68	63
12	120	115	110	107	105	102	100	97	95	92	90	85	80
15	145	140	135	132	130	127	125	122	120	117	115	110	105
20	186	181	176	174	171	169	166	164	161	159	156	151	146
25	228	223	218	215	213	210	208	205	203	200	198	193	188
30	270	265	260	257	255	252	250	247	245	242	240	235	230
35	311	306	301	299	296	294	291	289	286	284	281	276	271
40	353	348	343	340	338	335	333	330	328	325	323	318	313

Table 3-48

Hub Diameter, inch: 3

Leg Distance, inch: 30

SHIM REQUIRED FOR ALIGNMENT

Face - OD Method

FACE TIR MILS	OD TIR, MILS -40	-30	-20	-15	-10	-5	0	5	10	15	20	30	40
-40	-380	-385	-390	-392	-395	-397	-400	-402	-405	-407	-410	-415	-420
-35	-330	-335	-340	-342	-345	-347	-350	-352	-355	-357	-360	-365	-370
-30	-280	-285	-290	-292	-295	-297	-300	-302	-305	-307	-310	-315	-320
-25	-230	-235	-240	-242	-245	-247	-250	-252	-255	-257	-260	-265	-270
-20	-180	-185	-190	-192	-195	-197	-200	-202	-205	-207	-210	-215	-220
-15	-130	-135	-140	-142	-145	-147	-150	-152	-155	-157	-160	-165	-170
-12	-100	-105	-110	-112	-115	-117	-120	-122	-125	-127	-130	-135	-140
-10	-80	-85	-90	-92	-95	-97	-100	-102	-105	-107	-110	-115	-120
-9	-70	-75	-80	-82	-85	-87	-90	-92	-95	-97	-100	-105	-110
-8	-60	-65	-70	-72	-75	-77	-80	-82	-85	-87	-90	-95	-100
-7	-50	-55	-60	-62	-65	-67	-70	-72	-75	-77	-80	-85	-90
-6	-40	-45	-50	-52	-55	-57	-60	-62	-65	-67	-70	-75	-80
-5	-30	-35	-40	-42	-45	-47	-50	-52	-55	-57	-60	-65	-70
-4	-20	-25	-30	-32	-35	-37	-40	-42	-45	-47	-50	-55	-60
-3	-10	-15	-20	-22	-25	-27	-30	-32	-35	-37	-40	-45	-50
-2	0	-5	-10	-12	-15	-17	-20	-22	-25	-27	-30	-35	-40
-1	10	5	0	-2	-5	-7	-10	-12	-15	-17	-20	-25	-30
0	20	15	10	7	5	2	0	-2	-5	-7	-10	-15	-20
1	30	25	20	17	15	12	10	7	5	2	0	-5	-10
2	40	35	30	27	25	22	20	17	15	12	10	5	0
3	50	45	40	37	35	32	30	27	25	22	20	15	10
4	60	55	50	47	45	42	40	37	35	32	30	25	20
5	70	65	60	57	55	52	50	47	45	42	40	35	30
6	80	75	70	67	65	62	60	57	55	52	50	45	40
7	90	85	80	77	75	72	70	67	65	62	60	55	50
8	100	95	90	87	85	82	80	77	75	72	70	65	60
9	110	105	100	97	95	92	90	87	85	82	80	75	70
10	120	115	110	107	105	102	100	97	95	92	90	85	80
12	140	135	130	127	125	122	120	117	115	112	110	105	100
15	170	165	160	157	155	152	150	147	145	142	140	135	130
20	220	215	210	207	205	202	200	197	195	192	190	185	180
25	270	265	260	257	255	252	250	247	245	242	240	235	230
30	320	315	310	307	305	302	300	297	295	292	290	285	280
35	370	365	360	357	355	352	350	347	345	342	340	335	330
40	420	415	410	407	405	402	400	397	395	392	390	385	380

Table 3-49

Hub Diameter, inch: 3

Leg Distance, inch: 35

SHIM REQUIRED FOR ALIGNMENT

Face - OD Method

FACE TIR MILS	OD TIR, MILS -40	-30	-20	-15	-10	-5	0	5	10	15	20	30	40
-40	-446	-451	-456	-459	-461	-464	-466	-469	-471	-474	-476	-481	-486
-35	-388	-393	-398	-400	-403	-405	-408	-410	-413	-415	-418	-423	-428
-30	-330	-335	-340	-342	-345	-347	-350	-352	-355	-357	-360	-365	-370
-25	-271	-276	-281	-284	-286	-289	-291	-294	-296	-299	-301	-306	-311
-20	-213	-218	-223	-225	-228	-230	-233	-235	-238	-240	-243	-248	-253
-15	-155	-160	-165	-167	-170	-172	-175	-177	-180	-182	-185	-190	-195
-12	-120	-125	-130	-132	-135	-137	-140	-142	-145	-147	-150	-155	-160
-10	-96	-101	-106	-109	-111	-114	-116	-119	-121	-124	-126	-131	-136
-9	-85	-90	-95	-97	-100	-102	-105	-107	-110	-112	-115	-120	-125
-8	-73	-78	-83	-85	-88	-90	-93	-95	-98	-100	-103	-108	-113
-7	-61	-66	-71	-74	-76	-79	-81	-84	-86	-89	-91	-96	-101
-6	-50	-55	-60	-62	-65	-67	-70	-72	-75	-77	-80	-85	-90
-5	-38	-43	-48	-50	-53	-55	-58	-60	-63	-65	-68	-73	-78
-4	-26	-31	-36	-39	-41	-44	-46	-49	-51	-54	-56	-61	-66
-3	-15	-20	-25	-27	-30	-32	-35	-37	-40	-42	-45	-50	-55
-2	-3	-8	-13	-15	-18	-20	-23	-25	-28	-30	-33	-38	-43
-1	8	3	-1	-4	-6	-9	-11	-14	-16	-19	-21	-26	-31
0	20	15	10	7	5	2	0	-2	-5	-7	-10	-15	-20
1	31	26	21	19	16	14	11	9	6	4	1	-3	-8
2	43	38	33	30	28	25	23	20	18	15	13	8	3
3	55	50	45	42	40	37	35	32	30	27	25	20	15
4	66	61	56	54	51	49	46	44	41	39	36	31	26
5	78	73	68	65	63	60	58	55	53	50	48	43	38
6	90	85	80	77	75	72	70	67	65	62	60	55	50
7	101	96	91	89	86	84	81	79	76	74	71	66	61
8	113	108	103	100	98	95	93	90	88	85	83	78	73
9	125	120	115	112	110	107	105	102	100	97	95	90	85
10	136	131	126	124	121	119	116	114	111	109	106	101	96
12	160	155	150	147	145	142	140	137	135	132	130	125	120
15	195	190	185	182	180	177	175	172	170	167	165	160	155
20	253	248	243	240	238	235	233	230	228	225	223	218	213
25	311	306	301	299	296	294	291	289	286	284	281	276	271
30	370	365	360	357	355	352	350	347	345	342	340	335	330
35	428	423	418	415	413	410	408	405	403	400	398	393	388
40	486	481	476	474	471	469	466	464	461	459	456	451	446

Table 3-50

Hub Diameter, inch: 3

Leg Distance, inch: 40

SHIM REQUIRED FOR ALIGNMENT

Face - OD Method

FACE TIR MILS	OD TIR, MILS -40	-30	-20	-15	-10	-5	0	5	10	15	20	30	40
-40	-513	-518	-523	-525	-528	-530	-533	-535	-538	-540	-543	-548	-553
-35	-446	-451	-456	-459	-461	-464	-466	-469	-471	-474	-476	-481	-486
-30	-380	-385	-390	-392	-395	-397	-400	-402	-405	-407	-410	-415	-420
-25	-313	-318	-323	-325	-328	-330	-333	-335	-338	-340	-343	-348	-353
-20	-246	-251	-256	-259	-261	-264	-266	-269	-271	-274	-276	-281	-286
-15	-180	-185	-190	-192	-195	-197	-200	-202	-205	-207	-210	-215	-220
-12	-140	-145	-150	-152	-155	-157	-160	-162	-165	-167	-170	-175	-180
-10	-113	-118	-123	-125	-128	-130	-133	-135	-138	-140	-143	-148	-153
-9	-100	-105	-110	-112	-115	-117	-120	-122	-125	-127	-130	-135	-140
-8	-86	-91	-96	-99	-101	-104	-106	-109	-111	-114	-116	-121	-126
-7	-73	-78	-83	-85	-88	-90	-93	-95	-98	-100	-103	-108	-113
-6	-60	-65	-70	-72	-75	-77	-80	-82	-85	-87	-90	-95	-100
-5	-46	-51	-56	-59	-61	-64	-66	-69	-71	-74	-76	-81	-86
-4	-33	-38	-43	-45	-48	-50	-53	-55	-58	-60	-63	-68	-73
-3	-20	-25	-30	-32	-35	-37	-40	-42	-45	-47	-50	-55	-60
-2	-6	-11	-16	-19	-21	-24	-26	-29	-31	-34	-36	-41	-46
-1	6	1	-3	-5	-8	-10	-13	-15	-18	-20	-23	-28	-33
0	20	15	10	7	5	2	0	-2	-5	-7	-10	-15	-20
1	33	28	23	20	18	15	13	10	8	5	3	-1	-6
2	46	41	36	34	31	29	26	24	21	19	16	11	6
3	60	55	50	47	45	42	40	37	35	32	30	25	20
4	73	68	63	60	58	55	53	50	48	45	43	38	33
5	86	81	76	74	71	69	66	64	61	59	56	51	46
6	100	95	90	87	85	82	80	77	75	72	70	65	60
7	113	108	103	100	98	95	93	90	88	85	83	78	73
8	126	121	116	114	111	109	106	104	101	99	96	91	86
9	140	135	130	127	125	122	120	117	115	112	110	105	100
10	153	148	143	140	138	135	133	130	128	125	123	118	113
12	180	175	170	167	165	162	160	157	155	152	150	145	140
15	220	215	210	207	205	202	200	197	195	192	190	185	180
20	286	281	276	274	271	269	266	264	261	259	256	251	246
25	353	348	343	340	338	335	333	330	328	325	323	318	313
30	420	415	410	407	405	402	400	397	395	392	390	385	380
35	486	481	476	474	471	469	466	464	461	459	456	451	446
40	553	548	543	540	538	535	533	530	528	525	523	518	513

Table 3-51

Hub Diameter, inch: 3

Leg Distance, inch: 50

SHIM REQUIRED FOR ALIGNMENT

Face - OD Method

FACE TIR MILS	OD TIR, MILS -40	-30	-20	-15	-10	-5	0	5	10	15	20	30	40
-40	-646	-651	-656	-659	-661	-664	-666	-669	-671	-674	-676	-681	-686
-35	-563	-568	-573	-575	-578	-580	-583	-585	-588	-590	-593	-598	-603
-30	-480	-485	-490	-492	-495	-497	-500	-502	-505	-507	-510	-515	-520
-25	-396	-401	-406	-409	-411	-414	-416	-419	-421	-424	-426	-431	-436
-20	-313	-318	-323	-325	-328	-330	-333	-335	-338	-340	-343	-348	-353
-15	-230	-235	-240	-242	-245	-247	-250	-252	-255	-257	-260	-265	-270
-12	-180	-185	-190	-192	-195	-197	-200	-202	-205	-207	-210	-215	-220
-10	-146	-151	-156	-159	-161	-164	-166	-169	-171	-174	-176	-181	-186
-9	-130	-135	-140	-142	-145	-147	-150	-152	-155	-157	-160	-165	-170
-8	-113	-118	-123	-125	-128	-130	-133	-135	-138	-140	-143	-148	-153
-7	-96	-101	-106	-109	-111	-114	-116	-119	-121	-124	-126	-131	-136
-6	-80	-85	-90	-92	-95	-97	-100	-102	-105	-107	-110	-115	-120
-5	-63	-68	-73	-75	-78	-80	-83	-85	-88	-90	-93	-98	-103
-4	-46	-51	-56	-59	-61	-64	-66	-69	-71	-74	-76	-81	-86
-3	-30	-35	-40	-42	-45	-47	-50	-52	-55	-57	-60	-65	-70
-2	-13	-18	-23	-25	-28	-30	-33	-35	-38	-40	-43	-48	-53
-1	3	-1	-6	-9	-11	-14	-16	-19	-21	-24	-26	-31	-36
0	20	15	10	7	5	2	0	-2	-5	-7	-10	-15	-20
1	36	31	26	24	21	19	16	14	11	9	6	1	-3
2	53	48	43	40	38	35	33	30	28	25	23	18	13
3	70	65	60	57	55	52	50	47	45	42	40	35	30
4	86	81	76	74	71	69	66	64	61	59	56	51	46
5	103	98	93	90	88	85	83	80	78	75	73	68	63
6	120	115	110	107	105	102	100	97	95	92	90	85	80
7	136	131	126	124	121	119	116	114	111	109	106	101	96
8	153	148	143	140	138	135	133	130	128	125	123	118	113
9	170	165	160	157	155	152	150	147	145	142	140	135	130
10	186	181	176	174	171	169	166	164	161	159	156	151	146
12	220	215	210	207	205	202	200	197	195	192	190	185	180
15	270	265	260	257	255	252	250	247	245	242	240	235	230
20	353	348	343	340	338	335	333	330	328	325	323	318	313
25	436	431	426	424	421	419	416	414	411	409	406	401	396
30	520	515	510	507	505	502	500	497	495	492	490	485	480
35	603	598	593	590	588	585	583	580	578	575	573	568	563
40	686	681	676	674	671	669	666	664	661	659	656	651	646

Table 3-52

Hub Diameter, inch: 3

Leg Distance, inch: 60

SHIM REQUIRED FOR ALIGNMENT

Face - OD Method

FACE TIR MILS	OD TIR, MILS -40	-30	-20	-15	-10	-5	0	5	10	15	20	30	40
-40	-780	-785	-790	-792	-795	-797	-800	-802	-805	-807	-810	-815	-820
-35	-680	-685	-690	-692	-695	-697	-700	-702	-705	-707	-710	-715	-720
-30	-580	-585	-590	-592	-595	-597	-600	-602	-605	-607	-610	-615	-620
-25	-480	-485	-490	-492	-495	-497	-500	-502	-505	-507	-510	-515	-520
-20	-380	-385	-390	-392	-395	-397	-400	-402	-405	-407	-410	-415	-420
-15	-280	-285	-290	-292	-295	-297	-300	-302	-305	-307	-310	-315	-320
-12	-220	-225	-230	-232	-235	-237	-240	-242	-245	-247	-250	-255	-260
-10	-180	-185	-190	-192	-195	-197	-200	-202	-205	-207	-210	-215	-220
-9	-160	-165	-170	-172	-175	-177	-180	-182	-185	-187	-190	-195	-200
-8	-140	-145	-150	-152	-155	-157	-160	-162	-165	-167	-170	-175	-180
-7	-120	-125	-130	-132	-135	-137	-140	-142	-145	-147	-150	-155	-160
-6	-100	-105	-110	-112	-115	-117	-120	-122	-125	-127	-130	-135	-140
-5	-80	-85	-90	-92	-95	-97	-100	-102	-105	-107	-110	-115	-120
-4	-60	-65	-70	-72	-75	-77	-80	-82	-85	-87	-90	-95	-100
-3	-40	-45	-50	-52	-55	-57	-60	-62	-65	-67	-70	-75	-80
-2	-20	-25	-30	-32	-35	-37	-40	-42	-45	-47	-50	-55	-60
-1	0	-5	-10	-12	-15	-17	-20	-22	-25	-27	-30	-35	-40
0	20	15	10	7	5	2	0	-2	-5	-7	-10	-15	-20
1	40	35	30	27	25	22	20	17	15	12	10	5	0
2	60	55	50	47	45	42	40	37	35	32	30	25	20
3	80	75	70	67	65	62	60	57	55	52	50	45	40
4	100	95	90	87	85	82	80	77	75	72	70	65	60
5	120	115	110	107	105	102	100	97	95	92	90	85	80
6	140	135	130	127	125	122	120	117	115	112	110	105	100
7	160	155	150	147	145	142	140	137	135	132	130	125	120
8	180	175	170	167	165	162	160	157	155	152	150	145	140
9	200	195	190	187	185	182	180	177	175	172	170	165	160
10	220	215	210	207	205	202	200	197	195	192	190	185	180
12	260	255	250	247	245	242	240	237	235	232	230	225	220
15	320	315	310	307	305	302	300	297	295	292	290	285	280
20	420	415	410	407	405	402	400	397	395	392	390	385	380
25	520	515	510	507	505	502	500	497	495	492	490	485	480
30	620	615	610	607	605	602	600	597	595	592	590	585	580
35	720	715	710	707	705	702	700	697	695	692	690	685	680
40	820	815	810	807	805	802	800	797	795	792	790	785	780

Table 3-53

Hub Diameter, inch: 3
Leg Distance, inch: 70

SHIM REQUIRED FOR ALIGNMENT

Face - OD Method

FACE TIR MILS	OD TIR, MILS -40	-30	-20	-15	-10	-5	0	5	10	15	20	30	40
-40	-913	-918	-923	-925	-928	-930	-933	-935	-938	-940	-943	-948	-953
-35	-796	-801	-806	-809	-811	-814	-816	-819	-821	-824	-826	-831	-836
-30	-680	-685	-690	-692	-695	-697	-700	-702	-705	-707	-710	-715	-720
-25	-563	-568	-573	-575	-578	-580	-583	-585	-588	-590	-593	-598	-603
-20	-446	-451	-456	-459	-461	-464	-466	-469	-471	-474	-476	-481	-486
-15	-330	-335	-340	-342	-345	-347	-350	-352	-355	-357	-360	-365	-370
-12	-260	-265	-270	-272	-275	-277	-280	-282	-285	-287	-290	-295	-300
-10	-213	-218	-223	-225	-228	-230	-233	-235	-238	-240	-243	-248	-253
-9	-190	-195	-200	-202	-205	-207	-210	-212	-215	-217	-220	-225	-230
-8	-166	-171	-176	-179	-181	-184	-186	-189	-191	-194	-196	-201	-206
-7	-143	-148	-153	-155	-158	-160	-163	-165	-168	-170	-173	-178	-183
-6	-120	-125	-130	-132	-135	-137	-140	-142	-145	-147	-150	-155	-160
-5	-96	-101	-106	-109	-111	-114	-116	-119	-121	-124	-126	-131	-136
-4	-73	-78	-83	-85	-88	-90	-93	-95	-98	-100	-103	-108	-113
-3	-50	-55	-60	-62	-65	-67	-70	-72	-75	-77	-80	-85	-90
-2	-26	-31	-36	-39	-41	-44	-46	-49	-51	-54	-56	-61	-66
-1	-3	-8	-13	-15	-18	-20	-23	-25	-28	-30	-33	-38	-43
0	20	15	10	7	5	2	0	-2	-5	-7	-10	-15	-20
1	43	38	33	30	28	25	23	20	18	15	13	8	3
2	66	61	56	54	51	49	46	44	41	39	36	31	26
3	90	85	80	77	75	72	70	67	65	62	60	55	50
4	113	108	103	100	98	95	93	90	88	85	83	78	73
5	136	131	126	124	121	119	116	114	111	109	106	101	96
6	160	155	150	147	145	142	140	137	135	132	130	125	120
7	183	178	173	170	168	165	163	160	158	155	153	148	143
8	206	201	196	194	191	189	186	184	181	179	176	171	166
9	230	225	220	217	215	212	210	207	205	202	200	195	190
10	253	248	243	240	238	235	233	230	228	225	223	218	213
12	300	295	290	287	285	282	280	277	275	272	270	265	260
15	370	365	360	357	355	352	350	347	345	342	340	335	330
20	486	481	476	474	471	469	466	464	461	459	456	451	446
25	603	598	593	590	588	585	583	580	578	575	573	568	563
30	720	715	710	707	705	702	700	697	695	692	690	685	680
35	836	831	826	824	821	819	816	814	811	809	806	801	796
40	953	948	943	940	938	935	933	930	928	925	923	918	913

Table 3-54

Hub Diameter, inch: 3

Leg Distance, inch: 80

SHIM REQUIRED FOR ALIGNMENT

Face – OD Method

FACE TIR MILS	OD TIR, MILS -40	-30	-20	-15	-10	-5	0	5	10	15	20	30	40
-40	-1046	-1051	-1056	-1059	-1061	-1064	-1066	-1069	-1071	-1074	-1076	-1081	-1086
-35	-913	-918	-923	-925	-928	-930	-933	-935	-938	-940	-943	-948	-953
-30	-780	-785	-790	-792	-795	-797	-800	-802	-805	-807	-810	-815	-820
-25	-646	-651	-656	-659	-661	-664	-666	-669	-671	-674	-676	-681	-686
-20	-513	-518	-523	-525	-528	-530	-533	-535	-538	-540	-543	-548	-553
-15	-380	-385	-390	-392	-395	-397	-400	-402	-405	-407	-410	-415	-420
-12	-300	-305	-310	-312	-315	-317	-320	-322	-325	-327	-330	-335	-340
-10	-246	-251	-256	-259	-261	-264	-266	-269	-271	-274	-276	-281	-286
-9	-220	-225	-230	-232	-235	-237	-240	-242	-245	-247	-250	-255	-260
-8	-193	-198	-203	-205	-208	-210	-213	-215	-218	-220	-223	-228	-233
-7	-166	-171	-176	-179	-181	-184	-186	-189	-191	-194	-196	-201	-206
-6	-140	-145	-150	-152	-155	-157	-160	-162	-165	-167	-170	-175	-180
-5	-113	-118	-123	-125	-128	-130	-133	-135	-138	-140	-143	-148	-153
-4	-86	-91	-96	-99	-101	-104	-106	-109	-111	-114	-116	-121	-126
-3	-60	-65	-70	-72	-75	-77	-80	-82	-85	-87	-90	-95	-100
-2	-33	-38	-43	-45	-48	-50	-53	-55	-58	-60	-63	-68	-73
-1	-6	-11	-16	-19	-21	-24	-26	-29	-31	-34	-36	-41	-46
0	20	15	10	7	5	2	0	-2	-5	-7	-10	-15	-20
1	46	41	36	34	31	29	26	24	21	19	16	11	6
2	73	68	63	60	58	55	53	50	48	45	43	38	33
3	100	95	90	87	85	82	80	77	75	72	70	65	60
4	126	121	116	114	111	109	106	104	101	99	96	91	86
5	153	148	143	140	138	135	133	130	128	125	123	118	113
6	180	175	170	167	165	162	160	157	155	152	150	145	140
7	206	201	196	194	191	189	186	184	181	179	176	171	166
8	233	228	223	220	218	215	213	210	208	205	203	198	193
9	260	255	250	247	245	242	240	237	235	232	230	225	220
10	286	281	276	274	271	269	266	264	261	259	256	251	246
12	340	335	330	327	325	322	320	317	315	312	310	305	300
15	420	415	410	407	405	402	400	397	395	392	390	385	380
20	553	548	543	540	538	535	533	530	528	525	523	518	513
25	686	681	676	674	671	669	666	664	661	659	656	651	646
30	820	815	810	807	805	802	800	797	795	792	790	785	780
35	953	948	943	940	938	935	933	930	928	925	923	918	913
40	1086	1081	1076	1074	1071	1069	1066	1064	1061	1059	1056	1051	1046

Table 3-55

Hub Diameter, inch: 3.5

Leg Distance, inch: 15

SHIM REQUIRED FOR ALIGNMENT

Face - OD Method

FACE TIR MILS	OD TIR, MILS -40	-30	-20	-15	-10	-5	0	5	10	15	20	30	40
-40	-151	-156	-161	-163	-166	-168	-171	-173	-176	-178	-181	-186	-191
-35	-130	-135	-140	-142	-145	-147	-150	-152	-155	-157	-160	-165	-170
-30	-108	-113	-118	-121	-123	-126	-128	-131	-133	-136	-138	-143	-148
-25	-87	-92	-97	-99	-102	-104	-107	-109	-112	-114	-117	-122	-127
-20	-65	-70	-75	-78	-80	-83	-85	-88	-90	-93	-95	-100	-105
-15	-44	-49	-54	-56	-59	-61	-64	-66	-69	-71	-74	-79	-84
-12	-31	-36	-41	-43	-46	-48	-51	-53	-56	-58	-61	-66	-71
-10	-22	-27	-32	-35	-37	-40	-42	-45	-47	-50	-52	-57	-62
-9	-18	-23	-28	-31	-33	-36	-38	-41	-43	-46	-48	-53	-58
-8	-14	-19	-24	-26	-29	-31	-34	-36	-39	-41	-44	-49	-54
-7	-10	-15	-20	-22	-25	-27	-30	-32	-35	-37	-40	-45	-50
-6	-5	-10	-15	-18	-20	-23	-25	-28	-30	-33	-35	-40	-45
-5	-1	-6	-11	-13	-16	-18	-21	-23	-26	-28	-31	-36	-41
-4	2	-2	-7	-9	-12	-14	-17	-19	-22	-24	-27	-32	-37
-3	7	2	-2	-5	-7	-10	-12	-15	-17	-20	-22	-27	-32
-2	11	6	1	-1	-3	-6	-8	-11	-13	-16	-18	-23	-28
-1	15	10	5	3	0	-1	-4	-6	-9	-11	-14	-19	-24
0	20	15	10	7	5	2	0	-2	-5	-7	-10	-15	-20
1	24	19	14	11	9	6	4	1	0	-3	-5	-10	-15
2	28	23	18	16	13	11	8	6	3	1	-1	-6	-11
3	32	27	22	20	17	15	12	10	7	5	2	-2	-7
4	37	32	27	24	22	19	17	14	12	9	7	2	-2
5	41	36	31	28	26	23	21	18	16	13	11	6	1
6	45	40	35	33	30	28	25	23	20	18	15	10	5
7	50	45	40	37	35	32	30	27	25	22	20	15	10
8	54	49	44	41	39	36	34	31	29	26	24	19	14
9	58	53	48	46	43	41	38	36	33	31	28	23	18
10	62	57	52	50	47	45	42	40	37	35	32	27	22
12	71	66	61	58	56	53	51	48	46	43	41	36	31
15	84	79	74	71	69	66	64	61	59	56	54	49	44
20	105	100	95	93	90	88	85	83	80	78	75	70	65
25	127	122	117	114	112	109	107	104	102	99	97	92	87
30	148	143	138	136	133	131	128	126	123	121	118	113	108
35	170	165	160	157	155	152	150	147	145	142	140	135	130
40	191	186	181	178	176	173	171	168	166	163	161	156	151

Table 3-56

Hub Diameter, inch: 3.5

Leg Distance, inch: 20

SHIM REQUIRED FOR ALIGNMENT

Face - OD Method

FACE TIR MILS	OD TIR, MILS -40	-30	-20	-15	-10	-5	0	5	10	15	20	30	40
-40	-208	-213	-218	-221	-223	-226	-228	-231	-233	-236	-238	-243	-248
-35	-180	-185	-190	-192	-195	-197	-200	-202	-205	-207	-210	-215	-220
-30	-151	-156	-161	-163	-166	-168	-171	-173	-176	-178	-181	-186	-191
-25	-122	-127	-132	-135	-137	-140	-142	-145	-147	-150	-152	-157	-162
-20	-94	-99	-104	-106	-109	-111	-114	-116	-119	-121	-124	-129	-134
-15	-65	-70	-75	-78	-80	-83	-85	-88	-90	-93	-95	-100	-105
-12	-48	-53	-58	-61	-63	-66	-68	-71	-73	-76	-78	-83	-88
-10	-37	-42	-47	-49	-52	-54	-57	-59	-62	-64	-67	-72	-77
-9	-31	-36	-41	-43	-46	-48	-51	-53	-56	-58	-61	-66	-71
-8	-25	-30	-35	-38	-40	-43	-45	-48	-50	-53	-55	-60	-65
-7	-20	-25	-30	-32	-35	-37	-40	-42	-45	-47	-50	-55	-60
-6	-14	-19	-24	-26	-29	-31	-34	-36	-39	-41	-44	-49	-54
-5	-8	-13	-18	-21	-23	-26	-28	-31	-33	-36	-38	-43	-48
-4	-2	-7	-12	-15	-17	-20	-22	-25	-27	-30	-32	-37	-42
-3	2	-2	-7	-9	-12	-14	-17	-19	-22	-24	-27	-32	-37
-2	8	3	-1	-3	-6	-8	-11	-13	-16	-18	-21	-26	-31
-1	14	9	4	1	0	-3	-5	-8	-10	-13	-15	-20	-25
0	20	15	10	7	5	2	0	-2	-5	-7	-10	-15	-20
1	25	20	15	13	10	8	5	3	0	-1	-4	-9	-14
2	31	26	21	18	16	13	11	8	6	3	1	-3	-8
3	37	32	27	24	22	19	17	14	12	9	7	2	-2
4	42	37	32	30	27	25	22	20	17	15	12	7	2
5	48	43	38	36	33	31	28	26	23	21	18	13	8
6	54	49	44	41	39	36	34	31	29	26	24	19	14
7	60	55	50	47	45	42	40	37	35	32	30	25	20
8	65	60	55	53	50	48	45	43	40	38	35	30	25
9	71	66	61	58	56	53	51	48	46	43	41	36	31
10	77	72	67	64	62	59	57	54	52	49	47	42	37
12	88	83	78	76	73	71	68	66	63	61	58	53	48
15	105	100	95	93	90	88	85	83	80	78	75	70	65
20	134	129	124	121	119	116	114	111	109	106	104	99	94
25	162	157	152	150	147	145	142	140	137	135	132	127	122
30	191	186	181	178	176	173	171	168	166	163	161	156	151
35	220	215	210	207	205	202	200	197	195	192	190	185	180
40	248	243	238	236	233	231	228	226	223	221	218	213	208

Table 3-57

Hub Diameter, inch: 3.5
Leg Distance, inch: 25

SHIM REQUIRED FOR ALIGNMENT

Face – OD Method

FACE TIR MILS	OD TIR, MILS -40	-30	-20	-15	-10	-5	0	5	10	15	20	30	40
-40	-265	-270	-275	-278	-280	-283	-285	-288	-290	-293	-295	-300	-305
-35	-230	-235	-240	-242	-245	-247	-250	-252	-255	-257	-260	-265	-270
-30	-194	-199	-204	-206	-209	-211	-214	-216	-219	-221	-224	-229	-234
-25	-158	-163	-168	-171	-173	-176	-178	-181	-183	-186	-188	-193	-198
-20	-122	-127	-132	-135	-137	-140	-142	-145	-147	-150	-152	-157	-162
-15	-87	-92	-97	-99	-102	-104	-107	-109	-112	-114	-117	-122	-127
-12	-65	-70	-75	-78	-80	-83	-85	-88	-90	-93	-95	-100	-105
-10	-51	-56	-61	-63	-66	-68	-71	-73	-76	-78	-81	-86	-91
-9	-44	-49	-54	-56	-59	-61	-64	-66	-69	-71	-74	-79	-84
-8	-37	-42	-47	-49	-52	-54	-57	-59	-62	-64	-67	-72	-77
-7	-30	-35	-40	-42	-45	-47	-50	-52	-55	-57	-60	-65	-70
-6	-22	-27	-32	-35	-37	-40	-42	-45	-47	-50	-52	-57	-62
-5	-15	-20	-25	-28	-30	-33	-35	-38	-40	-43	-45	-50	-55
-4	-8	-13	-18	-21	-23	-26	-28	-31	-33	-36	-38	-43	-48
-3	-1	-6	-11	-13	-16	-18	-21	-23	-26	-28	-31	-36	-41
-2	5	0	-4	-6	-9	-11	-14	-16	-19	-21	-24	-29	-34
-1	12	7	2	0	-2	-4	-7	-9	-12	-14	-17	-22	-27
0	20	15	10	7	5	2	0	-2	-5	-7	-10	-15	-20
1	27	22	17	14	12	9	7	4	2	0	-2	-7	-12
2	34	29	24	21	19	16	14	11	9	6	4	0	-5
3	41	36	31	28	26	23	21	18	16	13	11	6	1
4	48	43	38	36	33	31	28	26	23	21	18	13	8
5	55	50	45	43	40	38	35	33	30	28	25	20	15
6	62	57	52	50	47	45	42	40	37	35	32	27	22
7	70	65	60	57	55	52	50	47	45	42	40	35	30
8	77	72	67	64	62	59	57	54	52	49	47	42	37
9	84	79	74	71	69	66	64	61	59	56	54	49	44
10	91	86	81	78	76	73	71	68	66	63	61	56	51
12	105	100	95	93	90	88	85	83	80	78	75	70	65
15	127	122	117	114	112	109	107	104	102	99	97	92	87
20	162	157	152	150	147	145	142	140	137	135	132	127	122
25	198	193	188	186	183	181	178	176	173	171	168	163	158
30	234	229	224	221	219	216	214	211	209	206	204	199	194
35	270	265	260	257	255	252	250	247	245	242	240	235	230
40	305	300	295	293	290	288	285	283	280	278	275	270	265

Table 3-58

Hub Diameter, inch: 3.5

Leg Distance, inch: 30

SHIM REQUIRED FOR ALIGNMENT

Face - OD Method

FACE TIR MILS	OD TIR, MILS -40	-30	-20	-15	-10	-5	0	5	10	15	20	30	40
-40	-322	-327	-332	-335	-337	-340	-342	-345	-347	-350	-352	-357	-362
-35	-280	-285	-290	-292	-295	-297	-300	-302	-305	-307	-310	-315	-320
-30	-237	-242	-247	-249	-252	-254	-257	-259	-262	-264	-267	-272	-277
-25	-194	-199	-204	-206	-209	-211	-214	-216	-219	-221	-224	-229	-234
-20	-151	-156	-161	-163	-166	-168	-171	-173	-176	-178	-181	-186	-191
-15	-108	-113	-118	-121	-123	-126	-128	-131	-133	-136	-138	-143	-148
-12	-82	-87	-92	-95	-97	-100	-102	-105	-107	-110	-112	-117	-122
-10	-65	-70	-75	-78	-80	-83	-85	-88	-90	-93	-95	-100	-105
-9	-57	-62	-67	-69	-72	-74	-77	-79	-82	-84	-87	-92	-97
-8	-48	-53	-58	-61	-63	-66	-68	-71	-73	-76	-78	-83	-88
-7	-40	-45	-50	-52	-55	-57	-60	-62	-65	-67	-70	-75	-80
-6	-31	-36	-41	-43	-46	-48	-51	-53	-56	-58	-61	-66	-71
-5	-22	-27	-32	-35	-37	-40	-42	-45	-47	-50	-52	-57	-62
-4	-14	-19	-24	-26	-29	-31	-34	-36	-39	-41	-44	-49	-54
-3	-5	-10	-15	-18	-20	-23	-25	-28	-30	-33	-35	-40	-45
-2	2	-2	-7	-9	-12	-14	-17	-19	-22	-24	-27	-32	-37
-1	11	6	1	-1	-3	-6	-8	-11	-13	-16	-18	-23	-28
0	20	15	10	7	5	2	0	-2	-5	-7	-10	-15	-20
1	28	23	18	16	13	11	8	6	3	1	-1	-6	-11
2	37	32	27	24	22	19	17	14	12	9	7	2	-2
3	45	40	35	33	30	28	25	23	20	18	15	10	5
4	54	49	44	41	39	36	34	31	29	26	24	19	14
5	62	57	52	50	47	45	42	40	37	35	32	27	22
6	71	66	61	58	56	53	51	48	46	43	41	36	31
7	80	75	70	67	65	62	60	57	55	52	50	45	40
8	88	83	78	76	73	71	68	66	63	61	58	53	48
9	97	92	87	84	82	79	77	74	72	69	67	62	57
10	105	100	95	93	90	88	85	83	80	78	75	70	65
12	122	117	112	110	107	105	102	100	97	95	92	87	82
15	148	143	138	136	133	131	128	126	123	121	118	113	108
20	191	186	181	178	176	173	171	168	166	163	161	156	151
25	234	229	224	221	219	216	214	211	209	206	204	199	194
30	277	272	267	264	262	259	257	254	252	249	247	242	237
35	320	315	310	307	305	302	300	297	295	292	290	285	280
40	362	357	352	350	347	345	342	340	337	335	332	327	322

Table 3-59

Hub Diameter, inch: 3.5

Leg Distance, inch: 35

SHIM REQUIRED FOR ALIGNMENT

Face - OD Method

FACE TIR MILS	OD TIR, MILS -40	-30	-20	-15	-10	-5	0	5	10	15	20	30	40
-40	-380	-385	-390	-392	-395	-397	-400	-402	-405	-407	-410	-415	-420
-35	-330	-335	-340	-342	-345	-347	-350	-352	-355	-357	-360	-365	-370
-30	-280	-285	-290	-292	-295	-297	-300	-302	-305	-307	-310	-315	-320
-25	-230	-235	-240	-242	-245	-247	-250	-252	-255	-257	-260	-265	-270
-20	-180	-185	-190	-192	-195	-197	-200	-202	-205	-207	-210	-215	-220
-15	-130	-135	-140	-142	-145	-147	-150	-152	-155	-157	-160	-165	-170
-12	-100	-105	-110	-112	-115	-117	-120	-122	-125	-127	-130	-135	-140
-10	-80	-85	-90	-92	-95	-97	-100	-102	-105	-107	-110	-115	-120
-9	-70	-75	-80	-82	-85	-87	-90	-92	-95	-97	-100	-105	-110
-8	-60	-65	-70	-72	-75	-77	-80	-82	-85	-87	-90	-95	-100
-7	-50	-55	-60	-62	-65	-67	-70	-72	-75	-77	-80	-85	-90
-6	-40	-45	-50	-52	-55	-57	-60	-62	-65	-67	-70	-75	-80
-5	-30	-35	-40	-42	-45	-47	-50	-52	-55	-57	-60	-65	-70
-4	-20	-25	-30	-32	-35	-37	-40	-42	-45	-47	-50	-55	-60
-3	-10	-15	-20	-22	-25	-27	-30	-32	-35	-37	-40	-45	-50
-2	0	-5	-10	-12	-15	-17	-20	-22	-25	-27	-30	-35	-40
-1	10	5	0	-2	-5	-7	-10	-12	-15	-17	-20	-25	-30
0	20	15	10	7	5	2	0	-2	-5	-7	-10	-15	-20
1	30	25	20	17	15	12	10	7	5	2	0	-5	-10
2	40	35	30	27	25	22	20	17	15	12	10	5	0
3	50	45	40	37	35	32	30	27	25	22	20	15	10
4	60	55	50	47	45	42	40	37	35	32	30	25	20
5	70	65	60	57	55	52	50	47	45	42	40	35	30
6	80	75	70	67	65	62	60	57	55	52	50	45	40
7	90	85	80	77	75	72	70	67	65	62	60	55	50
8	100	95	90	87	85	82	80	77	75	72	70	65	60
9	110	105	100	97	95	92	90	87	85	82	80	75	70
10	120	115	110	107	105	102	100	97	95	92	90	85	80
12	140	135	130	127	125	122	120	117	115	112	110	105	100
15	170	165	160	157	155	152	150	147	145	142	140	135	130
20	220	215	210	207	205	202	200	197	195	192	190	185	180
25	270	265	260	257	255	252	250	247	245	242	240	235	230
30	320	315	310	307	305	302	300	297	295	292	290	285	280
35	370	365	360	357	355	352	350	347	345	342	340	335	330
40	420	415	410	407	405	402	400	397	395	392	390	385	380

Table 3-60

Hub Diameter, inch: 3.5

Leg Distance, inch: 40

SHIM REQUIRED FOR ALIGNMENT

Face - OD Method

FACE TIR MILS	OD TIR, MILS -40	-30	-20	-15	-10	-5	0	5	10	15	20	30	40
-40	-437	-442	-447	-449	-452	-454	-457	-459	-462	-464	-467	-472	-477
-35	-380	-385	-390	-392	-395	-397	-400	-402	-405	-407	-410	-415	-420
-30	-322	-327	-332	-335	-337	-340	-342	-345	-347	-350	-352	-357	-362
-25	-265	-270	-275	-278	-280	-283	-285	-288	-290	-293	-295	-300	-305
-20	-208	-213	-218	-221	-223	-226	-228	-231	-233	-236	-238	-243	-248
-15	-151	-156	-161	-163	-166	-168	-171	-173	-176	-178	-181	-186	-191
-12	-117	-122	-127	-129	-132	-134	-137	-139	-142	-144	-147	-152	-157
-10	-94	-99	-104	-106	-109	-111	-114	-116	-119	-121	-124	-129	-134
-9	-82	-87	-92	-95	-97	-100	-102	-105	-107	-110	-112	-117	-122
-8	-71	-76	-81	-83	-86	-88	-91	-93	-96	-98	-101	-106	-111
-7	-60	-65	-70	-72	-75	-77	-80	-82	-85	-87	-90	-95	-100
-6	-48	-53	-58	-61	-63	-66	-68	-71	-73	-76	-78	-83	-88
-5	-37	-42	-47	-49	-52	-54	-57	-59	-62	-64	-67	-72	-77
-4	-25	-30	-35	-38	-40	-43	-45	-48	-50	-53	-55	-60	-65
-3	-14	-19	-24	-26	-29	-31	-34	-36	-39	-41	-44	-49	-54
-2	-2	-7	-12	-15	-17	-20	-22	-25	-27	-30	-32	-37	-42
-1	8	3	-1	-3	-6	-8	-11	-13	-16	-18	-21	-26	-31
0	20	15	10	7	5	2	0	-2	-5	-7	-10	-15	-20
1	31	26	21	18	16	13	11	8	6	3	1	-3	-8
2	42	37	32	30	27	25	22	20	17	15	12	7	2
3	54	49	44	41	39	36	34	31	29	26	24	19	14
4	65	60	55	53	50	48	45	43	40	38	35	30	25
5	77	72	67	64	62	59	57	54	52	49	47	42	37
6	88	83	78	76	73	71	68	66	63	61	58	53	48
7	100	95	90	87	85	82	80	77	75	72	70	65	60
8	111	106	101	98	96	93	91	88	86	83	81	76	71
9	122	117	112	110	107	105	102	100	97	95	92	87	82
10	134	129	124	121	119	116	114	111	109	106	104	99	94
12	157	152	147	144	142	139	137	134	132	129	127	122	117
15	191	186	181	178	176	173	171	168	166	163	161	156	151
20	248	243	238	236	233	231	228	226	223	221	218	213	208
25	305	300	295	293	290	288	285	283	280	278	275	270	265
30	362	357	352	350	347	345	342	340	337	335	332	327	322
35	420	415	410	407	405	402	400	397	395	392	390	385	380
40	477	472	467	464	462	459	457	454	452	449	447	442	437

Table 3-61

Hub Diameter, inch: 3.5

Leg Distance, inch: 50

SHIM REQUIRED FOR ALIGNMENT

Face - OD Method

FACE TIR MILS	OD TIR, MILS -40	-30	-20	-15	-10	-5	0	5	10	15	20	30	40
-40	-551	-556	-561	-563	-566	-568	-571	-573	-576	-578	-581	-586	-591
-35	-480	-485	-490	-492	-495	-497	-500	-502	-505	-507	-510	-515	-520
-30	-408	-413	-418	-421	-423	-426	-428	-431	-433	-436	-438	-443	-448
-25	-337	-342	-347	-349	-352	-354	-357	-359	-362	-364	-367	-372	-377
-20	-265	-270	-275	-278	-280	-283	-285	-288	-290	-293	-295	-300	-305
-15	-194	-199	-204	-206	-209	-211	-214	-216	-219	-221	-224	-229	-234
-12	-151	-156	-161	-163	-166	-168	-171	-173	-176	-178	-181	-186	-191
-10	-122	-127	-132	-135	-137	-140	-142	-145	-147	-150	-152	-157	-162
-9	-108	-113	-118	-121	-123	-126	-128	-131	-133	-136	-138	-143	-148
-8	-94	-99	-104	-106	-109	-111	-114	-116	-119	-121	-124	-129	-134
-7	-80	-85	-90	-92	-95	-97	-100	-102	-105	-107	-110	-115	-120
-6	-65	-70	-75	-78	-80	-83	-85	-88	-90	-93	-95	-100	-105
-5	-51	-56	-61	-63	-66	-68	-71	-73	-76	-78	-81	-86	-91
-4	-37	-42	-47	-49	-52	-54	-57	-59	-62	-64	-67	-72	-77
-3	-22	-27	-32	-35	-37	-40	-42	-45	-47	-50	-52	-57	-62
-2	-8	-13	-18	-21	-23	-26	-28	-31	-33	-36	-38	-43	-48
-1	5	0	-4	-6	-9	-11	-14	-16	-19	-21	-24	-29	-34
0	20	15	10	7	5	2	0	-2	-5	-7	-10	-15	-20
1	34	29	24	21	19	16	14	11	9	6	4	0	-5
2	48	43	38	36	33	31	28	26	23	21	18	13	8
3	62	57	52	50	47	45	42	40	37	35	32	27	22
4	77	72	67	64	62	59	57	54	52	49	47	42	37
5	91	86	81	78	76	73	71	68	66	63	61	56	51
6	105	100	95	93	90	88	85	83	80	78	75	70	65
7	120	115	110	107	105	102	100	97	95	92	90	85	80
8	134	129	124	121	119	116	114	111	109	106	104	99	94
9	148	143	138	136	133	131	128	126	123	121	118	113	108
10	162	157	152	150	147	145	142	140	137	135	132	127	122
12	191	186	181	178	176	173	171	168	166	163	161	156	151
15	234	229	224	221	219	216	214	211	209	206	204	199	194
20	305	300	295	293	290	288	285	283	280	278	275	270	265
25	377	372	367	364	362	359	357	354	352	349	347	342	337
30	448	443	438	436	433	431	428	426	423	421	418	413	408
35	520	515	510	507	505	502	500	497	495	492	490	485	480
40	591	586	581	578	576	573	571	568	566	563	561	556	551

Table 3-62

Hub Diameter, inch: 3.5

Leg Distance, inch: 60

SHIM REQUIRED FOR ALIGNMENT

Face - OD Method

FACE TIR MILS	OD TIR, MILS -40	-30	-20	-15	-10	-5	0	5	10	15	20	30	40
-40	-665	-670	-675	-678	-680	-683	-685	-688	-690	-693	-695	-700	-705
-35	-580	-585	-590	-592	-595	-597	-600	-602	-605	-607	-610	-615	-620
-30	-494	-499	-504	-506	-509	-511	-514	-516	-519	-521	-524	-529	-534
-25	-408	-413	-418	-421	-423	-426	-428	-431	-433	-436	-438	-443	-448
-20	-322	-327	-332	-335	-337	-340	-342	-345	-347	-350	-352	-357	-362
-15	-237	-242	-247	-249	-252	-254	-257	-259	-262	-264	-267	-272	-277
-12	-185	-190	-195	-198	-200	-203	-205	-208	-210	-213	-215	-220	-225
-10	-151	-156	-161	-163	-166	-168	-171	-173	-176	-178	-181	-186	-191
-9	-134	-139	-144	-146	-149	-151	-154	-156	-159	-161	-164	-169	-174
-8	-117	-122	-127	-129	-132	-134	-137	-139	-142	-144	-147	-152	-157
-7	-100	-105	-110	-112	-115	-117	-120	-122	-125	-127	-130	-135	-140
-6	-82	-87	-92	-95	-97	-100	-102	-105	-107	-110	-112	-117	-122
-5	-65	-70	-75	-78	-80	-83	-85	-88	-90	-93	-95	-100	-105
-4	-48	-53	-58	-61	-63	-66	-68	-71	-73	-76	-78	-83	-88
-3	-31	-36	-41	-43	-46	-48	-51	-53	-56	-58	-61	-66	-71
-2	-14	-19	-24	-26	-29	-31	-34	-36	-39	-41	-44	-49	-54
-1	2	-2	-7	-9	-12	-14	-17	-19	-22	-24	-27	-32	-37
0	20	15	10	7	5	2	0	-2	-5	-7	-10	-15	-20
1	37	32	27	24	22	19	17	14	12	9	7	2	-2
2	54	49	44	41	39	36	34	31	29	26	24	19	14
3	71	66	61	58	56	53	51	48	46	43	41	36	31
4	88	83	78	76	73	71	68	66	63	61	58	53	48
5	105	100	95	93	90	88	85	83	80	78	75	70	65
6	122	117	112	110	107	105	102	100	97	95	92	87	82
7	140	135	130	127	125	122	120	117	115	112	110	105	100
8	157	152	147	144	142	139	137	134	132	129	127	122	117
9	174	169	164	161	159	156	154	151	149	146	144	139	134
10	191	186	181	178	176	173	171	168	166	163	161	156	151
12	225	220	215	213	210	208	205	203	200	198	195	190	185
15	277	272	267	264	262	259	257	254	252	249	247	242	237
20	362	357	352	350	347	345	342	340	337	335	332	327	322
25	448	443	438	436	433	431	428	426	423	421	418	413	408
30	534	529	524	521	519	516	514	511	509	506	504	499	494
35	620	615	610	607	605	602	600	597	595	592	590	585	580
40	705	700	695	693	690	688	685	683	680	678	675	670	665

Table 3-63

Hub Diameter, inch: 3.5

Leg Distance, inch: 70

SHIM REQUIRED FOR ALIGNMENT

Face - OD Method

FACE TIR MILS	OD TIR, MILS -40	-30	-20	-15	-10	-5	0	5	10	15	20	30	40
-40	-780	-785	-790	-792	-795	-797	-800	-802	-805	-807	-810	-815	-820
-35	-680	-685	-690	-692	-695	-697	-700	-702	-705	-707	-710	-715	-720
-30	-580	-585	-590	-592	-595	-597	-600	-602	-605	-607	-610	-615	-620
-25	-480	-485	-490	-492	-495	-497	-500	-502	-505	-507	-510	-515	-520
-20	-380	-385	-390	-392	-395	-397	-400	-402	-405	-407	-410	-415	-420
-15	-280	-285	-290	-292	-295	-297	-300	-302	-305	-307	-310	-315	-320
-12	-220	-225	-230	-232	-235	-237	-240	-242	-245	-247	-250	-255	-260
-10	-180	-185	-190	-192	-195	-197	-200	-202	-205	-207	-210	-215	-220
-9	-160	-165	-170	-172	-175	-177	-180	-182	-185	-187	-190	-195	-200
-8	-140	-145	-150	-152	-155	-157	-160	-162	-165	-167	-170	-175	-180
-7	-120	-125	-130	-132	-135	-137	-140	-142	-145	-147	-150	-155	-160
-6	-100	-105	-110	-112	-115	-117	-120	-122	-125	-127	-130	-135	-140
-5	-80	-85	-90	-92	-95	-97	-100	-102	-105	-107	-110	-115	-120
-4	-60	-65	-70	-72	-75	-77	-80	-82	-85	-87	-90	-95	-100
-3	-40	-45	-50	-52	-55	-57	-60	-62	-65	-67	-70	-75	-80
-2	-20	-25	-30	-32	-35	-37	-40	-42	-45	-47	-50	-55	-60
-1	0	-5	-10	-12	-15	-17	-20	-22	-25	-27	-30	-35	-40
0	20	15	10	7	5	2	0	-2	-5	-7	-10	-15	-20
1	40	35	30	27	25	22	20	17	15	12	10	5	0
2	60	55	50	47	45	42	40	37	35	32	30	25	20
3	80	75	70	67	65	62	60	57	55	52	50	45	40
4	100	95	90	87	85	82	80	77	75	72	70	65	60
5	120	115	110	107	105	102	100	97	95	92	90	85	80
6	140	135	130	127	125	122	120	117	115	112	110	105	100
7	160	155	150	147	145	142	140	137	135	132	130	125	120
8	180	175	170	167	165	162	160	157	155	152	150	145	140
9	200	195	190	187	185	182	180	177	175	172	170	165	160
10	220	215	210	207	205	202	200	197	195	192	190	185	180
12	260	255	250	247	245	242	240	237	235	232	230	225	220
15	320	315	310	307	305	302	300	297	295	292	290	285	280
20	420	415	410	407	405	402	400	397	395	392	390	385	380
25	520	515	510	507	505	502	500	497	495	492	490	485	480
30	620	615	610	607	605	602	600	597	595	592	590	585	580
35	720	715	710	707	705	702	700	697	695	692	690	685	680
40	820	815	810	807	805	802	800	797	795	792	790	785	780

Table 3-64

Hub Diameter, inch: 3.5

Leg Distance, inch: 80

SHIM REQUIRED FOR ALIGNMENT

Face - OD Method

FACE TIR MILS	OD TIR, MILS												
	-40	-30	-20	-15	-10	-5	0	5	10	15	20	30	40
-40	-894	-899	-904	-906	-909	-911	-914	-916	-919	-921	-924	-929	-934
-35	-780	-785	-790	-792	-795	-797	-800	-802	-805	-807	-810	-815	-820
-30	-665	-670	-675	-678	-680	-683	-685	-688	-690	-693	-695	-700	-705
-25	-551	-556	-561	-563	-566	-568	-571	-573	-576	-578	-581	-586	-591
-20	-437	-442	-447	-449	-452	-454	-457	-459	-462	-464	-467	-472	-477
-15	-322	-327	-332	-335	-337	-340	-342	-345	-347	-350	-352	-357	-362
-12	-254	-259	-264	-266	-269	-271	-274	-276	-279	-281	-284	-289	-294
-10	-208	-213	-218	-221	-223	-226	-228	-231	-233	-236	-238	-243	-248
-9	-185	-190	-195	-198	-200	-203	-205	-208	-210	-213	-215	-220	-225
-8	-162	-167	-172	-175	-177	-180	-182	-185	-187	-190	-192	-197	-202
-7	-140	-145	-150	-152	-155	-157	-160	-162	-165	-167	-170	-175	-180
-6	-117	-122	-127	-129	-132	-134	-137	-139	-142	-144	-147	-152	-157
-5	-94	-99	-104	-106	-109	-111	-114	-116	-119	-121	-124	-129	-134
-4	-71	-76	-81	-83	-86	-88	-91	-93	-96	-98	-101	-106	-111
-3	-48	-53	-58	-61	-63	-66	-68	-71	-73	-76	-78	-83	-88
-2	-25	-30	-35	-38	-40	-43	-45	-48	-50	-53	-55	-60	-65
-1	-2	-7	-12	-15	-17	-20	-22	-25	-27	-30	-32	-37	-42
0	20	15	10	7	5	2	0	-2	-5	-7	-10	-15	-20
1	42	37	32	30	27	25	22	20	17	15	12	7	2
2	65	60	55	53	50	48	45	43	40	38	35	30	25
3	88	83	78	76	73	71	68	66	63	61	58	53	48
4	111	106	101	98	96	93	91	88	86	83	81	76	71
5	134	129	124	121	119	116	114	111	109	106	104	99	94
6	157	152	147	144	142	139	137	134	132	129	127	122	117
7	180	175	170	167	165	162	160	157	155	152	150	145	140
8	202	197	192	190	187	185	182	180	177	175	172	167	162
9	225	220	215	213	210	208	205	203	200	198	195	190	185
10	248	243	238	236	233	231	228	226	223	221	218	213	208
12	294	289	284	281	279	276	274	271	269	266	264	259	254
15	362	357	352	350	347	345	342	340	337	335	332	327	322
20	477	472	467	464	462	459	457	454	452	449	447	442	437
25	591	586	581	578	576	573	571	568	566	563	561	556	551
30	705	700	695	693	690	688	685	683	680	678	675	670	665
35	820	815	810	807	805	802	800	797	795	792	790	785	780
40	934	929	924	921	919	916	914	911	909	906	904	899	894

Table 3-65

Hub Diameter, inch: 4

Leg Distance, inch: 15

SHIM REQUIRED FOR ALIGNMENT

Face - OD Method

FACE TIR MILS	OD TIR, MILS -40	-30	-20	-15	-10	-5	0	5	10	15	20	30	40
-40	-130	-135	-140	-142	-145	-147	-150	-152	-155	-157	-160	-165	-170
-35	-111	-116	-121	-123	-126	-128	-131	-133	-136	-138	-141	-146	-151
-30	-92	-97	-102	-105	-107	-110	-112	-115	-117	-120	-122	-127	-132
-25	-73	-78	-83	-86	-88	-91	-93	-96	-98	-101	-103	-108	-113
-20	-55	-60	-65	-67	-70	-72	-75	-77	-80	-82	-85	-90	-95
-15	-36	-41	-46	-48	-51	-53	-56	-58	-61	-63	-66	-71	-76
-12	-25	-30	-35	-37	-40	-42	-45	-47	-50	-52	-55	-60	-65
-10	-17	-22	-27	-30	-32	-35	-37	-40	-42	-45	-47	-52	-57
-9	-13	-18	-23	-26	-28	-31	-33	-36	-38	-41	-43	-48	-53
-8	-10	-15	-20	-22	-25	-27	-30	-32	-35	-37	-40	-45	-50
-7	-6	-11	-16	-18	-21	-23	-26	-28	-31	-33	-36	-41	-46
-6	-2	-7	-12	-15	-17	-20	-22	-25	-27	-30	-32	-37	-42
-5	1	-3	-8	-11	-13	-16	-18	-21	-23	-26	-28	-33	-38
-4	5	0	-5	-7	-10	-12	-15	-17	-20	-22	-25	-30	-35
-3	8	3	-1	-3	-6	-8	-11	-13	-16	-18	-21	-26	-31
-2	12	7	2	0	-2	-5	-7	-10	-12	-15	-17	-22	-27
-1	16	11	6	3	1	-1	-3	-6	-8	-11	-13	-18	-23
0	20	15	10	7	5	2	0	-2	-5	-7	-10	-15	-20
1	23	18	13	11	8	6	3	1	-1	-3	-6	-11	-16
2	27	22	17	15	12	10	7	5	2	0	-2	-7	-12
3	31	26	21	18	16	13	11	8	6	3	1	-3	-8
4	35	30	25	22	20	17	15	12	10	7	5	0	-5
5	38	33	28	26	23	21	18	16	13	11	8	3	-1
6	42	37	32	30	27	25	22	20	17	15	12	7	2
7	46	41	36	33	31	28	26	23	21	18	16	11	6
8	50	45	40	37	35	32	30	27	25	22	20	15	10
9	53	48	43	41	38	36	33	31	28	26	23	18	13
10	57	52	47	45	42	40	37	35	32	30	27	22	17
12	65	60	55	52	50	47	45	42	40	37	35	30	25
15	76	71	66	63	61	58	56	53	51	48	46	41	36
20	95	90	85	82	80	77	75	72	70	67	65	60	55
25	113	108	103	101	98	96	93	91	88	86	83	78	73
30	132	127	122	120	117	115	112	110	107	105	102	97	92
35	151	146	141	138	136	133	131	128	126	123	121	116	111
40	170	165	160	157	155	152	150	147	145	142	140	135	130

Table 3-66

Hub Diameter, inch: 4

Leg Distance, inch: 20

SHIM REQUIRED FOR ALIGNMENT

Face - OD Method

FACE TIR MILS	OD TIR, MILS -40	-30	-20	-15	-10	-5	0	5	10	15	20	30	40
-40	-180	-185	-190	-192	-195	-197	-200	-202	-205	-207	-210	-215	-220
-35	-155	-160	-165	-167	-170	-172	-175	-177	-180	-182	-185	-190	-195
-30	-130	-135	-140	-142	-145	-147	-150	-152	-155	-157	-160	-165	-170
-25	-105	-110	-115	-117	-120	-122	-125	-127	-130	-132	-135	-140	-145
-20	-80	-85	-90	-92	-95	-97	-100	-102	-105	-107	-110	-115	-120
-15	-55	-60	-65	-67	-70	-72	-75	-77	-80	-82	-85	-90	-95
-12	-40	-45	-50	-52	-55	-57	-60	-62	-65	-67	-70	-75	-80
-10	-30	-35	-40	-42	-45	-47	-50	-52	-55	-57	-60	-65	-70
-9	-25	-30	-35	-37	-40	-42	-45	-47	-50	-52	-55	-60	-65
-8	-20	-25	-30	-32	-35	-37	-40	-42	-45	-47	-50	-55	-60
-7	-15	-20	-25	-27	-30	-32	-35	-37	-40	-42	-45	-50	-55
-6	-10	-15	-20	-22	-25	-27	-30	-32	-35	-37	-40	-45	-50
-5	-5	-10	-15	-17	-20	-22	-25	-27	-30	-32	-35	-40	-45
-4	0	-5	-10	-12	-15	-17	-20	-22	-25	-27	-30	-35	-40
-3	5	0	-5	-7	-10	-12	-15	-17	-20	-22	-25	-30	-35
-2	10	5	0	-2	-5	-7	-10	-12	-15	-17	-20	-25	-30
-1	15	10	5	2	0	-2	-5	-7	-10	-12	-15	-20	-25
0	20	15	10	7	5	2	0	-2	-5	-7	-10	-15	-20
1	25	20	15	12	10	7	5	2	0	-2	-5	-10	-15
2	30	25	20	17	15	12	10	7	5	2	0	-5	-10
3	35	30	25	22	20	17	15	12	10	7	5	0	-5
4	40	35	30	27	25	22	20	17	15	12	10	5	0
5	45	40	35	32	30	27	25	22	20	17	15	10	5
6	50	45	40	37	35	32	30	27	25	22	20	15	10
7	55	50	45	42	40	37	35	32	30	27	25	20	15
8	60	55	50	47	45	42	40	37	35	32	30	25	20
9	65	60	55	52	50	47	45	42	40	37	35	30	25
10	70	65	60	57	55	52	50	47	45	42	40	35	30
12	80	75	70	67	65	62	60	57	55	52	50	45	40
15	95	90	85	82	80	77	75	72	70	67	65	60	55
20	120	115	110	107	105	102	100	97	95	92	90	85	80
25	145	140	135	132	130	127	125	122	120	117	115	110	105
30	170	165	160	157	155	152	150	147	145	142	140	135	130
35	195	190	185	182	180	177	175	172	170	167	165	160	155
40	220	215	210	207	205	202	200	197	195	192	190	185	180

Table 3-67

Hub Diameter, inch: 4

Leg Distance, inch: 25

SHIM REQUIRED FOR ALIGNMENT

Face - OD Method

FACE TIR MILS	OD TIR, MILS -40	-30	-20	-15	-10	-5	0	5	10	15	20	30	40
-40	-230	-235	-240	-242	-245	-247	-250	-252	-255	-257	-260	-265	-270
-35	-198	-203	-208	-211	-213	-216	-218	-221	-223	-226	-228	-233	-238
-30	-167	-172	-177	-180	-182	-185	-187	-190	-192	-195	-197	-202	-207
-25	-136	-141	-146	-148	-151	-153	-156	-158	-161	-163	-166	-171	-176
-20	-105	-110	-115	-117	-120	-122	-125	-127	-130	-132	-135	-140	-145
-15	-73	-78	-83	-86	-88	-91	-93	-96	-98	-101	-103	-108	-113
-12	-55	-60	-65	-67	-70	-72	-75	-77	-80	-82	-85	-90	-95
-10	-42	-47	-52	-55	-57	-60	-62	-65	-67	-70	-72	-77	-82
-9	-36	-41	-46	-48	-51	-53	-56	-58	-61	-63	-66	-71	-76
-8	-30	-35	-40	-42	-45	-47	-50	-52	-55	-57	-60	-65	-70
-7	-23	-28	-33	-36	-38	-41	-43	-46	-48	-51	-53	-58	-63
-6	-17	-22	-27	-30	-32	-35	-37	-40	-42	-45	-47	-52	-57
-5	-11	-16	-21	-23	-26	-28	-31	-33	-36	-38	-41	-46	-51
-4	-5	-10	-15	-17	-20	-22	-25	-27	-30	-32	-35	-40	-45
-3	1	-3	-8	-11	-13	-16	-18	-21	-23	-26	-28	-33	-38
-2	7	2	-2	-5	-7	-10	-12	-15	-17	-20	-22	-27	-32
-1	13	8	3	1	-1	-3	-6	-8	-11	-13	-16	-21	-26
0	20	15	10	7	5	2	0	-2	-5	-7	-10	-15	-20
1	26	21	16	13	11	8	6	3	1	-1	-3	-8	-13
2	32	27	22	20	17	15	12	10	7	5	2	-2	-7
3	38	33	28	26	23	21	18	16	13	11	8	3	-1
4	45	40	35	32	30	27	25	22	20	17	15	10	5
5	51	46	41	38	36	33	31	28	26	23	21	16	11
6	57	52	47	45	42	40	37	35	32	30	27	22	17
7	63	58	53	51	48	46	43	41	38	36	33	28	23
8	70	65	60	57	55	52	50	47	45	42	40	35	30
9	76	71	66	63	61	58	56	53	51	48	46	41	36
10	82	77	72	70	67	65	62	60	57	55	52	47	42
12	95	90	85	82	80	77	75	72	70	67	65	60	55
15	113	108	103	101	98	96	93	91	88	86	83	78	73
20	145	140	135	132	130	127	125	122	120	117	115	110	105
25	176	171	166	163	161	158	156	153	151	148	146	141	136
30	207	202	197	195	192	190	187	185	182	180	177	172	167
35	238	233	228	226	223	221	218	216	213	211	208	203	198
40	270	265	260	257	255	252	250	247	245	242	240	235	230

Table 3-68

Hub Diameter, inch: 4

Leg Distance, inch: 30

SHIM REQUIRED FOR ALIGNMENT

Face - OD Method

FACE TIR MILS	OD TIR, MILS -40	-30	-20	-15	-10	-5	0	5	10	15	20	30	40
-40	-280	-285	-290	-292	-295	-297	-300	-302	-305	-307	-310	-315	-320
-35	-242	-247	-252	-255	-257	-260	-262	-265	-267	-270	-272	-277	-282
-30	-205	-210	-215	-217	-220	-222	-225	-227	-230	-232	-235	-240	-245
-25	-167	-172	-177	-180	-182	-185	-187	-190	-192	-195	-197	-202	-207
-20	-130	-135	-140	-142	-145	-147	-150	-152	-155	-157	-160	-165	-170
-15	-92	-97	-102	-105	-107	-110	-112	-115	-117	-120	-122	-127	-132
-12	-70	-75	-80	-82	-85	-87	-90	-92	-95	-97	-100	-105	-110
-10	-55	-60	-65	-67	-70	-72	-75	-77	-80	-82	-85	-90	-95
-9	-47	-52	-57	-60	-62	-65	-67	-70	-72	-75	-77	-82	-87
-8	-40	-45	-50	-52	-55	-57	-60	-62	-65	-67	-70	-75	-80
-7	-32	-37	-42	-45	-47	-50	-52	-55	-57	-60	-62	-67	-72
-6	-25	-30	-35	-37	-40	-42	-45	-47	-50	-52	-55	-60	-65
-5	-17	-22	-27	-30	-32	-35	-37	-40	-42	-45	-47	-52	-57
-4	-10	-15	-20	-22	-25	-27	-30	-32	-35	-37	-40	-45	-50
-3	-2	-7	-12	-15	-17	-20	-22	-25	-27	-30	-32	-37	-42
-2	5	0	-5	-7	-10	-12	-15	-17	-20	-22	-25	-30	-35
-1	12	7	2	0	-2	-5	-7	-10	-12	-15	-17	-22	-27
0	20	15	10	7	5	2	0	-2	-5	-7	-10	-15	-20
1	27	22	17	15	12	10	7	5	2	0	-2	-7	-12
2	35	30	25	22	20	17	15	12	10	7	5	0	-5
3	42	37	32	30	27	25	22	20	17	15	12	7	2
4	50	45	40	37	35	32	30	27	25	22	20	15	10
5	57	52	47	45	42	40	37	35	32	30	27	22	17
6	65	60	55	52	50	47	45	42	40	37	35	30	25
7	72	67	62	60	57	55	52	50	47	45	42	37	32
8	80	75	70	67	65	62	60	57	55	52	50	45	40
9	87	82	77	75	72	70	67	65	62	60	57	52	47
10	95	90	85	82	80	77	75	72	70	67	65	60	55
12	110	105	100	97	95	92	90	87	85	82	80	75	70
15	132	127	122	120	117	115	112	110	107	105	102	97	92
20	170	165	160	157	155	152	150	147	145	142	140	135	130
25	207	202	197	195	192	190	187	185	182	180	177	172	167
30	245	240	235	232	230	227	225	222	220	217	215	210	205
35	282	277	272	270	267	265	262	260	257	255	252	247	242
40	320	315	310	307	305	302	300	297	295	292	290	285	280

Table 3-69

Hub Diameter, inch: 4

Leg Distance, inch: 35

SHIM REQUIRED FOR ALIGNMENT

Face - OD Method

FACE TIR MILS	OD TIR, MILS -40	-30	-20	-15	-10	-5	0	5	10	15	20	30	40
-40	-330	-335	-340	-342	-345	-347	-350	-352	-355	-357	-360	-365	-370
-35	-286	-291	-296	-298	-301	-303	-306	-308	-311	-313	-316	-321	-326
-30	-242	-247	-252	-255	-257	-260	-262	-265	-267	-270	-272	-277	-282
-25	-198	-203	-208	-211	-213	-216	-218	-221	-223	-226	-228	-233	-238
-20	-155	-160	-165	-167	-170	-172	-175	-177	-180	-182	-185	-190	-195
-15	-111	-116	-121	-123	-126	-128	-131	-133	-136	-138	-141	-146	-151
-12	-85	-90	-95	-97	-100	-102	-105	-107	-110	-112	-115	-120	-125
-10	-67	-72	-77	-80	-82	-85	-87	-90	-92	-95	-97	-102	-107
-9	-58	-63	-68	-71	-73	-76	-78	-81	-83	-86	-88	-93	-98
-8	-50	-55	-60	-62	-65	-67	-70	-72	-75	-77	-80	-85	-90
-7	-41	-46	-51	-53	-56	-58	-61	-63	-66	-68	-71	-76	-81
-6	-32	-37	-42	-45	-47	-50	-52	-55	-57	-60	-62	-67	-72
-5	-23	-28	-33	-36	-38	-41	-43	-46	-48	-51	-53	-58	-63
-4	-15	-20	-25	-27	-30	-32	-35	-37	-40	-42	-45	-50	-55
-3	-6	-11	-16	-18	-21	-23	-26	-28	-31	-33	-36	-41	-46
-2	2	-2	-7	-10	-12	-15	-17	-20	-22	-25	-27	-32	-37
-1	11	6	1	-1	-3	-6	-8	-11	-13	-16	-18	-23	-28
0	20	15	10	7	5	2	0	-2	-5	-7	-10	-15	-20
1	28	23	18	16	13	11	8	6	3	1	-1	-6	-11
2	37	32	27	25	22	20	17	15	12	10	7	2	-2
3	46	41	36	33	31	28	26	23	21	18	16	11	6
4	55	50	45	42	40	37	35	32	30	27	25	20	15
5	63	58	53	51	48	46	43	41	38	36	33	28	23
6	72	67	62	60	57	55	52	50	47	45	42	37	32
7	81	76	71	68	66	63	61	58	56	53	51	46	41
8	90	85	80	77	75	72	70	67	65	62	60	55	50
9	98	93	88	86	83	81	78	76	73	71	68	63	58
10	107	102	97	95	92	90	87	85	82	80	77	72	67
12	125	120	115	112	110	107	105	102	100	97	95	90	85
15	151	146	141	138	136	133	131	128	126	123	121	116	111
20	195	190	185	182	180	177	175	172	170	167	165	160	155
25	238	233	228	226	223	221	218	216	213	211	208	203	198
30	282	277	272	270	267	265	262	260	257	255	252	247	242
35	326	321	316	313	311	308	306	303	301	298	296	291	286
40	370	365	360	357	355	352	350	347	345	342	340	335	330

Table 3-70

Hub Diameter, inch: 4

Leg Distance, inch: 40

SHIM REQUIRED FOR ALIGNMENT

Face - OD Method

FACE TIR MILS	OD TIR, MILS												
	-40	-30	-20	-15	-10	-5	0	5	10	15	20	30	40
-40	-380	-385	-390	-392	-395	-397	-400	-402	-405	-407	-410	-415	-420
-35	-330	-335	-340	-342	-345	-347	-350	-352	-355	-357	-360	-365	-370
-30	-280	-285	-290	-292	-295	-297	-300	-302	-305	-307	-310	-315	-320
-25	-230	-235	-240	-242	-245	-247	-250	-252	-255	-257	-260	-265	-270
-20	-180	-185	-190	-192	-195	-197	-200	-202	-205	-207	-210	-215	-220
-15	-130	-135	-140	-142	-145	-147	-150	-152	-155	-157	-160	-165	-170
-12	-100	-105	-110	-112	-115	-117	-120	-122	-125	-127	-130	-135	-140
-10	-80	-85	-90	-92	-95	-97	-100	-102	-105	-107	-110	-115	-120
-9	-70	-75	-80	-82	-85	-87	-90	-92	-95	-97	-100	-105	-110
-8	-60	-65	-70	-72	-75	-77	-80	-82	-85	-87	-90	-95	-100
-7	-50	-55	-60	-62	-65	-67	-70	-72	-75	-77	-80	-85	-90
-6	-40	-45	-50	-52	-55	-57	-60	-62	-65	-67	-70	-75	-80
-5	-30	-35	-40	-42	-45	-47	-50	-52	-55	-57	-60	-65	-70
-4	-20	-25	-30	-32	-35	-37	-40	-42	-45	-47	-50	-55	-60
-3	-10	-15	-20	-22	-25	-27	-30	-32	-35	-37	-40	-45	-50
-2	0	-5	-10	-12	-15	-17	-20	-22	-25	-27	-30	-35	-40
-1	10	5	0	-2	-5	-7	-10	-12	-15	-17	-20	-25	-30
0	20	15	10	7	5	2	0	-2	-5	-7	-10	-15	-20
1	30	25	20	17	15	12	10	7	5	2	0	-5	-10
2	40	35	30	27	25	22	20	17	15	12	10	5	0
3	50	45	40	37	35	32	30	27	25	22	20	15	10
4	60	55	50	47	45	42	40	37	35	32	30	25	20
5	70	65	60	57	55	52	50	47	45	42	40	35	30
6	80	75	70	67	65	62	60	57	55	52	50	45	40
7	90	85	80	77	75	72	70	67	65	62	60	55	50
8	100	95	90	87	85	82	80	77	75	72	70	65	60
9	110	105	100	97	95	92	90	87	85	82	80	75	70
10	120	115	110	107	105	102	100	97	95	92	90	85	80
12	140	135	130	127	125	122	120	117	115	112	110	105	100
15	170	165	160	157	155	152	150	147	145	142	140	135	130
20	220	215	210	207	205	202	200	197	195	192	190	185	180
25	270	265	260	257	255	252	250	247	245	242	240	235	230
30	320	315	310	307	305	302	300	297	295	292	290	285	280
35	370	365	360	357	355	352	350	347	345	342	340	335	330
40	420	415	410	407	405	402	400	397	395	392	390	385	380

Table 3-71

Hub Diameter, inch: 4

Leg Distance, inch: 50

SHIM REQUIRED FOR ALIGNMENT

Face - OD Method

FACE TIR MILS	OD TIR, MILS -40	-30	-20	-15	-10	-5	0	5	10	15	20	30	40
-40	-480	-485	-490	-492	-495	-497	-500	-502	-505	-507	-510	-515	-520
-35	-417	-422	-427	-430	-432	-435	-437	-440	-442	-445	-447	-452	-457
-30	-355	-360	-365	-367	-370	-372	-375	-377	-380	-382	-385	-390	-395
-25	-292	-297	-302	-305	-307	-310	-312	-315	-317	-320	-322	-327	-332
-20	-230	-235	-240	-242	-245	-247	-250	-252	-255	-257	-260	-265	-270
-15	-167	-172	-177	-180	-182	-185	-187	-190	-192	-195	-197	-202	-207
-12	-130	-135	-140	-142	-145	-147	-150	-152	-155	-157	-160	-165	-170
-10	-105	-110	-115	-117	-120	-122	-125	-127	-130	-132	-135	-140	-145
-9	-92	-97	-102	-105	-107	-110	-112	-115	-117	-120	-122	-127	-132
-8	-80	-85	-90	-92	-95	-97	-100	-102	-105	-107	-110	-115	-120
-7	-67	-72	-77	-80	-82	-85	-87	-90	-92	-95	-97	-102	-107
-6	-55	-60	-65	-67	-70	-72	-75	-77	-80	-82	-85	-90	-95
-5	-42	-47	-52	-55	-57	-60	-62	-65	-67	-70	-72	-77	-82
-4	-30	-35	-40	-42	-45	-47	-50	-52	-55	-57	-60	-65	-70
-3	-17	-22	-27	-30	-32	-35	-37	-40	-42	-45	-47	-52	-57
-2	-5	-10	-15	-17	-20	-22	-25	-27	-30	-32	-35	-40	-45
-1	7	2	-2	-5	-7	-10	-12	-15	-17	-20	-22	-27	-32
0	20	15	10	7	5	2	0	-2	-5	-7	-10	-15	-20
1	32	27	22	20	17	15	12	10	7	5	2	-2	-7
2	45	40	35	32	30	27	25	22	20	17	15	10	5
3	57	52	47	45	42	40	37	35	32	30	27	22	17
4	70	65	60	57	55	52	50	47	45	42	40	35	30
5	82	77	72	70	67	65	62	60	57	55	52	47	42
6	95	90	85	82	80	77	75	72	70	67	65	60	55
7	107	102	97	95	92	90	87	85	82	80	77	72	67
8	120	115	110	107	105	102	100	97	95	92	90	85	80
9	132	127	122	120	117	115	112	110	107	105	102	97	92
10	145	140	135	132	130	127	125	122	120	117	115	110	105
12	170	165	160	157	155	152	150	147	145	142	140	135	130
15	207	202	197	195	192	190	187	185	182	180	177	172	167
20	270	265	260	257	255	252	250	247	245	242	240	235	230
25	332	327	322	320	317	315	312	310	307	305	302	297	292
30	395	390	385	382	380	377	375	372	370	367	365	360	355
35	457	452	447	445	442	440	437	435	432	430	427	422	417
40	520	515	510	507	505	502	500	497	495	492	490	485	480

Table 3-72

Hub Diameter, inch: 4

Leg Distance, inch: 60

SHIM REQUIRED FOR ALIGNMENT

Face - OD Method

FACE TIR MILS	OD TIR, MILS -40	-30	-20	-15	-10	-5	0	5	10	15	20	30	40
-40	-580	-585	-590	-592	-595	-597	-600	-602	-605	-607	-610	-615	-620
-35	-505	-510	-515	-517	-520	-522	-525	-527	-530	-532	-535	-540	-545
-30	-430	-435	-440	-442	-445	-447	-450	-452	-455	-457	-460	-465	-470
-25	-355	-360	-365	-367	-370	-372	-375	-377	-380	-382	-385	-390	-395
-20	-280	-285	-290	-292	-295	-297	-300	-302	-305	-307	-310	-315	-320
-15	-205	-210	-215	-217	-220	-222	-225	-227	-230	-232	-235	-240	-245
-12	-160	-165	-170	-172	-175	-177	-180	-182	-185	-187	-190	-195	-200
-10	-130	-135	-140	-142	-145	-147	-150	-152	-155	-157	-160	-165	-170
-9	-115	-120	-125	-127	-130	-132	-135	-137	-140	-142	-145	-150	-155
-8	-100	-105	-110	-112	-115	-117	-120	-122	-125	-127	-130	-135	-140
-7	-85	-90	-95	-97	-100	-102	-105	-107	-110	-112	-115	-120	-125
-6	-70	-75	-80	-82	-85	-87	-90	-92	-95	-97	-100	-105	-110
-5	-55	-60	-65	-67	-70	-72	-75	-77	-80	-82	-85	-90	-95
-4	-40	-45	-50	-52	-55	-57	-60	-62	-65	-67	-70	-75	-80
-3	-25	-30	-35	-37	-40	-42	-45	-47	-50	-52	-55	-60	-65
-2	-10	-15	-20	-22	-25	-27	-30	-32	-35	-37	-40	-45	-50
-1	5	0	-5	-7	-10	-12	-15	-17	-20	-22	-25	-30	-35
0	20	15	10	7	5	2	0	-2	-5	-7	-10	-15	-20
1	35	30	25	22	20	17	15	12	10	7	5	0	-5
2	50	45	40	37	35	32	30	27	25	22	20	15	10
3	65	60	55	52	50	47	45	42	40	37	35	30	25
4	80	75	70	67	65	62	60	57	55	52	50	45	40
5	95	90	85	82	80	77	75	72	70	67	65	60	55
6	110	105	100	97	95	92	90	87	85	82	80	75	70
7	125	120	115	112	110	107	105	102	100	97	95	90	85
8	140	135	130	127	125	122	120	117	115	112	110	105	100
9	155	150	145	142	140	137	135	132	130	127	125	120	115
10	170	165	160	157	155	152	150	147	145	142	140	135	130
12	200	195	190	187	185	182	180	177	175	172	170	165	160
15	245	240	235	232	230	227	225	222	220	217	215	210	205
20	320	315	310	307	305	302	300	297	295	292	290	285	280
25	395	390	385	382	380	377	375	372	370	367	365	360	355
30	470	465	460	457	455	452	450	447	445	442	440	435	430
35	545	540	535	532	530	527	525	522	520	517	515	510	505
40	620	615	610	607	605	602	600	597	595	592	590	585	580

Table 3-73

Hub Diameter, inch: 4

Leg Distance, inch: 70

SHIM REQUIRED FOR ALIGNMENT

Face - OD Method

FACE TIR MILS	OD TIR, MILS -40	-30	-20	-15	-10	-5	0	5	10	15	20	30	40
-40	-680	-685	-690	-692	-695	-697	-700	-702	-705	-707	-710	-715	-720
-35	-592	-597	-602	-605	-607	-610	-612	-615	-617	-620	-622	-627	-632
-30	-505	-510	-515	-517	-520	-522	-525	-527	-530	-532	-535	-540	-545
-25	-417	-422	-427	-430	-432	-435	-437	-440	-442	-445	-447	-452	-457
-20	-330	-335	-340	-342	-345	-347	-350	-352	-355	-357	-360	-365	-370
-15	-242	-247	-252	-255	-257	-260	-262	-265	-267	-270	-272	-277	-282
-12	-190	-195	-200	-202	-205	-207	-210	-212	-215	-217	-220	-225	-230
-10	-155	-160	-165	-167	-170	-172	-175	-177	-180	-182	-185	-190	-195
-9	-137	-142	-147	-150	-152	-155	-157	-160	-162	-165	-167	-172	-177
-8	-120	-125	-130	-132	-135	-137	-140	-142	-145	-147	-150	-155	-160
-7	-102	-107	-112	-115	-117	-120	-122	-125	-127	-130	-132	-137	-142
-6	-85	-90	-95	-97	-100	-102	-105	-107	-110	-112	-115	-120	-125
-5	-67	-72	-77	-80	-82	-85	-87	-90	-92	-95	-97	-102	-107
-4	-50	-55	-60	-62	-65	-67	-70	-72	-75	-77	-80	-85	-90
-3	-32	-37	-42	-45	-47	-50	-52	-55	-57	-60	-62	-67	-72
-2	-15	-20	-25	-27	-30	-32	-35	-37	-40	-42	-45	-50	-55
-1	2	-2	-7	-10	-12	-15	-17	-20	-22	-25	-27	-32	-37
0	20	15	10	7	5	2	0	-2	-5	-7	-10	-15	-20
1	37	32	27	25	22	20	17	15	12	10	7	2	-2
2	55	50	45	42	40	37	35	32	30	27	25	20	15
3	72	67	62	60	57	55	52	50	47	45	42	37	32
4	90	85	80	77	75	72	70	67	65	62	60	55	50
5	107	102	97	95	92	90	87	85	82	80	77	72	67
6	125	120	115	112	110	107	105	102	100	97	95	90	85
7	142	137	132	130	127	125	122	120	117	115	112	107	102
8	160	155	150	147	145	142	140	137	135	132	130	125	120
9	177	172	167	165	162	160	157	155	152	150	147	142	137
10	195	190	185	182	180	177	175	172	170	167	165	160	155
12	230	225	220	217	215	212	210	207	205	202	200	195	190
15	282	277	272	270	267	265	262	260	257	255	252	247	242
20	370	365	360	357	355	352	350	347	345	342	340	335	330
25	457	452	447	445	442	440	437	435	432	430	427	422	417
30	545	540	535	532	530	527	525	522	520	517	515	510	505
35	632	627	622	620	617	615	612	610	607	605	602	597	592
40	720	715	710	707	705	702	700	697	695	692	690	685	680

Table 3-74

Hub Diameter, inch: 4

Leg Distance, inch: 80

SHIM REQUIRED FOR ALIGNMENT

Face - OD Method

FACE TIR MILS	OD TIR, MILS -40	-30	-20	-15	-10	-5	0	5	10	15	20	30	40
-40	-780	-785	-790	-792	-795	-797	-800	-802	-805	-807	-810	-815	-820
-35	-680	-685	-690	-692	-695	-697	-700	-702	-705	-707	-710	-715	-720
-30	-580	-585	-590	-592	-595	-597	-600	-602	-605	-607	-610	-615	-620
-25	-480	-485	-490	-492	-495	-497	-500	-502	-505	-507	-510	-515	-520
-20	-380	-385	-390	-392	-395	-397	-400	-402	-405	-407	-410	-415	-420
-15	-280	-285	-290	-292	-295	-297	-300	-302	-305	-307	-310	-315	-320
-12	-220	-225	-230	-232	-235	-237	-240	-242	-245	-247	-250	-255	-260
-10	-180	-185	-190	-192	-195	-197	-200	-202	-205	-207	-210	-215	-220
-9	-160	-165	-170	-172	-175	-177	-180	-182	-185	-187	-190	-195	-200
-8	-140	-145	-150	-152	-155	-157	-160	-162	-165	-167	-170	-175	-180
-7	-120	-125	-130	-132	-135	-137	-140	-142	-145	-147	-150	-155	-160
-6	-100	-105	-110	-112	-115	-117	-120	-122	-125	-127	-130	-135	-140
-5	-80	-85	-90	-92	-95	-97	-100	-102	-105	-107	-110	-115	-120
-4	-60	-65	-70	-72	-75	-77	-80	-82	-85	-87	-90	-95	-100
-3	-40	-45	-50	-52	-55	-57	-60	-62	-65	-67	-70	-75	-80
-2	-20	-25	-30	-32	-35	-37	-40	-42	-45	-47	-50	-55	-60
-1	0	-5	-10	-12	-15	-17	-20	-22	-25	-27	-30	-35	-40
0	20	15	10	7	5	2	0	-2	-5	-7	-10	-15	-20
1	40	35	30	27	25	22	20	17	15	12	10	5	0
2	60	55	50	47	45	42	40	37	35	32	30	25	20
3	80	75	70	67	65	62	60	57	55	52	50	45	40
4	100	95	90	87	85	82	80	77	75	72	70	65	60
5	120	115	110	107	105	102	100	97	95	92	90	85	80
6	140	135	130	127	125	122	120	117	115	112	110	105	100
7	160	155	150	147	145	142	140	137	135	132	130	125	120
8	180	175	170	167	165	162	160	157	155	152	150	145	140
9	200	195	190	187	185	182	180	177	175	172	170	165	160
10	220	215	210	207	205	202	200	197	195	192	190	185	180
12	260	255	250	247	245	242	240	237	235	232	230	225	220
15	320	315	310	307	305	302	300	297	295	292	290	285	280
20	420	415	410	407	405	402	400	397	395	392	390	385	380
25	520	515	510	507	505	502	500	497	495	492	490	485	480
30	620	615	610	607	605	602	600	597	595	592	590	585	580
35	720	715	710	707	705	702	700	697	695	692	690	685	680
40	820	815	810	807	805	802	800	797	795	792	790	785	780

Table 3-75

Hub Diameter, inch: 5

Leg Distance, inch: 15

SHIM REQUIRED FOR ALIGNMENT

Face - OD Method

FACE TIR MILS	OD TIR, MILS -40	-30	-20	-15	-10	-5	0	5	10	15	20	30	40
-40	-100	-105	-110	-112	-115	-117	-120	-122	-125	-127	-130	-135	-140
-35	-85	-90	-95	-97	-100	-102	-105	-107	-110	-112	-115	-120	-125
-30	-70	-75	-80	-82	-85	-87	-90	-92	-95	-97	-100	-105	-110
-25	-55	-60	-65	-67	-70	-72	-75	-77	-80	-82	-85	-90	-95
-20	-40	-45	-50	-52	-55	-57	-60	-62	-65	-67	-70	-75	-80
-15	-25	-30	-35	-37	-40	-42	-45	-47	-50	-52	-55	-60	-65
-12	-16	-21	-26	-28	-31	-33	-36	-38	-41	-43	-46	-51	-56
-10	-10	-15	-20	-22	-25	-27	-30	-32	-35	-37	-40	-45	-50
-9	-7	-12	-17	-19	-22	-24	-27	-29	-32	-34	-37	-42	-47
-8	-4	-9	-14	-16	-19	-21	-24	-26	-29	-31	-34	-39	-44
-7	-1	-6	-11	-13	-16	-18	-21	-23	-26	-28	-31	-36	-41
-6	2	-3	-8	-10	-13	-15	-18	-20	-23	-25	-28	-33	-38
-5	5	0	-5	-7	-10	-12	-15	-17	-20	-22	-25	-30	-35
-4	8	3	-2	-4	-7	-9	-12	-14	-17	-19	-22	-27	-32
-3	11	6	1	-1	-4	-6	-9	-11	-14	-16	-19	-24	-29
-2	14	9	4	1	-1	-3	-6	-8	-11	-13	-16	-21	-26
-1	17	12	7	4	2	0	-3	-5	-8	-10	-13	-18	-23
0	20	15	10	7	5	2	0	-2	-5	-7	-10	-15	-20
1	23	18	13	10	8	5	3	0	-2	-4	-7	-12	-17
2	26	21	16	13	11	8	6	3	1	-1	-4	-9	-14
3	29	24	19	16	14	11	9	6	4	1	-1	-6	-11
4	32	27	22	19	17	14	12	9	7	4	2	-3	-8
5	35	30	25	22	20	17	15	12	10	7	5	0	-5
6	38	33	28	25	23	20	18	15	13	10	8	3	-2
7	41	36	31	28	26	23	21	18	16	13	11	6	1
8	44	39	34	31	29	26	24	21	19	16	14	9	4
9	47	42	37	34	32	29	27	24	22	19	17	12	7
10	50	45	40	37	35	32	30	27	25	22	20	15	10
12	56	51	46	43	41	38	36	33	31	28	26	21	16
15	65	60	55	52	50	47	45	42	40	37	35	30	25
20	80	75	70	67	65	62	60	57	55	52	50	45	40
25	95	90	85	82	80	77	75	72	70	67	65	60	55
30	110	105	100	97	95	92	90	87	85	82	80	75	70
35	125	120	115	112	110	107	105	102	100	97	95	90	85
40	140	135	130	127	125	122	120	117	115	112	110	105	100

Table 3-76

Hub Diameter, inch: 5

Leg Distance, inch: 20

SHIM REQUIRED FOR ALIGNMENT

Face - OD Method

FACE TIR MILS	OD TIR, MILS -40	-30	-20	-15	-10	-5	0	5	10	15	20	30	40
-40	-140	-145	-150	-152	-155	-157	-160	-162	-165	-167	-170	-175	-180
-35	-120	-125	-130	-132	-135	-137	-140	-142	-145	-147	-150	-155	-160
-30	-100	-105	-110	-112	-115	-117	-120	-122	-125	-127	-130	-135	-140
-25	-80	-85	-90	-92	-95	-97	-100	-102	-105	-107	-110	-115	-120
-20	-60	-65	-70	-72	-75	-77	-80	-82	-85	-87	-90	-95	-100
-15	-40	-45	-50	-52	-55	-57	-60	-62	-65	-67	-70	-75	-80
-12	-28	-33	-38	-40	-43	-45	-48	-50	-53	-55	-58	-63	-68
-10	-20	-25	-30	-32	-35	-37	-40	-42	-45	-47	-50	-55	-60
-9	-16	-21	-26	-28	-31	-33	-36	-38	-41	-43	-46	-51	-56
-8	-12	-17	-22	-24	-27	-29	-32	-34	-37	-39	-42	-47	-52
-7	-8	-13	-18	-20	-23	-25	-28	-30	-33	-35	-38	-43	-48
-6	-4	-9	-14	-16	-19	-21	-24	-26	-29	-31	-34	-39	-44
-5	0	-5	-10	-12	-15	-17	-20	-22	-25	-27	-30	-35	-40
-4	4	-1	-6	-8	-11	-13	-16	-18	-21	-23	-26	-31	-36
-3	8	3	-2	-4	-7	-9	-12	-14	-17	-19	-22	-27	-32
-2	12	7	2	0	-3	-5	-8	-10	-13	-15	-18	-23	-28
-1	16	11	6	3	1	-1	-4	-6	-9	-11	-14	-19	-24
0	20	15	10	7	5	2	0	-2	-5	-7	-10	-15	-20
1	24	19	14	11	9	6	4	1	-1	-3	-6	-11	-16
2	28	23	18	15	13	10	8	5	3	0	-2	-7	-12
3	32	27	22	19	17	14	12	9	7	4	2	-3	-8
4	36	31	26	23	21	18	16	13	11	8	6	1	-4
5	40	35	30	27	25	22	20	17	15	12	10	5	0
6	44	39	34	31	29	26	24	21	19	16	14	9	4
7	48	43	38	35	33	30	28	25	23	20	18	13	8
8	52	47	42	39	37	34	32	29	27	24	22	17	12
9	56	51	46	43	41	38	36	33	31	28	26	21	16
10	60	55	50	47	45	42	40	37	35	32	30	25	20
12	68	63	58	55	53	50	48	45	43	40	38	33	28
15	80	75	70	67	65	62	60	57	55	52	50	45	40
20	100	95	90	87	85	82	80	77	75	72	70	65	60
25	120	115	110	107	105	102	100	97	95	92	90	85	80
30	140	135	130	127	125	122	120	117	115	112	110	105	100
35	160	155	150	147	145	142	140	137	135	132	130	125	120
40	180	175	170	167	165	162	160	157	155	152	150	145	140

Table 3-77

Hub Diameter, inch: 5

Leg Distance, inch: 25

SHIM REQUIRED FOR ALIGNMENT

Face – OD Method

FACE TIR MILS	OD TIR, MILS -40	-30	-20	-15	-10	-5	0	5	10	15	20	30	40
-40	-180	-185	-190	-192	-195	-197	-200	-202	-205	-207	-210	-215	-220
-35	-155	-160	-165	-167	-170	-172	-175	-177	-180	-182	-185	-190	-195
-30	-130	-135	-140	-142	-145	-147	-150	-152	-155	-157	-160	-165	-170
-25	-105	-110	-115	-117	-120	-122	-125	-127	-130	-132	-135	-140	-145
-20	-80	-85	-90	-92	-95	-97	-100	-102	-105	-107	-110	-115	-120
-15	-55	-60	-65	-67	-70	-72	-75	-77	-80	-82	-85	-90	-95
-12	-40	-45	-50	-52	-55	-57	-60	-62	-65	-67	-70	-75	-80
-10	-30	-35	-40	-42	-45	-47	-50	-52	-55	-57	-60	-65	-70
-9	-25	-30	-35	-37	-40	-42	-45	-47	-50	-52	-55	-60	-65
-8	-20	-25	-30	-32	-35	-37	-40	-42	-45	-47	-50	-55	-60
-7	-15	-20	-25	-27	-30	-32	-35	-37	-40	-42	-45	-50	-55
-6	-10	-15	-20	-22	-25	-27	-30	-32	-35	-37	-40	-45	-50
-5	-5	-10	-15	-17	-20	-22	-25	-27	-30	-32	-35	-40	-45
-4	0	-5	-10	-12	-15	-17	-20	-22	-25	-27	-30	-35	-40
-3	5	0	-5	-7	-10	-12	-15	-17	-20	-22	-25	-30	-35
-2	10	5	0	-2	-5	-7	-10	-12	-15	-17	-20	-25	-30
-1	15	10	5	2	0	-2	-5	-7	-10	-12	-15	-20	-25
0	20	15	10	7	5	2	0	-2	-5	-7	-10	-15	-20
1	25	20	15	12	10	7	5	2	0	-2	-5	-10	-15
2	30	25	20	17	15	12	10	7	5	2	0	-5	-10
3	35	30	25	22	20	17	15	12	10	7	5	0	-5
4	40	35	30	27	25	22	20	17	15	12	10	5	0
5	45	40	35	32	30	27	25	22	20	17	15	10	5
6	50	45	40	37	35	32	30	27	25	22	20	15	10
7	55	50	45	42	40	37	35	32	30	27	25	20	15
8	60	55	50	47	45	42	40	37	35	32	30	25	20
9	65	60	55	52	50	47	45	42	40	37	35	30	25
10	70	65	60	57	55	52	50	47	45	42	40	35	30
12	80	75	70	67	65	62	60	57	55	52	50	45	40
15	95	90	85	82	80	77	75	72	70	67	65	60	55
20	120	115	110	107	105	102	100	97	95	92	90	85	80
25	145	140	135	132	130	127	125	122	120	117	115	110	105
30	170	165	160	157	155	152	150	147	145	142	140	135	130
35	195	190	185	182	180	177	175	172	170	167	165	160	155
40	220	215	210	207	205	202	200	197	195	192	190	185	180

Table 3-78

Hub Diameter, inch: 5

Leg Distance, inch: 30

SHIM REQUIRED FOR ALIGNMENT

Face - OD Method

FACE TIR MILS	OD TIR, MILS -40	-30	-20	-15	-10	-5	0	5	10	15	20	30	40
-40	-220	-225	-230	-232	-235	-237	-240	-242	-245	-247	-250	-255	-260
-35	-190	-195	-200	-202	-205	-207	-210	-212	-215	-217	-220	-225	-230
-30	-160	-165	-170	-172	-175	-177	-180	-182	-185	-187	-190	-195	-200
-25	-130	-135	-140	-142	-145	-147	-150	-152	-155	-157	-160	-165	-170
-20	-100	-105	-110	-112	-115	-117	-120	-122	-125	-127	-130	-135	-140
-15	-70	-75	-80	-82	-85	-87	-90	-92	-95	-97	-100	-105	-110
-12	-52	-57	-62	-64	-67	-69	-72	-74	-77	-79	-82	-87	-92
-10	-40	-45	-50	-52	-55	-57	-60	-62	-65	-67	-70	-75	-80
-9	-34	-39	-44	-46	-49	-51	-54	-56	-59	-61	-64	-69	-74
-8	-28	-33	-38	-40	-43	-45	-48	-50	-53	-55	-58	-63	-68
-7	-22	-27	-32	-34	-37	-39	-42	-44	-47	-49	-52	-57	-62
-6	-16	-21	-26	-28	-31	-33	-36	-38	-41	-43	-46	-51	-56
-5	-10	-15	-20	-22	-25	-27	-30	-32	-35	-37	-40	-45	-50
-4	-4	-9	-14	-16	-19	-21	-24	-26	-29	-31	-34	-39	-44
-3	2	-3	-8	-10	-13	-15	-18	-20	-23	-25	-28	-33	-38
-2	8	3	-2	-4	-7	-9	-12	-14	-17	-19	-22	-27	-32
-1	14	9	4	1	-1	-3	-6	-8	-11	-13	-16	-21	-26
0	20	15	10	7	5	2	0	-2	-5	-7	-10	-15	-20
1	26	21	16	13	11	8	6	3	1	-1	-4	-9	-14
2	32	27	22	19	17	14	12	9	7	4	2	-3	-8
3	38	33	28	25	23	20	18	15	13	10	8	3	-2
4	44	39	34	31	29	26	24	21	19	16	14	9	4
5	50	45	40	37	35	32	30	27	25	22	20	15	10
6	56	51	46	43	41	38	36	33	31	28	26	21	16
7	62	57	52	49	47	44	42	39	37	34	32	27	22
8	68	63	58	55	53	50	48	45	43	40	38	33	28
9	74	69	64	61	59	56	54	51	49	46	44	39	34
10	80	75	70	67	65	62	60	57	55	52	50	45	40
12	92	87	82	79	77	74	72	69	67	64	62	57	52
15	110	105	100	97	95	92	90	87	85	82	80	75	70
20	140	135	130	127	125	122	120	117	115	112	110	105	100
25	170	165	160	157	155	152	150	147	145	142	140	135	130
30	200	195	190	187	185	182	180	177	175	172	170	165	160
35	230	225	220	217	215	212	210	207	205	202	200	195	190
40	260	255	250	247	245	242	240	237	235	232	230	225	220

Table 3-79

Hub Diameter, inch: 5

Leg Distance, inch: 35

SHIM REQUIRED FOR ALIGNMENT

Face - OD Method

FACE TIR MILS	OD TIR, MILS -40	-30	-20	-15	-10	-5	0	5	10	15	20	30	40
-40	-260	-265	-270	-272	-275	-277	-280	-282	-285	-287	-290	-295	-300
-35	-225	-230	-235	-237	-240	-242	-245	-247	-250	-252	-255	-260	-265
-30	-190	-195	-200	-202	-205	-207	-210	-212	-215	-217	-220	-225	-230
-25	-155	-160	-165	-167	-170	-172	-175	-177	-180	-182	-185	-190	-195
-20	-120	-125	-130	-132	-135	-137	-140	-142	-145	-147	-150	-155	-160
-15	-85	-90	-95	-97	-100	-102	-105	-107	-110	-112	-115	-120	-125
-12	-64	-69	-74	-76	-79	-81	-84	-86	-89	-91	-94	-99	-104
-10	-50	-55	-60	-62	-65	-67	-70	-72	-75	-77	-80	-85	-90
-9	-43	-48	-53	-55	-58	-60	-63	-65	-68	-70	-73	-78	-83
-8	-36	-41	-46	-48	-51	-53	-56	-58	-61	-63	-66	-71	-76
-7	-29	-34	-39	-41	-44	-46	-49	-51	-54	-56	-59	-64	-69
-6	-22	-27	-32	-34	-37	-39	-42	-44	-47	-49	-52	-57	-62
-5	-15	-20	-25	-27	-30	-32	-35	-37	-40	-42	-45	-50	-55
-4	-8	-13	-18	-20	-23	-25	-28	-30	-33	-35	-38	-43	-48
-3	-1	-6	-11	-13	-16	-18	-21	-23	-26	-28	-31	-36	-41
-2	6	1	-4	-6	-9	-11	-14	-16	-19	-21	-24	-29	-34
-1	13	8	3	0	-2	-4	-7	-9	-12	-14	-17	-22	-27
0	20	15	10	7	5	2	0	-2	-5	-7	-10	-15	-20
1	27	22	17	14	12	9	7	4	2	0	-3	-8	-13
2	34	29	24	21	19	16	14	11	9	6	4	-1	-6
3	41	36	31	28	26	23	21	18	16	13	11	6	1
4	48	43	38	35	33	30	28	25	23	20	18	13	8
5	55	50	45	42	40	37	35	32	30	27	25	20	15
6	62	57	52	49	47	44	42	39	37	34	32	27	22
7	69	64	59	56	54	51	49	46	44	41	39	34	29
8	76	71	66	63	61	58	56	53	51	48	46	41	36
9	83	78	73	70	68	65	63	60	58	55	53	48	43
10	90	85	80	77	75	72	70	67	65	62	60	55	50
12	104	99	94	91	89	86	84	81	79	76	74	69	64
15	125	120	115	112	110	107	105	102	100	97	95	90	85
20	160	155	150	147	145	142	140	137	135	132	130	125	120
25	195	190	185	182	180	177	175	172	170	167	165	160	155
30	230	225	220	217	215	212	210	207	205	202	200	195	190
35	265	260	255	252	250	247	245	242	240	237	235	230	225
40	300	295	290	287	285	282	280	277	275	272	270	265	260

Table 3-80

Hub Diameter, inch: 5

Leg Distance, inch: 40

SHIM REQUIRED FOR ALIGNMENT

Face - OD Method

FACE TIR MILS	OD TIR, MILS -40	-30	-20	-15	-10	-5	0	5	10	15	20	30	40
-40	-300	-305	-310	-312	-315	-317	-320	-322	-325	-327	-330	-335	-340
-35	-260	-265	-270	-272	-275	-277	-280	-282	-285	-287	-290	-295	-300
-30	-220	-225	-230	-232	-235	-237	-240	-242	-245	-247	-250	-255	-260
-25	-180	-185	-190	-192	-195	-197	-200	-202	-205	-207	-210	-215	-220
-20	-140	-145	-150	-152	-155	-157	-160	-162	-165	-167	-170	-175	-180
-15	-100	-105	-110	-112	-115	-117	-120	-122	-125	-127	-130	-135	-140
-12	-76	-81	-86	-88	-91	-93	-96	-98	-101	-103	-106	-111	-116
-10	-60	-65	-70	-72	-75	-77	-80	-82	-85	-87	-90	-95	-100
-9	-52	-57	-62	-64	-67	-69	-72	-74	-77	-79	-82	-87	-92
-8	-44	-49	-54	-56	-59	-61	-64	-66	-69	-71	-74	-79	-84
-7	-36	-41	-46	-48	-51	-53	-56	-58	-61	-63	-66	-71	-76
-6	-28	-33	-38	-40	-43	-45	-48	-50	-53	-55	-58	-63	-68
-5	-20	-25	-30	-32	-35	-37	-40	-42	-45	-47	-50	-55	-60
-4	-12	-17	-22	-24	-27	-29	-32	-34	-37	-39	-42	-47	-52
-3	-4	-9	-14	-16	-19	-21	-24	-26	-29	-31	-34	-39	-44
-2	4	-1	-6	-8	-11	-13	-16	-18	-21	-23	-26	-31	-36
-1	12	7	2	0	-3	-5	-8	-10	-13	-15	-18	-23	-28
0	20	15	10	7	5	2	0	-2	-5	-7	-10	-15	-20
1	28	23	18	15	13	10	8	5	3	0	-2	-7	-12
2	36	31	26	23	21	18	16	13	11	8	6	1	-4
3	44	39	34	31	29	26	24	21	19	16	14	9	4
4	52	47	42	39	37	34	32	29	27	24	22	17	12
5	60	55	50	47	45	42	40	37	35	32	30	25	20
6	68	63	58	55	53	50	48	45	43	40	38	33	28
7	76	71	66	63	61	58	56	53	51	48	46	41	36
8	84	79	74	71	69	66	64	61	59	56	54	49	44
9	92	87	82	79	77	74	72	69	67	64	62	57	52
10	100	95	90	87	85	82	80	77	75	72	70	65	60
12	116	111	106	103	101	98	96	93	91	88	86	81	76
15	140	135	130	127	125	122	120	117	115	112	110	105	100
20	180	175	170	167	165	162	160	157	155	152	150	145	140
25	220	215	210	207	205	202	200	197	195	192	190	185	180
30	260	255	250	247	245	242	240	237	235	232	230	225	220
35	300	295	290	287	285	282	280	277	275	272	270	265	260
40	340	335	330	327	325	322	320	317	315	312	310	305	300

Table 3-81

Hub Diameter, inch: 5

Leg Distance, inch: 50

SHIM REQUIRED FOR ALIGNMENT

Face - OD Method

FACE TIR MILS	OD TIR, MILS -40	-30	-20	-15	-10	-5	0	5	10	15	20	30	40
-40	-380	-385	-390	-392	-395	-397	-400	-402	-405	-407	-410	-415	-420
-35	-330	-335	-340	-342	-345	-347	-350	-352	-355	-357	-360	-365	-370
-30	-280	-285	-290	-292	-295	-297	-300	-302	-305	-307	-310	-315	-320
-25	-230	-235	-240	-242	-245	-247	-250	-252	-255	-257	-260	-265	-270
-20	-180	-185	-190	-192	-195	-197	-200	-202	-205	-207	-210	-215	-220
-15	-130	-135	-140	-142	-145	-147	-150	-152	-155	-157	-160	-165	-170
-12	-100	-105	-110	-112	-115	-117	-120	-122	-125	-127	-130	-135	-140
-10	-80	-85	-90	-92	-95	-97	-100	-102	-105	-107	-110	-115	-120
-9	-70	-75	-80	-82	-85	-87	-90	-92	-95	-97	-100	-105	-110
-8	-60	-65	-70	-72	-75	-77	-80	-82	-85	-87	-90	-95	-100
-7	-50	-55	-60	-62	-65	-67	-70	-72	-75	-77	-80	-85	-90
-6	-40	-45	-50	-52	-55	-57	-60	-62	-65	-67	-70	-75	-80
-5	-30	-35	-40	-42	-45	-47	-50	-52	-55	-57	-60	-65	-70
-4	-20	-25	-30	-32	-35	-37	-40	-42	-45	-47	-50	-55	-60
-3	-10	-15	-20	-22	-25	-27	-30	-32	-35	-37	-40	-45	-50
-2	0	-5	-10	-12	-15	-17	-20	-22	-25	-27	-30	-35	-40
-1	10	5	0	-2	-5	-7	-10	-12	-15	-17	-20	-25	-30
0	20	15	10	7	5	2	0	-2	-5	-7	-10	-15	-20
1	30	25	20	17	15	12	10	7	5	2	0	-5	-10
2	40	35	30	27	25	22	20	17	15	12	10	5	0
3	50	45	40	37	35	32	30	27	25	22	20	15	10
4	60	55	50	47	45	42	40	37	35	32	30	25	20
5	70	65	60	57	55	52	50	47	45	42	40	35	30
6	80	75	70	67	65	62	60	57	55	52	50	45	40
7	90	85	80	77	75	72	70	67	65	62	60	55	50
8	100	95	90	87	85	82	80	77	75	72	70	65	60
9	110	105	100	97	95	92	90	87	85	82	80	75	70
10	120	115	110	107	105	102	100	97	95	92	90	85	80
12	140	135	130	127	125	122	120	117	115	112	110	105	100
15	170	165	160	157	155	152	150	147	145	142	140	135	130
20	220	215	210	207	205	202	200	197	195	192	190	185	180
25	270	265	260	257	255	252	250	247	245	242	240	235	230
30	320	315	310	307	305	302	300	297	295	292	290	285	280
35	370	365	360	357	355	352	350	347	345	342	340	335	330
40	420	415	410	407	405	402	400	397	395	392	390	385	380

Table 3-82

Hub Diameter, inch: 5

Leg Distance, inch: 60

SHIM REQUIRED FOR ALIGNMENT

Face - OD Method

FACE TIR MILS	OD TIR, MILS -40	-30	-20	-15	-10	-5	0	5	10	15	20	30	40
-40	-460	-465	-470	-472	-475	-477	-480	-482	-485	-487	-490	-495	-500
-35	-400	-405	-410	-412	-415	-417	-420	-422	-425	-427	-430	-435	-440
-30	-340	-345	-350	-352	-355	-357	-360	-362	-365	-367	-370	-375	-380
-25	-280	-285	-290	-292	-295	-297	-300	-302	-305	-307	-310	-315	-320
-20	-220	-225	-230	-232	-235	-237	-240	-242	-245	-247	-250	-255	-260
-15	-160	-165	-170	-172	-175	-177	-180	-182	-185	-187	-190	-195	-200
-12	-124	-129	-134	-136	-139	-141	-144	-146	-149	-151	-154	-159	-164
-10	-100	-105	-110	-112	-115	-117	-120	-122	-125	-127	-130	-135	-140
-9	-88	-93	-98	-100	-103	-105	-108	-110	-113	-115	-118	-123	-128
-8	-76	-81	-86	-88	-91	-93	-96	-98	-101	-103	-106	-111	-116
-7	-64	-69	-74	-76	-79	-81	-84	-86	-89	-91	-94	-99	-104
-6	-52	-57	-62	-64	-67	-69	-72	-74	-77	-79	-82	-87	-92
-5	-40	-45	-50	-52	-55	-57	-60	-62	-65	-67	-70	-75	-80
-4	-28	-33	-38	-40	-43	-45	-48	-50	-53	-55	-58	-63	-68
-3	-16	-21	-26	-28	-31	-33	-36	-38	-41	-43	-46	-51	-56
-2	-4	-9	-14	-16	-19	-21	-24	-26	-29	-31	-34	-39	-44
-1	8	3	-2	-4	-7	-9	-12	-14	-17	-19	-22	-27	-32
0	20	15	10	7	5	2	0	-2	-5	-7	-10	-15	-20
1	32	27	22	19	17	14	12	9	7	4	2	-3	-8
2	44	39	34	31	29	26	24	21	19	16	14	9	4
3	56	51	46	43	41	38	36	33	31	28	26	21	16
4	68	63	58	55	53	50	48	45	43	40	38	33	28
5	80	75	70	67	65	62	60	57	55	52	50	45	40
6	92	87	82	79	77	74	72	69	67	64	62	57	52
7	104	99	94	91	89	86	84	81	79	76	74	69	64
8	116	111	106	103	101	98	96	93	91	88	86	81	76
9	128	123	118	115	113	110	108	105	103	100	98	93	88
10	140	135	130	127	125	122	120	117	115	112	110	105	100
12	164	159	154	151	149	146	144	141	139	136	134	129	124
15	200	195	190	187	185	182	180	177	175	172	170	165	160
20	260	255	250	247	245	242	240	237	235	232	230	225	220
25	320	315	310	307	305	302	300	297	295	292	290	285	280
30	380	375	370	367	365	362	360	357	355	352	350	345	340
35	440	435	430	427	425	422	420	417	415	412	410	405	400
40	500	495	490	487	485	482	480	477	475	472	470	465	460

Table 3-83

Hub Diameter, inch: 5

Leg Distance, inch: 70

SHIM REQUIRED FOR ALIGNMENT

Face - OD Method

FACE TIR MILS	OD TIR, MILS -40	-30	-20	-15	-10	-5	0	5	10	15	20	30	40
-40	-540	-545	-550	-552	-555	-557	-560	-562	-565	-567	-570	-575	-580
-35	-470	-475	-480	-482	-485	-487	-490	-492	-495	-497	-500	-505	-510
-30	-400	-405	-410	-412	-415	-417	-420	-422	-425	-427	-430	-435	-440
-25	-330	-335	-340	-342	-345	-347	-350	-352	-355	-357	-360	-365	-370
-20	-260	-265	-270	-272	-275	-277	-280	-282	-285	-287	-290	-295	-300
-15	-190	-195	-200	-202	-205	-207	-210	-212	-215	-217	-220	-225	-230
-12	-148	-153	-158	-160	-163	-165	-168	-170	-173	-175	-178	-183	-188
-10	-120	-125	-130	-132	-135	-137	-140	-142	-145	-147	-150	-155	-160
-9	-106	-111	-116	-118	-121	-123	-126	-128	-131	-133	-136	-141	-146
-8	-92	-97	-102	-104	-107	-109	-112	-114	-117	-119	-122	-127	-132
-7	-78	-83	-88	-90	-93	-95	-98	-100	-103	-105	-108	-113	-118
-6	-64	-69	-74	-76	-79	-81	-84	-86	-89	-91	-94	-99	-104
-5	-50	-55	-60	-62	-65	-67	-70	-72	-75	-77	-80	-85	-90
-4	-36	-41	-46	-48	-51	-53	-56	-58	-61	-63	-66	-71	-76
-3	-22	-27	-32	-34	-37	-39	-42	-44	-47	-49	-52	-57	-62
-2	-8	-13	-18	-20	-23	-25	-28	-30	-33	-35	-38	-43	-48
-1	6	1	-4	-6	-9	-11	-14	-16	-19	-21	-24	-29	-34
0	20	15	10	7	5	2	0	-2	-5	-7	-10	-15	-20
1	34	29	24	21	19	16	14	11	9	6	4	-1	-6
2	48	43	38	35	33	30	28	25	23	20	18	13	8
3	62	57	52	49	47	44	42	39	37	34	32	27	22
4	76	71	66	63	61	58	56	53	51	48	46	41	36
5	90	85	80	77	75	72	70	67	65	62	60	55	50
6	104	99	94	91	89	86	84	81	79	76	74	69	64
7	118	113	108	105	103	100	98	95	93	90	88	83	78
8	132	127	122	119	117	114	112	109	107	104	102	97	92
9	146	141	136	133	131	128	126	123	121	118	116	111	106
10	160	155	150	147	145	142	140	137	135	132	130	125	120
12	188	183	178	175	173	170	168	165	163	160	158	153	148
15	230	225	220	217	215	212	210	207	205	202	200	195	190
20	300	295	290	287	285	282	280	277	275	272	270	265	260
25	370	365	360	357	355	352	350	347	345	342	340	335	330
30	440	435	430	427	425	422	420	417	415	412	410	405	400
35	510	505	500	497	495	492	490	487	485	482	480	475	470
40	580	575	570	567	565	562	560	557	555	552	550	545	540

Table 3-84

Hub Diameter, inch: 5

Leg Distance, inch: 80

SHIM REQUIRED FOR ALIGNMENT

Face - OD Method

FACE TIR MILS	OD TIR, MILS -40	-30	-20	-15	-10	-5	0	5	10	15	20	30	40
-40	-620	-625	-630	-632	-635	-637	-640	-642	-645	-647	-650	-655	-660
-35	-540	-545	-550	-552	-555	-557	-560	-562	-565	-567	-570	-575	-580
-30	-460	-465	-470	-472	-475	-477	-480	-482	-485	-487	-490	-495	-500
-25	-380	-385	-390	-392	-395	-397	-400	-402	-405	-407	-410	-415	-420
-20	-300	-305	-310	-312	-315	-317	-320	-322	-325	-327	-330	-335	-340
-15	-220	-225	-230	-232	-235	-237	-240	-242	-245	-247	-250	-255	-260
-12	-172	-177	-182	-184	-187	-189	-192	-194	-197	-199	-202	-207	-212
-10	-140	-145	-150	-152	-155	-157	-160	-162	-165	-167	-170	-175	-180
-9	-124	-129	-134	-136	-139	-141	-144	-146	-149	-151	-154	-159	-164
-8	-108	-113	-118	-120	-123	-125	-128	-130	-133	-135	-138	-143	-148
-7	-92	-97	-102	-104	-107	-109	-112	-114	-117	-119	-122	-127	-132
-6	-76	-81	-86	-88	-91	-93	-96	-98	-101	-103	-106	-111	-116
-5	-60	-65	-70	-72	-75	-77	-80	-82	-85	-87	-90	-95	-100
-4	-44	-49	-54	-56	-59	-61	-64	-66	-69	-71	-74	-79	-84
-3	-28	-33	-38	-40	-43	-45	-48	-50	-53	-55	-58	-63	-68
-2	-12	-17	-22	-24	-27	-29	-32	-34	-37	-39	-42	-47	-52
-1	4	-1	-6	-8	-11	-13	-16	-18	-21	-23	-26	-31	-36
0	20	15	10	7	5	2	0	-2	-5	-7	-10	-15	-20
1	36	31	26	23	21	18	16	13	11	8	6	1	-4
2	52	47	42	39	37	34	32	29	27	24	22	17	12
3	68	63	58	55	53	50	48	45	43	40	38	33	28
4	84	79	74	71	69	66	64	61	59	56	54	49	44
5	100	95	90	87	85	82	80	77	75	72	70	65	60
6	116	111	106	103	101	98	96	93	91	88	86	81	76
7	132	127	122	119	117	114	112	109	107	104	102	97	92
8	148	143	138	135	133	130	128	125	123	120	118	113	108
9	164	159	154	151	149	146	144	141	139	136	134	129	124
10	180	175	170	167	165	162	160	157	155	152	150	145	140
12	212	207	202	199	197	194	192	189	187	184	182	177	172
15	260	255	250	247	245	242	240	237	235	232	230	225	220
20	340	335	330	327	325	322	320	317	315	312	310	305	300
25	420	415	410	407	405	402	400	397	395	392	390	385	380
30	500	495	490	487	485	482	480	477	475	472	470	465	460
35	580	575	570	567	565	562	560	557	555	552	550	545	540
40	660	655	650	647	645	642	640	637	635	632	630	625	620

Table 3-85

Hub Diameter, inch: 6

Leg Distance, inch: 25

SHIM REQUIRED FOR ALIGNMENT

Face - OD Method

FACE TIR MILS	OD TIR, MILS -40	-30	-20	-15	-10	-5	0	5	10	15	20	30	40
-40	-146	-151	-156	-159	-161	-164	-166	-169	-171	-174	-176	-181	-186
-35	-125	-130	-135	-138	-140	-143	-145	-148	-150	-153	-155	-160	-165
-30	-105	-110	-115	-117	-120	-122	-125	-127	-130	-132	-135	-140	-145
-25	-84	-89	-94	-96	-99	-101	-104	-106	-109	-111	-114	-119	-124
-20	-63	-68	-73	-75	-78	-80	-83	-85	-88	-90	-93	-98	-103
-15	-42	-47	-52	-55	-57	-60	-62	-65	-67	-70	-72	-77	-82
-12	-30	-35	-40	-42	-45	-47	-50	-52	-55	-57	-60	-65	-70
-10	-21	-26	-31	-34	-36	-39	-41	-44	-46	-49	-51	-56	-61
-9	-17	-22	-27	-30	-32	-35	-37	-40	-42	-45	-47	-52	-57
-8	-13	-18	-23	-25	-28	-30	-33	-35	-38	-40	-43	-48	-53
-7	-9	-14	-19	-21	-24	-26	-29	-31	-34	-36	-39	-44	-49
-6	-5	-10	-15	-17	-20	-22	-25	-27	-30	-32	-35	-40	-45
-5	0	-5	-10	-13	-15	-18	-20	-23	-25	-28	-30	-35	-40
-4	3	-1	-6	-9	-11	-14	-16	-19	-21	-24	-26	-31	-36
-3	7	2	-2	-5	-7	-10	-12	-15	-17	-20	-22	-27	-32
-2	11	6	1	0	-3	-5	-8	-10	-13	-15	-18	-23	-28
-1	15	10	5	3	0	-1	-4	-6	-9	-11	-14	-19	-24
0	20	15	10	7	5	2	0	-2	-5	-7	-10	-15	-20
1	24	19	14	11	9	6	4	1	0	-3	-5	-10	-15
2	28	23	18	15	13	10	8	5	3	0	-1	-6	-11
3	32	27	22	20	17	15	12	10	7	5	2	-2	-7
4	36	31	26	24	21	19	16	14	11	9	6	1	-3
5	40	35	30	28	25	23	20	18	15	13	10	5	0
6	45	40	35	32	30	27	25	22	20	17	15	10	5
7	49	44	39	36	34	31	29	26	24	21	19	14	9
8	53	48	43	40	38	35	33	30	28	25	23	18	13
9	57	52	47	45	42	40	37	35	32	30	27	22	17
10	61	56	51	49	46	44	41	39	36	34	31	26	21
12	70	65	60	57	55	52	50	47	45	42	40	35	30
15	82	77	72	70	67	65	62	60	57	55	52	47	42
20	103	98	93	90	88	85	83	80	78	75	73	68	63
25	124	119	114	111	109	106	104	101	99	96	94	89	84
30	145	140	135	132	130	127	125	122	120	117	115	110	105
35	165	160	155	153	150	148	145	143	140	138	135	130	125
40	186	181	176	174	171	169	166	164	161	159	156	151	146

Table 3-86

Hub Diameter, inch: 6

Leg Distance, inch: 30

SHIM REQUIRED FOR ALIGNMENT

Face - OD Method

FACE TIR MILS	OD TIR, MILS -40	-30	-20	-15	-10	-5	0	5	10	15	20	30	40
-40	-180	-185	-190	-192	-195	-197	-200	-202	-205	-207	-210	-215	-220
-35	-155	-160	-165	-167	-170	-172	-175	-177	-180	-182	-185	-190	-195
-30	-130	-135	-140	-142	-145	-147	-150	-152	-155	-157	-160	-165	-170
-25	-105	-110	-115	-117	-120	-122	-125	-127	-130	-132	-135	-140	-145
-20	-80	-85	-90	-92	-95	-97	-100	-102	-105	-107	-110	-115	-120
-15	-55	-60	-65	-67	-70	-72	-75	-77	-80	-82	-85	-90	-95
-12	-40	-45	-50	-52	-55	-57	-60	-62	-65	-67	-70	-75	-80
-10	-30	-35	-40	-42	-45	-47	-50	-52	-55	-57	-60	-65	-70
-9	-25	-30	-35	-37	-40	-42	-45	-47	-50	-52	-55	-60	-65
-8	-20	-25	-30	-32	-35	-37	-40	-42	-45	-47	-50	-55	-60
-7	-15	-20	-25	-27	-30	-32	-35	-37	-40	-42	-45	-50	-55
-6	-10	-15	-20	-22	-25	-27	-30	-32	-35	-37	-40	-45	-50
-5	-5	-10	-15	-17	-20	-22	-25	-27	-30	-32	-35	-40	-45
-4	0	-5	-10	-12	-15	-17	-20	-22	-25	-27	-30	-35	-40
-3	5	0	-5	-7	-10	-12	-15	-17	-20	-22	-25	-30	-35
-2	10	5	0	-2	-5	-7	-10	-12	-15	-17	-20	-25	-30
-1	15	10	5	2	0	-2	-5	-7	-10	-12	-15	-20	-25
0	20	15	10	7	5	2	0	-2	-5	-7	-10	-15	-20
1	25	20	15	12	10	7	5	2	0	-2	-5	-10	-15
2	30	25	20	17	15	12	10	7	5	2	0	-5	-10
3	35	30	25	22	20	17	15	12	10	7	5	0	-5
4	40	35	30	27	25	22	20	17	15	12	10	5	0
5	45	40	35	32	30	27	25	22	20	17	15	10	5
6	50	45	40	37	35	32	30	27	25	22	20	15	10
7	55	50	45	42	40	37	35	32	30	27	25	20	15
8	60	55	50	47	45	42	40	37	35	32	30	25	20
9	65	60	55	52	50	47	45	42	40	37	35	30	25
10	70	65	60	57	55	52	50	47	45	42	40	35	30
12	80	75	70	67	65	62	60	57	55	52	50	45	40
15	95	90	85	82	80	77	75	72	70	67	65	60	55
20	120	115	110	107	105	102	100	97	95	92	90	85	80
25	145	140	135	132	130	127	125	122	120	117	115	110	105
30	170	165	160	157	155	152	150	147	145	142	140	135	130
35	195	190	185	182	180	177	175	172	170	167	165	160	155
40	220	215	210	207	205	202	200	197	195	192	190	185	180

Table 3-87

Hub Diameter, inch: 6
Leg Distance, inch: 35

SHIM REQUIRED FOR ALIGNMENT

Face - OD Method

FACE TIR MILS	OD TIR, MILS -40	-30	-20	-15	-10	-5	0	5	10	15	20	30	40
-40	-213	-218	-223	-225	-228	-230	-233	-235	-238	-240	-243	-248	-253
-35	-184	-189	-194	-196	-199	-201	-204	-206	-209	-211	-214	-219	-224
-30	-155	-160	-165	-167	-170	-172	-175	-177	-180	-182	-185	-190	-195
-25	-125	-130	-135	-138	-140	-143	-145	-148	-150	-153	-155	-160	-165
-20	-96	-101	-106	-109	-111	-114	-116	-119	-121	-124	-126	-131	-136
-15	-67	-72	-77	-80	-82	-85	-87	-90	-92	-95	-97	-102	-107
-12	-50	-55	-60	-62	-65	-67	-70	-72	-75	-77	-80	-85	-90
-10	-38	-43	-48	-50	-53	-55	-58	-60	-63	-65	-68	-73	-78
-9	-32	-37	-42	-45	-47	-50	-52	-55	-57	-60	-62	-67	-72
-8	-26	-31	-36	-39	-41	-44	-46	-49	-51	-54	-56	-61	-66
-7	-20	-25	-30	-33	-35	-38	-40	-43	-45	-48	-50	-55	-60
-6	-15	-20	-25	-27	-30	-32	-35	-37	-40	-42	-45	-50	-55
-5	-9	-14	-19	-21	-24	-26	-29	-31	-34	-36	-39	-44	-49
-4	-3	-8	-13	-15	-18	-20	-23	-25	-28	-30	-33	-38	-43
-3	2	-2	-7	-10	-12	-15	-17	-20	-22	-25	-27	-32	-37
-2	8	3	-1	-4	-6	-9	-11	-14	-16	-19	-21	-26	-31
-1	14	9	4	1	0	-3	-5	-8	-10	-13	-15	-20	-25
0	20	15	10	7	5	2	0	-2	-5	-7	-10	-15	-20
1	25	20	15	13	10	8	5	3	0	-1	-4	-9	-14
2	31	26	21	19	16	14	11	9	6	4	1	-3	-8
3	37	32	27	25	22	20	17	15	12	10	7	2	-2
4	43	38	33	30	28	25	23	20	18	15	13	8	3
5	49	44	39	36	34	31	29	26	24	21	19	14	9
6	55	50	45	42	40	37	35	32	30	27	25	20	15
7	60	55	50	48	45	43	40	38	35	33	30	25	20
8	66	61	56	54	51	49	46	44	41	39	36	31	26
9	72	67	62	60	57	55	52	50	47	45	42	37	32
10	78	73	68	65	63	60	58	55	53	50	48	43	38
12	90	85	80	77	75	72	70	67	65	62	60	55	50
15	107	102	97	95	92	90	87	85	82	80	77	72	67
20	136	131	126	124	121	119	116	114	111	109	106	101	96
25	165	160	155	153	150	148	145	143	140	138	135	130	125
30	195	190	185	182	180	177	175	172	170	167	165	160	155
35	224	219	214	211	209	206	204	201	199	196	194	189	184
40	253	248	243	240	238	235	233	230	228	225	223	218	213

Table 3-88

Hub Diameter, inch: 6

Leg Distance, inch: 40

SHIM REQUIRED FOR ALIGNMENT

Face - OD Method

FACE TIR MILS	OD TIR, MILS -40	-30	-20	-15	-10	-5	0	5	10	15	20	30	40
-40	-246	-251	-256	-259	-261	-264	-266	-269	-271	-274	-276	-281	-286
-35	-213	-218	-223	-225	-228	-230	-233	-235	-238	-240	-243	-248	-253
-30	-180	-185	-190	-192	-195	-197	-200	-202	-205	-207	-210	-215	-220
-25	-146	-151	-156	-159	-161	-164	-166	-169	-171	-174	-176	-181	-186
-20	-113	-118	-123	-125	-128	-130	-133	-135	-138	-140	-143	-148	-153
-15	-80	-85	-90	-92	-95	-97	-100	-102	-105	-107	-110	-115	-120
-12	-60	-65	-70	-72	-75	-77	-80	-82	-85	-87	-90	-95	-100
-10	-46	-51	-56	-59	-61	-64	-66	-69	-71	-74	-76	-81	-86
-9	-40	-45	-50	-52	-55	-57	-60	-62	-65	-67	-70	-75	-80
-8	-33	-38	-43	-45	-48	-50	-53	-55	-58	-60	-63	-68	-73
-7	-26	-31	-36	-39	-41	-44	-46	-49	-51	-54	-56	-61	-66
-6	-20	-25	-30	-32	-35	-37	-40	-42	-45	-47	-50	-55	-60
-5	-13	-18	-23	-25	-28	-30	-33	-35	-38	-40	-43	-48	-53
-4	-6	-11	-16	-19	-21	-24	-26	-29	-31	-34	-36	-41	-46
-3	0	-5	-10	-12	-15	-17	-20	-22	-25	-27	-30	-35	-40
-2	6	1	-3	-5	-8	-10	-13	-15	-18	-20	-23	-28	-33
-1	13	8	3	0	-1	-4	-6	-9	-11	-14	-16	-21	-26
0	20	15	10	7	5	2	0	-2	-5	-7	-10	-15	-20
1	26	21	16	14	11	9	6	4	1	0	-3	-8	-13
2	33	28	23	20	18	15	13	10	8	5	3	-1	-6
3	40	35	30	27	25	22	20	17	15	12	10	5	0
4	46	41	36	34	31	29	26	24	21	19	16	11	6
5	53	48	43	40	38	35	33	30	28	25	23	18	13
6	60	55	50	47	45	42	40	37	35	32	30	25	20
7	66	61	56	54	51	49	46	44	41	39	36	31	26
8	73	68	63	60	58	55	53	50	48	45	43	38	33
9	80	75	70	67	65	62	60	57	55	52	50	45	40
10	86	81	76	74	71	69	66	64	61	59	56	51	46
12	100	95	90	87	85	82	80	77	75	72	70	65	60
15	120	115	110	107	105	102	100	97	95	92	90	85	80
20	153	148	143	140	138	135	133	130	128	125	123	118	113
25	186	181	176	174	171	169	166	164	161	159	156	151	146
30	220	215	210	207	205	202	200	197	195	192	190	185	180
35	253	248	243	240	238	235	233	230	228	225	223	218	213
40	286	281	276	274	271	269	266	264	261	259	256	251	246

Table 3-89

Hub Diameter, inch: 6
Leg Distance, inch: 50

SHIM REQUIRED FOR ALIGNMENT

Face - OD Method

FACE TIR MILS	OD TIR, MILS -40	-30	-20	-15	-10	-5	0	5	10	15	20	30	40
-40	-313	-318	-323	-325	-328	-330	-333	-335	-338	-340	-343	-348	-353
-35	-271	-276	-281	-284	-286	-289	-291	-294	-296	-299	-301	-306	-311
-30	-230	-235	-240	-242	-245	-247	-250	-252	-255	-257	-260	-265	-270
-25	-188	-193	-198	-200	-203	-205	-208	-210	-213	-215	-218	-223	-228
-20	-146	-151	-156	-159	-161	-164	-166	-169	-171	-174	-176	-181	-186
-15	-105	-110	-115	-117	-120	-122	-125	-127	-130	-132	-135	-140	-145
-12	-80	-85	-90	-92	-95	-97	-100	-102	-105	-107	-110	-115	-120
-10	-63	-68	-73	-75	-78	-80	-83	-85	-88	-90	-93	-98	-103
-9	-55	-60	-65	-67	-70	-72	-75	-77	-80	-82	-85	-90	-95
-8	-46	-51	-56	-59	-61	-64	-66	-69	-71	-74	-76	-81	-86
-7	-38	-43	-48	-50	-53	-55	-58	-60	-63	-65	-68	-73	-78
-6	-30	-35	-40	-42	-45	-47	-50	-52	-55	-57	-60	-65	-70
-5	-21	-26	-31	-34	-36	-39	-41	-44	-46	-49	-51	-56	-61
-4	-13	-18	-23	-25	-28	-30	-33	-35	-38	-40	-43	-48	-53
-3	-5	-10	-15	-17	-20	-22	-25	-27	-30	-32	-35	-40	-45
-2	3	-1	-6	-9	-11	-14	-16	-19	-21	-24	-26	-31	-36
-1	11	6	1	0	-3	-5	-8	-10	-13	-15	-18	-23	-28
0	20	15	10	7	5	2	0	-2	-5	-7	-10	-15	-20
1	28	23	18	15	13	10	8	5	3	0	-1	-6	-11
2	36	31	26	24	21	19	16	14	11	9	6	1	-3
3	45	40	35	32	30	27	25	22	20	17	15	10	5
4	53	48	43	40	38	35	33	30	28	25	23	18	13
5	61	56	51	49	46	44	41	39	36	34	31	26	21
6	70	65	60	57	55	52	50	47	45	42	40	35	30
7	78	73	68	65	63	60	58	55	53	50	48	43	38
8	86	81	76	74	71	69	66	64	61	59	56	51	46
9	95	90	85	82	80	77	75	72	70	67	65	60	55
10	103	98	93	90	88	85	83	80	78	75	73	68	63
12	120	115	110	107	105	102	100	97	95	92	90	85	80
15	145	140	135	132	130	127	125	122	120	117	115	110	105
20	186	181	176	174	171	169	166	164	161	159	156	151	146
25	228	223	218	215	213	210	208	205	203	200	198	193	188
30	270	265	260	257	255	252	250	247	245	242	240	235	230
35	311	306	301	299	296	294	291	289	286	284	281	276	271
40	353	348	343	340	338	335	333	330	328	325	323	318	313

Table 3-90

Hub Diameter, inch: 6

Leg Distance, inch: 60

SHIM REQUIRED FOR ALIGNMENT

Face - OD Method

FACE TIR MILS	OD TIR, MILS -40	-30	-20	-15	-10	-5	0	5	10	15	20	30	40
-40	-380	-385	-390	-392	-395	-397	-400	-402	-405	-407	-410	-415	-420
-35	-330	-335	-340	-342	-345	-347	-350	-352	-355	-357	-360	-365	-370
-30	-280	-285	-290	-292	-295	-297	-300	-302	-305	-307	-310	-315	-320
-25	-230	-235	-240	-242	-245	-247	-250	-252	-255	-257	-260	-265	-270
-20	-180	-185	-190	-192	-195	-197	-200	-202	-205	-207	-210	-215	-220
-15	-130	-135	-140	-142	-145	-147	-150	-152	-155	-157	-160	-165	-170
-12	-100	-105	-110	-112	-115	-117	-120	-122	-125	-127	-130	-135	-140
-10	-80	-85	-90	-92	-95	-97	-100	-102	-105	-107	-110	-115	-120
-9	-70	-75	-80	-82	-85	-87	-90	-92	-95	-97	-100	-105	-110
-8	-60	-65	-70	-72	-75	-77	-80	-82	-85	-87	-90	-95	-100
-7	-50	-55	-60	-62	-65	-67	-70	-72	-75	-77	-80	-85	-90
-6	-40	-45	-50	-52	-55	-57	-60	-62	-65	-67	-70	-75	-80
-5	-30	-35	-40	-42	-45	-47	-50	-52	-55	-57	-60	-65	-70
-4	-20	-25	-30	-32	-35	-37	-40	-42	-45	-47	-50	-55	-60
-3	-10	-15	-20	-22	-25	-27	-30	-32	-35	-37	-40	-45	-50
-2	0	-5	-10	-12	-15	-17	-20	-22	-25	-27	-30	-35	-40
-1	10	5	0	-2	-5	-7	-10	-12	-15	-17	-20	-25	-30
0	20	15	10	7	5	2	0	-2	-5	-7	-10	-15	-20
1	30	25	20	17	15	12	10	7	5	2	0	-5	-10
2	40	35	30	27	25	22	20	17	15	12	10	5	0
3	50	45	40	37	35	32	30	27	25	22	20	15	10
4	60	55	50	47	45	42	40	37	35	32	30	25	20
5	70	65	60	57	55	52	50	47	45	42	40	35	30
6	80	75	70	67	65	62	60	57	55	52	50	45	40
7	90	85	80	77	75	72	70	67	65	62	60	55	50
8	100	95	90	87	85	82	80	77	75	72	70	65	60
9	110	105	100	97	95	92	90	87	85	82	80	75	70
10	120	115	110	107	105	102	100	97	95	92	90	85	80
12	140	135	130	127	125	122	120	117	115	112	110	105	100
15	170	165	160	157	155	152	150	147	145	142	140	135	130
20	220	215	210	207	205	202	200	197	195	192	190	185	180
25	270	265	260	257	255	252	250	247	245	242	240	235	230
30	320	315	310	307	305	302	300	297	295	292	290	285	280
35	370	365	360	357	355	352	350	347	345	342	340	335	330
40	420	415	410	407	405	402	400	397	395	392	390	385	380

Table 3-91

Hub Diameter, inch: 6
Leg Distance, inch: 70

SHIM REQUIRED FOR ALIGNMENT

Face - OD Method

FACE TIR MILS	OD TIR, MILS -40	-30	-20	-15	-10	-5	0	5	10	15	20	30	40
-40	-446	-451	-456	-459	-461	-464	-466	-469	-471	-474	-476	-481	-486
-35	-388	-393	-398	-400	-403	-405	-408	-410	-413	-415	-418	-423	-428
-30	-330	-335	-340	-342	-345	-347	-350	-352	-355	-357	-360	-365	-370
-25	-271	-276	-281	-284	-286	-289	-291	-294	-296	-299	-301	-306	-311
-20	-213	-218	-223	-225	-228	-230	-233	-235	-238	-240	-243	-248	-253
-15	-155	-160	-165	-167	-170	-172	-175	-177	-180	-182	-185	-190	-195
-12	-120	-125	-130	-132	-135	-137	-140	-142	-145	-147	-150	-155	-160
-10	-96	-101	-106	-109	-111	-114	-116	-119	-121	-124	-126	-131	-136
-9	-85	-90	-95	-97	-100	-102	-105	-107	-110	-112	-115	-120	-125
-8	-73	-78	-83	-85	-88	-90	-93	-95	-98	-100	-103	-108	-113
-7	-61	-66	-71	-74	-76	-79	-81	-84	-86	-89	-91	-96	-101
-6	-50	-55	-60	-62	-65	-67	-70	-72	-75	-77	-80	-85	-90
-5	-38	-43	-48	-50	-53	-55	-58	-60	-63	-65	-68	-73	-78
-4	-26	-31	-36	-39	-41	-44	-46	-49	-51	-54	-56	-61	-66
-3	-15	-20	-25	-27	-30	-32	-35	-37	-40	-42	-45	-50	-55
-2	-3	-8	-13	-15	-18	-20	-23	-25	-28	-30	-33	-38	-43
-1	8	3	-1	-4	-6	-9	-11	-14	-16	-19	-21	-26	-31
0	20	15	10	7	5	2	0	-2	-5	-7	-10	-15	-20
1	31	26	21	19	16	14	11	9	6	4	1	-3	-8
2	43	38	33	30	28	25	23	20	18	15	13	8	3
3	55	50	45	42	40	37	35	32	30	27	25	20	15
4	66	61	56	54	51	49	46	44	41	39	36	31	26
5	78	73	68	65	63	60	58	55	53	50	48	43	38
6	90	85	80	77	75	72	70	67	65	62	60	55	50
7	101	96	91	89	86	84	81	79	76	74	71	66	61
8	113	108	103	100	98	95	93	90	88	85	83	78	73
9	125	120	115	112	110	107	105	102	100	97	95	90	85
10	136	131	126	124	121	119	116	114	111	109	106	101	96
12	160	155	150	147	145	142	140	137	135	132	130	125	120
15	195	190	185	182	180	177	175	172	170	167	165	160	155
20	253	248	243	240	238	235	233	230	228	225	223	218	213
25	311	306	301	299	296	294	291	289	286	284	281	276	271
30	370	365	360	357	355	352	350	347	345	342	340	335	330
35	428	423	418	415	413	410	408	405	403	400	398	393	388
40	486	481	476	474	471	469	466	464	461	459	456	451	446

Table 3-92

Hub Diameter, inch: 6

Leg Distance, inch: 80

SHIM REQUIRED FOR ALIGNMENT

Face - OD Method

FACE TIR MILS	OD TIR, MILS -40	-30	-20	-15	-10	-5	0	5	10	15	20	30	40
-40	-513	-518	-523	-525	-528	-530	-533	-535	-538	-540	-543	-548	-553
-35	-446	-451	-456	-459	-461	-464	-466	-469	-471	-474	-476	-481	-486
-30	-380	-385	-390	-392	-395	-397	-400	-402	-405	-407	-410	-415	-420
-25	-313	-318	-323	-325	-328	-330	-333	-335	-338	-340	-343	-348	-353
-20	-246	-251	-256	-259	-261	-264	-266	-269	-271	-274	-276	-281	-286
-15	-180	-185	-190	-192	-195	-197	-200	-202	-205	-207	-210	-215	-220
-12	-140	-145	-150	-152	-155	-157	-160	-162	-165	-167	-170	-175	-180
-10	-113	-118	-123	-125	-128	-130	-133	-135	-138	-140	-143	-148	-153
-9	-100	-105	-110	-112	-115	-117	-120	-122	-125	-127	-130	-135	-140
-8	-86	-91	-96	-99	-101	-104	-106	-109	-111	-114	-116	-121	-126
-7	-73	-78	-83	-85	-88	-90	-93	-95	-98	-100	-103	-108	-113
-6	-60	-65	-70	-72	-75	-77	-80	-82	-85	-87	-90	-95	-100
-5	-46	-51	-56	-59	-61	-64	-66	-69	-71	-74	-76	-81	-86
-4	-33	-38	-43	-45	-48	-50	-53	-55	-58	-60	-63	-68	-73
-3	-20	-25	-30	-32	-35	-37	-40	-42	-45	-47	-50	-55	-60
-2	-6	-11	-16	-19	-21	-24	-26	-29	-31	-34	-36	-41	-46
-1	6	1	-3	-5	-8	-10	-13	-15	-18	-20	-23	-28	-33
0	20	15	10	7	5	2	0	-2	-5	-7	-10	-15	-20
1	33	28	23	20	18	15	13	10	8	5	3	-1	-6
2	46	41	36	34	31	29	26	24	21	19	16	11	6
3	60	55	50	47	45	42	40	37	35	32	30	25	20
4	73	68	63	60	58	55	53	50	48	45	43	38	33
5	86	81	76	74	71	69	66	64	61	59	56	51	46
6	100	95	90	87	85	82	80	77	75	72	70	65	60
7	113	108	103	100	98	95	93	90	88	85	83	78	73
8	126	121	116	114	111	109	106	104	101	99	96	91	86
9	140	135	130	127	125	122	120	117	115	112	110	105	100
10	153	148	143	140	138	135	133	130	128	125	123	118	113
12	180	175	170	167	165	162	160	157	155	152	150	145	140
15	220	215	210	207	205	202	200	197	195	192	190	185	180
20	286	281	276	274	271	269	266	264	261	259	256	251	246
25	353	348	343	340	338	335	333	330	328	325	323	318	313
30	420	415	410	407	405	402	400	397	395	392	390	385	380
35	486	481	476	474	471	469	466	464	461	459	456	451	446
40	553	548	543	540	538	535	533	530	528	525	523	518	513

Table 3-93

Hub Diameter, inch: 6

Leg Distance, inch: 90

SHIM REQUIRED FOR ALIGNMENT

Face - OD Method

FACE TIR MILS	OD TIR, MILS -40	-30	-20	-15	-10	-5	0	5	10	15	20	30	40
-40	-580	-585	-590	-592	-595	-597	-600	-602	-605	-607	-610	-615	-620
-35	-505	-510	-515	-517	-520	-522	-525	-527	-530	-532	-535	-540	-545
-30	-430	-435	-440	-442	-445	-447	-450	-452	-455	-457	-460	-465	-470
-25	-355	-360	-365	-367	-370	-372	-375	-377	-380	-382	-385	-390	-395
-20	-280	-285	-290	-292	-295	-297	-300	-302	-305	-307	-310	-315	-320
-15	-205	-210	-215	-217	-220	-222	-225	-227	-230	-232	-235	-240	-245
-12	-160	-165	-170	-172	-175	-177	-180	-182	-185	-187	-190	-195	-200
-10	-130	-135	-140	-142	-145	-147	-150	-152	-155	-157	-160	-165	-170
-9	-115	-120	-125	-127	-130	-132	-135	-137	-140	-142	-145	-150	-155
-8	-100	-105	-110	-112	-115	-117	-120	-122	-125	-127	-130	-135	-140
-7	-85	-90	-95	-97	-100	-102	-105	-107	-110	-112	-115	-120	-125
-6	-70	-75	-80	-82	-85	-87	-90	-92	-95	-97	-100	-105	-110
-5	-55	-60	-65	-67	-70	-72	-75	-77	-80	-82	-85	-90	-95
-4	-40	-45	-50	-52	-55	-57	-60	-62	-65	-67	-70	-75	-80
-3	-25	-30	-35	-37	-40	-42	-45	-47	-50	-52	-55	-60	-65
-2	-10	-15	-20	-22	-25	-27	-30	-32	-35	-37	-40	-45	-50
-1	5	0	-5	-7	-10	-12	-15	-17	-20	-22	-25	-30	-35
0	20	15	10	7	5	2	0	-2	-5	-7	-10	-15	-20
1	35	30	25	22	20	17	15	12	10	7	5	0	-5
2	50	45	40	37	35	32	30	27	25	22	20	15	10
3	65	60	55	52	50	47	45	42	40	37	35	30	25
4	80	75	70	67	65	62	60	57	55	52	50	45	40
5	95	90	85	82	80	77	75	72	70	67	65	60	55
6	110	105	100	97	95	92	90	87	85	82	80	75	70
7	125	120	115	112	110	107	105	102	100	97	95	90	85
8	140	135	130	127	125	122	120	117	115	112	110	105	100
9	155	150	145	142	140	137	135	132	130	127	125	120	115
10	170	165	160	157	155	152	150	147	145	142	140	135	130
12	200	195	190	187	185	182	180	177	175	172	170	165	160
15	245	240	235	232	230	227	225	222	220	217	215	210	205
20	320	315	310	307	305	302	300	297	295	292	290	285	280
25	395	390	385	382	380	377	375	372	370	367	365	360	355
30	470	465	460	457	455	452	450	447	445	442	440	435	430
35	545	540	535	532	530	527	525	522	520	517	515	510	505
40	620	615	610	607	605	602	600	597	595	592	590	585	580

Table 3-94

Hub Diameter, inch: 6

Leg Distance, inch: 100

SHIM REQUIRED FOR ALIGNMENT

Face - OD Method

FACE TIR MILS	OD TIR, MILS -40	-30	-20	-15	-10	-5	0	5	10	15	20	30	40
-40	-646	-651	-656	-659	-661	-664	-666	-669	-671	-674	-676	-681	-686
-35	-563	-568	-573	-575	-578	-580	-583	-585	-588	-590	-593	-598	-603
-30	-480	-485	-490	-492	-495	-497	-500	-502	-505	-507	-510	-515	-520
-25	-396	-401	-406	-409	-411	-414	-416	-419	-421	-424	-426	-431	-436
-20	-313	-318	-323	-325	-328	-330	-333	-335	-338	-340	-343	-348	-353
-15	-230	-235	-240	-242	-245	-247	-250	-252	-255	-257	-260	-265	-270
-12	-180	-185	-190	-192	-195	-197	-200	-202	-205	-207	-210	-215	-220
-10	-146	-151	-156	-159	-161	-164	-166	-169	-171	-174	-176	-181	-186
-9	-130	-135	-140	-142	-145	-147	-150	-152	-155	-157	-160	-165	-170
-8	-113	-118	-123	-125	-128	-130	-133	-135	-138	-140	-143	-148	-153
-7	-96	-101	-106	-109	-111	-114	-116	-119	-121	-124	-126	-131	-136
-6	-80	-85	-90	-92	-95	-97	-100	-102	-105	-107	-110	-115	-120
-5	-63	-68	-73	-75	-78	-80	-83	-85	-88	-90	-93	-98	-103
-4	-46	-51	-56	-59	-61	-64	-66	-69	-71	-74	-76	-81	-86
-3	-30	-35	-40	-42	-45	-47	-50	-52	-55	-57	-60	-65	-70
-2	-13	-18	-23	-25	-28	-30	-33	-35	-38	-40	-43	-48	-53
-1	3	-1	-6	-9	-11	-14	-16	-19	-21	-24	-26	-31	-36
0	20	15	10	7	5	2	0	-2	-5	-7	-10	-15	-20
1	36	31	26	24	21	19	16	14	11	9	6	1	-3
2	53	48	43	40	38	35	33	30	28	25	23	18	13
3	70	65	60	57	55	52	50	47	45	42	40	35	30
4	86	81	76	74	71	69	66	64	61	59	56	51	46
5	103	98	93	90	88	85	83	80	78	75	73	68	63
6	120	115	110	107	105	102	100	97	95	92	90	85	80
7	136	131	126	124	121	119	116	114	111	109	106	101	96
8	153	148	143	140	138	135	133	130	128	125	123	118	113
9	170	165	160	157	155	152	150	147	145	142	140	135	130
10	186	181	176	174	171	169	166	164	161	159	156	151	146
12	220	215	210	207	205	202	200	197	195	192	190	185	180
15	270	265	260	257	255	252	250	247	245	242	240	235	230
20	353	348	343	340	338	335	333	330	328	325	323	318	313
25	436	431	426	424	421	419	416	414	411	409	406	401	396
30	520	515	510	507	505	502	500	497	495	492	490	485	480
35	603	598	593	590	588	585	583	580	578	575	573	568	563
40	686	681	676	674	671	669	666	664	661	659	656	651	646

Table 3-95

Hub Diameter, inch: 7

Leg Distance, inch: 25

SHIM REQUIRED FOR ALIGNMENT

Face - OD Method

FACE TIR MILS	OD TIR, MILS -40	-30	-20	-15	-10	-5	0	5	10	15	20	30	40
-40	-122	-127	-132	-135	-137	-140	-142	-145	-147	-150	-152	-157	-162
-35	-105	-110	-115	-117	-120	-122	-125	-127	-130	-132	-135	-140	-145
-30	-87	-92	-97	-99	-102	-104	-107	-109	-112	-114	-117	-122	-127
-25	-69	-74	-79	-81	-84	-86	-89	-91	-94	-96	-99	-104	-109
-20	-51	-56	-61	-63	-66	-68	-71	-73	-76	-78	-81	-86	-91
-15	-33	-38	-43	-46	-48	-51	-53	-56	-58	-61	-63	-68	-73
-12	-22	-27	-32	-35	-37	-40	-42	-45	-47	-50	-52	-57	-62
-10	-15	-20	-25	-28	-30	-33	-35	-38	-40	-43	-45	-50	-55
-9	-12	-17	-22	-24	-27	-29	-32	-34	-37	-39	-42	-47	-52
-8	-8	-13	-18	-21	-23	-26	-28	-31	-33	-36	-38	-43	-48
-7	-5	-10	-15	-17	-20	-22	-25	-27	-30	-32	-35	-40	-45
-6	-1	-6	-11	-13	-16	-18	-21	-23	-26	-28	-31	-36	-41
-5	2	-2	-7	-10	-12	-15	-17	-20	-22	-25	-27	-32	-37
-4	5	0	-4	-6	-9	-11	-14	-16	-19	-21	-24	-29	-34
-3	9	4	0	-3	-5	-8	-10	-13	-15	-18	-20	-25	-30
-2	12	7	2	0	-2	-4	-7	-9	-12	-14	-17	-22	-27
-1	16	11	6	3	1	-1	-3	-6	-8	-11	-13	-18	-23
0	20	15	10	7	5	2	0	-2	-5	-7	-10	-15	-20
1	23	18	13	11	8	6	3	1	-1	-3	-6	-11	-16
2	27	22	17	14	12	9	7	4	2	0	-2	-7	-12
3	30	25	20	18	15	13	10	8	5	3	0	-4	-9
4	34	29	24	21	19	16	14	11	9	6	4	0	-5
5	37	32	27	25	22	20	17	15	12	10	7	2	-2
6	41	36	31	28	26	23	21	18	16	13	11	6	1
7	45	40	35	32	30	27	25	22	20	17	15	10	5
8	48	43	38	36	33	31	28	26	23	21	18	13	8
9	52	47	42	39	37	34	32	29	27	24	22	17	12
10	55	50	45	43	40	38	35	33	30	28	25	20	15
12	62	57	52	50	47	45	42	40	37	35	32	27	22
15	73	68	63	61	58	56	53	51	48	46	43	38	33
20	91	86	81	78	76	73	71	68	66	63	61	56	51
25	109	104	99	96	94	91	89	86	84	81	79	74	69
30	127	122	117	114	112	109	107	104	102	99	97	92	87
35	145	140	135	132	130	127	125	122	120	117	115	110	105
40	162	157	152	150	147	145	142	140	137	135	132	127	122

Table 3-96

Hub Diameter, inch: 7

Leg Distance, inch: 30

SHIM REQUIRED FOR ALIGNMENT

Face - OD Method

FACE TIR MILS	OD TIR, MILS -40	-30	-20	-15	-10	-5	0	5	10	15	20	30	40
-40	-151	-156	-161	-163	-166	-168	-171	-173	-176	-178	-181	-186	-191
-35	-130	-135	-140	-142	-145	-147	-150	-152	-155	-157	-160	-165	-170
-30	-108	-113	-118	-121	-123	-126	-128	-131	-133	-136	-138	-143	-148
-25	-87	-92	-97	-99	-102	-104	-107	-109	-112	-114	-117	-122	-127
-20	-65	-70	-75	-78	-80	-83	-85	-88	-90	-93	-95	-100	-105
-15	-44	-49	-54	-56	-59	-61	-64	-66	-69	-71	-74	-79	-84
-12	-31	-36	-41	-43	-46	-48	-51	-53	-56	-58	-61	-66	-71
-10	-22	-27	-32	-35	-37	-40	-42	-45	-47	-50	-52	-57	-62
-9	-18	-23	-28	-31	-33	-36	-38	-41	-43	-46	-48	-53	-58
-8	-14	-19	-24	-26	-29	-31	-34	-36	-39	-41	-44	-49	-54
-7	-10	-15	-20	-22	-25	-27	-30	-32	-35	-37	-40	-45	-50
-6	-5	-10	-15	-18	-20	-23	-25	-28	-30	-33	-35	-40	-45
-5	-1	-6	-11	-13	-16	-18	-21	-23	-26	-28	-31	-36	-41
-4	2	-2	-7	-9	-12	-14	-17	-19	-22	-24	-27	-32	-37
-3	7	2	-2	-5	-7	-10	-12	-15	-17	-20	-22	-27	-32
-2	11	6	1	-1	-3	-6	-8	-11	-13	-16	-18	-23	-28
-1	15	10	5	3	0	-1	-4	-6	-9	-11	-14	-19	-24
0	20	15	10	7	5	2	0	-2	-5	-7	-10	-15	-20
1	24	19	14	11	9	6	4	1	0	-3	-5	-10	-15
2	28	23	18	16	13	11	8	6	3	1	-1	-6	-11
3	32	27	22	20	17	15	12	10	7	5	2	-2	-7
4	37	32	27	24	22	19	17	14	12	9	7	2	-2
5	41	36	31	28	26	23	21	18	16	13	11	6	1
6	45	40	35	33	30	28	25	23	20	18	15	10	5
7	50	45	40	37	35	32	30	27	25	22	20	15	10
8	54	49	44	41	39	36	34	31	29	26	24	19	14
9	58	53	48	46	43	41	38	36	33	31	28	23	18
10	62	57	52	50	47	45	42	40	37	35	32	27	22
12	71	66	61	58	56	53	51	48	46	43	41	36	31
15	84	79	74	71	69	66	64	61	59	56	54	49	44
20	105	100	95	93	90	88	85	83	80	78	75	70	65
25	127	122	117	114	112	109	107	104	102	99	97	92	87
30	148	143	138	136	133	131	128	126	123	121	118	113	108
35	170	165	160	157	155	152	150	147	145	142	140	135	130
40	191	186	181	178	176	173	171	168	166	163	161	156	151

Table 3-97

Hub Diameter, inch: 7

Leg Distance, inch: 35

SHIM REQUIRED FOR ALIGNMENT

Face - OD Method

FACE TIR MILS	OD TIR, MILS -40	-30	-20	-15	-10	-5	0	5	10	15	20	30	40
-40	-180	-185	-190	-192	-195	-197	-200	-202	-205	-207	-210	-215	-220
-35	-155	-160	-165	-167	-170	-172	-175	-177	-180	-182	-185	-190	-195
-30	-130	-135	-140	-142	-145	-147	-150	-152	-155	-157	-160	-165	-170
-25	-105	-110	-115	-117	-120	-122	-125	-127	-130	-132	-135	-140	-145
-20	-80	-85	-90	-92	-95	-97	-100	-102	-105	-107	-110	-115	-120
-15	-55	-60	-65	-67	-70	-72	-75	-77	-80	-82	-85	-90	-95
-12	-40	-45	-50	-52	-55	-57	-60	-62	-65	-67	-70	-75	-80
-10	-30	-35	-40	-42	-45	-47	-50	-52	-55	-57	-60	-65	-70
-9	-25	-30	-35	-37	-40	-42	-45	-47	-50	-52	-55	-60	-65
-8	-20	-25	-30	-32	-35	-37	-40	-42	-45	-47	-50	-55	-60
-7	-15	-20	-25	-27	-30	-32	-35	-37	-40	-42	-45	-50	-55
-6	-10	-15	-20	-22	-25	-27	-30	-32	-35	-37	-40	-45	-50
-5	-5	-10	-15	-17	-20	-22	-25	-27	-30	-32	-35	-40	-45
-4	0	-5	-10	-12	-15	-17	-20	-22	-25	-27	-30	-35	-40
-3	5	0	-5	-7	-10	-12	-15	-17	-20	-22	-25	-30	-35
-2	10	5	0	-2	-5	-7	-10	-12	-15	-17	-20	-25	-30
-1	15	10	5	2	0	-2	-5	-7	-10	-12	-15	-20	-25
0	20	15	10	7	5	2	0	-2	-5	-7	-10	-15	-20
1	25	20	15	12	10	7	5	2	0	-2	-5	-10	-15
2	30	25	20	17	15	12	10	7	5	2	0	-5	-10
3	35	30	25	22	20	17	15	12	10	7	5	0	-5
4	40	35	30	27	25	22	20	17	15	12	10	5	0
5	45	40	35	32	30	27	25	22	20	17	15	10	5
6	50	45	40	37	35	32	30	27	25	22	20	15	10
7	55	50	45	42	40	37	35	32	30	27	25	20	15
8	60	55	50	47	45	42	40	37	35	32	30	25	20
9	65	60	55	52	50	47	45	42	40	37	35	30	25
10	70	65	60	57	55	52	50	47	45	42	40	35	30
12	80	75	70	67	65	62	60	57	55	52	50	45	40
15	95	90	85	82	80	77	75	72	70	67	65	60	55
20	120	115	110	107	105	102	100	97	95	92	90	85	80
25	145	140	135	132	130	127	125	122	120	117	115	110	105
30	170	165	160	157	155	152	150	147	145	142	140	135	130
35	195	190	185	182	180	177	175	172	170	167	165	160	155
40	220	215	210	207	205	202	200	197	195	192	190	185	180

Table 3-98

Hub Diameter, inch: 7

Leg Distance, inch: 40

SHIM REQUIRED FOR ALIGNMENT

Face - OD Method

FACE TIR MILS	OD TIR, MILS -40	-30	-20	-15	-10	-5	0	5	10	15	20	30	40
-40	-208	-213	-218	-221	-223	-226	-228	-231	-233	-236	-238	-243	-248
-35	-180	-185	-190	-192	-195	-197	-200	-202	-205	-207	-210	-215	-220
-30	-151	-156	-161	-163	-166	-168	-171	-173	-176	-178	-181	-186	-191
-25	-122	-127	-132	-135	-137	-140	-142	-145	-147	-150	-152	-157	-162
-20	-94	-99	-104	-106	-109	-111	-114	-116	-119	-121	-124	-129	-134
-15	-65	-70	-75	-78	-80	-83	-85	-88	-90	-93	-95	-100	-105
-12	-48	-53	-58	-61	-63	-66	-68	-71	-73	-76	-78	-83	-88
-10	-37	-42	-47	-49	-52	-54	-57	-59	-62	-64	-67	-72	-77
-9	-31	-36	-41	-43	-46	-48	-51	-53	-56	-58	-61	-66	-71
-8	-25	-30	-35	-38	-40	-43	-45	-48	-50	-53	-55	-60	-65
-7	-20	-25	-30	-32	-35	-37	-40	-42	-45	-47	-50	-55	-60
-6	-14	-19	-24	-26	-29	-31	-34	-36	-39	-41	-44	-49	-54
-5	-8	-13	-18	-21	-23	-26	-28	-31	-33	-36	-38	-43	-48
-4	-2	-7	-12	-15	-17	-20	-22	-25	-27	-30	-32	-37	-42
-3	2	-2	-7	-9	-12	-14	-17	-19	-22	-24	-27	-32	-37
-2	8	3	-1	-3	-6	-8	-11	-13	-16	-18	-21	-26	-31
-1	14	9	4	1	0	-3	-5	-8	-10	-13	-15	-20	-25
0	20	15	10	7	5	2	0	-2	-5	-7	-10	-15	-20
1	25	20	15	13	10	8	5	3	0	-1	-4	-9	-14
2	31	26	21	18	16	13	11	8	6	3	1	-3	-8
3	37	32	27	24	22	19	17	14	12	9	7	2	-2
4	42	37	32	30	27	25	22	20	17	15	12	7	2
5	48	43	38	36	33	31	28	26	23	21	18	13	8
6	54	49	44	41	39	36	34	31	29	26	24	19	14
7	60	55	50	47	45	42	40	37	35	32	30	25	20
8	65	60	55	53	50	48	45	43	40	38	35	30	25
9	71	66	61	58	56	53	51	48	46	43	41	36	31
10	77	72	67	64	62	59	57	54	52	49	47	42	37
12	88	83	78	76	73	71	68	66	63	61	58	53	48
15	105	100	95	93	90	88	85	83	80	78	75	70	65
20	134	129	124	121	119	116	114	111	109	106	104	99	94
25	162	157	152	150	147	145	142	140	137	135	132	127	122
30	191	186	181	178	176	173	171	168	166	163	161	156	151
35	220	215	210	207	205	202	200	197	195	192	190	185	180
40	248	243	238	236	233	231	228	226	223	221	218	213	208

Table 3-99

Hub Diameter, inch: 7

Leg Distance, inch: 50

SHIM REQUIRED FOR ALIGNMENT

Face - OD Method

FACE TIR MILS	OD TIR, MILS -40	-30	-20	-15	-10	-5	0	5	10	15	20	30	40
-40	-265	-270	-275	-278	-280	-283	-285	-288	-290	-293	-295	-300	-305
-35	-230	-235	-240	-242	-245	-247	-250	-252	-255	-257	-260	-265	-270
-30	-194	-199	-204	-206	-209	-211	-214	-216	-219	-221	-224	-229	-234
-25	-158	-163	-168	-171	-173	-176	-178	-181	-183	-186	-188	-193	-198
-20	-122	-127	-132	-135	-137	-140	-142	-145	-147	-150	-152	-157	-162
-15	-87	-92	-97	-99	-102	-104	-107	-109	-112	-114	-117	-122	-127
-12	-65	-70	-75	-78	-80	-83	-85	-88	-90	-93	-95	-100	-105
-10	-51	-56	-61	-63	-66	-68	-71	-73	-76	-78	-81	-86	-91
-9	-44	-49	-54	-56	-59	-61	-64	-66	-69	-71	-74	-79	-84
-8	-37	-42	-47	-49	-52	-54	-57	-59	-62	-64	-67	-72	-77
-7	-30	-35	-40	-42	-45	-47	-50	-52	-55	-57	-60	-65	-70
-6	-22	-27	-32	-35	-37	-40	-42	-45	-47	-50	-52	-57	-62
-5	-15	-20	-25	-28	-30	-33	-35	-38	-40	-43	-45	-50	-55
-4	-8	-13	-18	-21	-23	-26	-28	-31	-33	-36	-38	-43	-48
-3	-1	-6	-11	-13	-16	-18	-21	-23	-26	-28	-31	-36	-41
-2	5	0	-4	-6	-9	-11	-14	-16	-19	-21	-24	-29	-34
-1	12	7	2	0	-2	-4	-7	-9	-12	-14	-17	-22	-27
0	20	15	10	7	5	2	0	-2	-5	-7	-10	-15	-20
1	27	22	17	14	12	9	7	4	2	0	-2	-7	-12
2	34	29	24	21	19	16	14	11	9	6	4	0	-5
3	41	36	31	28	26	23	21	18	16	13	11	6	1
4	48	43	38	36	33	31	28	26	23	21	18	13	8
5	55	50	45	43	40	38	35	33	30	28	25	20	15
6	62	57	52	50	47	45	42	40	37	35	32	27	22
7	70	65	60	57	55	52	50	47	45	42	40	35	30
8	77	72	67	64	62	59	57	54	52	49	47	42	37
9	84	79	74	71	69	66	64	61	59	56	54	49	44
10	91	86	81	78	76	73	71	68	66	63	61	56	51
12	105	100	95	93	90	88	85	83	80	78	75	70	65
15	127	122	117	114	112	109	107	104	102	99	97	92	87
20	162	157	152	150	147	145	142	140	137	135	132	127	122
25	198	193	188	186	183	181	178	176	173	171	168	163	158
30	234	229	224	221	219	216	214	211	209	206	204	199	194
35	270	265	260	257	255	252	250	247	245	242	240	235	230
40	305	300	295	293	290	288	285	283	280	278	275	270	265

Table 3-100

Hub Diameter, inch: 7

Leg Distance, inch: 60

SHIM REQUIRED FOR ALIGNMENT

Face – OD Method

FACE TIR MILS	OD TIR, MILS -40	-30	-20	-15	-10	-5	0	5	10	15	20	30	40
-40	-322	-327	-332	-335	-337	-340	-342	-345	-347	-350	-352	-357	-362
-35	-280	-285	-290	-292	-295	-297	-300	-302	-305	-307	-310	-315	-320
-30	-237	-242	-247	-249	-252	-254	-257	-259	-262	-264	-267	-272	-277
-25	-194	-199	-204	-206	-209	-211	-214	-216	-219	-221	-224	-229	-234
-20	-151	-156	-161	-163	-166	-168	-171	-173	-176	-178	-181	-186	-191
-15	-108	-113	-118	-121	-123	-126	-128	-131	-133	-136	-138	-143	-148
-12	-82	-87	-92	-95	-97	-100	-102	-105	-107	-110	-112	-117	-122
-10	-65	-70	-75	-78	-80	-83	-85	-88	-90	-93	-95	-100	-105
-9	-57	-62	-67	-69	-72	-74	-77	-79	-82	-84	-87	-92	-97
-8	-48	-53	-58	-61	-63	-66	-68	-71	-73	-76	-78	-83	-88
-7	-40	-45	-50	-52	-55	-57	-60	-62	-65	-67	-70	-75	-80
-6	-31	-36	-41	-43	-46	-48	-51	-53	-56	-58	-61	-66	-71
-5	-22	-27	-32	-35	-37	-40	-42	-45	-47	-50	-52	-57	-62
-4	-14	-19	-24	-26	-29	-31	-34	-36	-39	-41	-44	-49	-54
-3	-5	-10	-15	-18	-20	-23	-25	-28	-30	-33	-35	-40	-45
-2	2	-2	-7	-9	-12	-14	-17	-19	-22	-24	-27	-32	-37
-1	11	6	1	-1	-3	-6	-8	-11	-13	-16	-18	-23	-28
0	20	15	10	7	5	2	0	-2	-5	-7	-10	-15	-20
1	28	23	18	16	13	11	8	6	3	1	-1	-6	-11
2	37	32	27	24	22	19	17	14	12	9	7	2	-2
3	45	40	35	33	30	28	25	23	20	18	15	10	5
4	54	49	44	41	39	36	34	31	29	26	24	19	14
5	62	57	52	50	47	45	42	40	37	35	32	27	22
6	71	66	61	58	56	53	51	48	46	43	41	36	31
7	80	75	70	67	65	62	60	57	55	52	50	45	40
8	88	83	78	76	73	71	68	66	63	61	58	53	48
9	97	92	87	84	82	79	77	74	72	69	67	62	57
10	105	100	95	93	90	88	85	83	80	78	75	70	65
12	122	117	112	110	107	105	102	100	97	95	92	87	82
15	148	143	138	136	133	131	128	126	123	121	118	113	108
20	191	186	181	178	176	173	171	168	166	163	161	156	151
25	234	229	224	221	219	216	214	211	209	206	204	199	194
30	277	272	267	264	262	259	257	254	252	249	247	242	237
35	320	315	310	307	305	302	300	297	295	292	290	285	280
40	362	357	352	350	347	345	342	340	337	335	332	327	322

Table 3-101

Hub Diameter, inch: 7

Leg Distance, inch: 70

SHIM REQUIRED FOR ALIGNMENT

Face - OD Method

FACE TIR MILS	OD TIR, MILS -40	-30	-20	-15	-10	-5	0	5	10	15	20	30	40
-40	-380	-385	-390	-392	-395	-397	-400	-402	-405	-407	-410	-415	-420
-35	-330	-335	-340	-342	-345	-347	-350	-352	-355	-357	-360	-365	-370
-30	-280	-285	-290	-292	-295	-297	-300	-302	-305	-307	-310	-315	-320
-25	-230	-235	-240	-242	-245	-247	-250	-252	-255	-257	-260	-265	-270
-20	-180	-185	-190	-192	-195	-197	-200	-202	-205	-207	-210	-215	-220
-15	-130	-135	-140	-142	-145	-147	-150	-152	-155	-157	-160	-165	-170
-12	-100	-105	-110	-112	-115	-117	-120	-122	-125	-127	-130	-135	-140
-10	-80	-85	-90	-92	-95	-97	-100	-102	-105	-107	-110	-115	-120
-9	-70	-75	-80	-82	-85	-87	-90	-92	-95	-97	-100	-105	-110
-8	-60	-65	-70	-72	-75	-77	-80	-82	-85	-87	-90	-95	-100
-7	-50	-55	-60	-62	-65	-67	-70	-72	-75	-77	-80	-85	-90
-6	-40	-45	-50	-52	-55	-57	-60	-62	-65	-67	-70	-75	-80
-5	-30	-35	-40	-42	-45	-47	-50	-52	-55	-57	-60	-65	-70
-4	-20	-25	-30	-32	-35	-37	-40	-42	-45	-47	-50	-55	-60
-3	-10	-15	-20	-22	-25	-27	-30	-32	-35	-37	-40	-45	-50
-2	0	-5	-10	-12	-15	-17	-20	-22	-25	-27	-30	-35	-40
-1	10	5	0	-2	-5	-7	-10	-12	-15	-17	-20	-25	-30
0	20	15	10	7	5	2	0	-2	-5	-7	-10	-15	-20
1	30	25	20	17	15	12	10	7	5	2	0	-5	-10
2	40	35	30	27	25	22	20	17	15	12	10	5	0
3	50	45	40	37	35	32	30	27	25	22	20	15	10
4	60	55	50	47	45	42	40	37	35	32	30	25	20
5	70	65	60	57	55	52	50	47	45	42	40	35	30
6	80	75	70	67	65	62	60	57	55	52	50	45	40
7	90	85	80	77	75	72	70	67	65	62	60	55	50
8	100	95	90	87	85	82	80	77	75	72	70	65	60
9	110	105	100	97	95	92	90	87	85	82	80	75	70
10	120	115	110	107	105	102	100	97	95	92	90	85	80
12	140	135	130	127	125	122	120	117	115	112	110	105	100
15	170	165	160	157	155	152	150	147	145	142	140	135	130
20	220	215	210	207	205	202	200	197	195	192	190	185	180
25	270	265	260	257	255	252	250	247	245	242	240	235	230
30	320	315	310	307	305	302	300	297	295	292	290	285	280
35	370	365	360	357	355	352	350	347	345	342	340	335	330
40	420	415	410	407	405	402	400	397	395	392	390	385	380

Table 3-102

Hub Diameter, inch: 7

Leg Distance, inch: 80

SHIM REQUIRED FOR ALIGNMENT

Face - OD Method

FACE TIR MILS	OD TIR, MILS -40	-30	-20	-15	-10	-5	0	5	10	15	20	30	40
-40	-437	-442	-447	-449	-452	-454	-457	-459	-462	-464	-467	-472	-477
-35	-380	-385	-390	-392	-395	-397	-400	-402	-405	-407	-410	-415	-420
-30	-322	-327	-332	-335	-337	-340	-342	-345	-347	-350	-352	-357	-362
-25	-265	-270	-275	-278	-280	-283	-285	-288	-290	-293	-295	-300	-305
-20	-208	-213	-218	-221	-223	-226	-228	-231	-233	-236	-238	-243	-248
-15	-151	-156	-161	-163	-166	-168	-171	-173	-176	-178	-181	-186	-191
-12	-117	-122	-127	-129	-132	-134	-137	-139	-142	-144	-147	-152	-157
-10	-94	-99	-104	-106	-109	-111	-114	-116	-119	-121	-124	-129	-134
-9	-82	-87	-92	-95	-97	-100	-102	-105	-107	-110	-112	-117	-122
-8	-71	-76	-81	-83	-86	-88	-91	-93	-96	-98	-101	-106	-111
-7	-60	-65	-70	-72	-75	-77	-80	-82	-85	-87	-90	-95	-100
-6	-48	-53	-58	-61	-63	-66	-68	-71	-73	-76	-78	-83	-88
-5	-37	-42	-47	-49	-52	-54	-57	-59	-62	-64	-67	-72	-77
-4	-25	-30	-35	-38	-40	-43	-45	-48	-50	-53	-55	-60	-65
-3	-14	-19	-24	-26	-29	-31	-34	-36	-39	-41	-44	-49	-54
-2	-2	-7	-12	-15	-17	-20	-22	-25	-27	-30	-32	-37	-42
-1	8	3	-1	-3	-6	-8	-11	-13	-16	-18	-21	-26	-31
0	20	15	10	7	5	2	0	-2	-5	-7	-10	-15	-20
1	31	26	21	18	16	13	11	8	6	3	1	-3	-8
2	42	37	32	30	27	25	22	20	17	15	12	7	2
3	54	49	44	41	39	36	34	31	29	26	24	19	14
4	65	60	55	53	50	48	45	43	40	38	35	30	25
5	77	72	67	64	62	59	57	54	52	49	47	42	37
6	88	83	78	76	73	71	68	66	63	61	58	53	48
7	100	95	90	87	85	82	80	77	75	72	70	65	60
8	111	106	101	98	96	93	91	88	86	83	81	76	71
9	122	117	112	110	107	105	102	100	97	95	92	87	82
10	134	129	124	121	119	116	114	111	109	106	104	99	94
12	157	152	147	144	142	139	137	134	132	129	127	122	117
15	191	186	181	178	176	173	171	168	166	163	161	156	151
20	248	243	238	236	233	231	228	226	223	221	218	213	208
25	305	300	295	293	290	288	285	283	280	278	275	270	265
30	362	357	352	350	347	345	342	340	337	335	332	327	322
35	420	415	410	407	405	402	400	397	395	392	390	385	380
40	477	472	467	464	462	459	457	454	452	449	447	442	437

Table 3-103

Hub Diameter, inch: 7

Leg Distance, inch: 90

SHIM REQUIRED FOR ALIGNMENT

Face - OD Method

FACE TIR MILS	OD TIR, MILS -40	-30	-20	-15	-10	-5	0	5	10	15	20	30	40
-40	-494	-499	-504	-506	-509	-511	-514	-516	-519	-521	-524	-529	-534
-35	-430	-435	-440	-442	-445	-447	-450	-452	-455	-457	-460	-465	-470
-30	-365	-370	-375	-378	-380	-383	-385	-388	-390	-393	-395	-400	-405
-25	-301	-306	-311	-313	-316	-318	-321	-323	-326	-328	-331	-336	-341
-20	-237	-242	-247	-249	-252	-254	-257	-259	-262	-264	-267	-272	-277
-15	-172	-177	-182	-185	-187	-190	-192	-195	-197	-200	-202	-207	-212
-12	-134	-139	-144	-146	-149	-151	-154	-156	-159	-161	-164	-169	-174
-10	-108	-113	-118	-121	-123	-126	-128	-131	-133	-136	-138	-143	-148
-9	-95	-100	-105	-108	-110	-113	-115	-118	-120	-123	-125	-130	-135
-8	-82	-87	-92	-95	-97	-100	-102	-105	-107	-110	-112	-117	-122
-7	-70	-75	-80	-82	-85	-87	-90	-92	-95	-97	-100	-105	-110
-6	-57	-62	-67	-69	-72	-74	-77	-79	-82	-84	-87	-92	-97
-5	-44	-49	-54	-56	-59	-61	-64	-66	-69	-71	-74	-79	-84
-4	-31	-36	-41	-43	-46	-48	-51	-53	-56	-58	-61	-66	-71
-3	-18	-23	-28	-31	-33	-36	-38	-41	-43	-46	-48	-53	-58
-2	-5	-10	-15	-18	-20	-23	-25	-28	-30	-33	-35	-40	-45
-1	7	2	-2	-5	-7	-10	-12	-15	-17	-20	-22	-27	-32
0	20	15	10	7	5	2	0	-2	-5	-7	-10	-15	-20
1	32	27	22	20	17	15	12	10	7	5	2	-2	-7
2	45	40	35	33	30	28	25	23	20	18	15	10	5
3	58	53	48	46	43	41	38	36	33	31	28	23	18
4	71	66	61	58	56	53	51	48	46	43	41	36	31
5	84	79	74	71	69	66	64	61	59	56	54	49	44
6	97	92	87	84	82	79	77	74	72	69	67	62	57
7	110	105	100	97	95	92	90	87	85	82	80	75	70
8	122	117	112	110	107	105	102	100	97	95	92	87	82
9	135	130	125	123	120	118	115	113	110	108	105	100	95
10	148	143	138	136	133	131	128	126	123	121	118	113	108
12	174	169	164	161	159	156	154	151	149	146	144	139	134
15	212	207	202	200	197	195	192	190	187	185	182	177	172
20	277	272	267	264	262	259	257	254	252	249	247	242	237
25	341	336	331	328	326	323	321	318	316	313	311	306	301
30	405	400	395	393	390	388	385	383	380	378	375	370	365
35	470	465	460	457	455	452	450	447	445	442	440	435	430
40	534	529	524	521	519	516	514	511	509	506	504	499	494

Table 3-104

Hub Diameter, inch: 7

Leg Distance, inch: 100

SHIM REQUIRED FOR ALIGNMENT

Face - OD Method

FACE TIR MILS	OD TIR, MILS -40	-30	-20	-15	-10	-5	0	5	10	15	20	30	40
-40	-551	-556	-561	-563	-566	-568	-571	-573	-576	-578	-581	-586	-591
-35	-480	-485	-490	-492	-495	-497	-500	-502	-505	-507	-510	-515	-520
-30	-408	-413	-418	-421	-423	-426	-428	-431	-433	-436	-438	-443	-448
-25	-337	-342	-347	-349	-352	-354	-357	-359	-362	-364	-367	-372	-377
-20	-265	-270	-275	-278	-280	-283	-285	-288	-290	-293	-295	-300	-305
-15	-194	-199	-204	-206	-209	-211	-214	-216	-219	-221	-224	-229	-234
-12	-151	-156	-161	-163	-166	-168	-171	-173	-176	-178	-181	-186	-191
-10	-122	-127	-132	-135	-137	-140	-142	-145	-147	-150	-152	-157	-162
-9	-108	-113	-118	-121	-123	-126	-128	-131	-133	-136	-138	-143	-148
-8	-94	-99	-104	-106	-109	-111	-114	-116	-119	-121	-124	-129	-134
-7	-80	-85	-90	-92	-95	-97	-100	-102	-105	-107	-110	-115	-120
-6	-65	-70	-75	-78	-80	-83	-85	-88	-90	-93	-95	-100	-105
-5	-51	-56	-61	-63	-66	-68	-71	-73	-76	-78	-81	-86	-91
-4	-37	-42	-47	-49	-52	-54	-57	-59	-62	-64	-67	-72	-77
-3	-22	-27	-32	-35	-37	-40	-42	-45	-47	-50	-52	-57	-62
-2	-8	-13	-18	-21	-23	-26	-28	-31	-33	-36	-38	-43	-48
-1	5	0	-4	-6	-9	-11	-14	-16	-19	-21	-24	-29	-34
0	20	15	10	7	5	2	0	-2	-5	-7	-10	-15	-20
1	34	29	24	21	19	16	14	11	9	6	4	0	-5
2	48	43	38	36	33	31	28	26	23	21	18	13	8
3	62	57	52	50	47	45	42	40	37	35	32	27	22
4	77	72	67	64	62	59	57	54	52	49	47	42	37
5	91	86	81	78	76	73	71	68	66	63	61	56	51
6	105	100	95	93	90	88	85	83	80	78	75	70	65
7	120	115	110	107	105	102	100	97	95	92	90	85	80
8	134	129	124	121	119	116	114	111	109	106	104	99	94
9	148	143	138	136	133	131	128	126	123	121	118	113	108
10	162	157	152	150	147	145	142	140	137	135	132	127	122
12	191	186	181	178	176	173	171	168	166	163	161	156	151
15	234	229	224	221	219	216	214	211	209	206	204	199	194
20	305	300	295	293	290	288	285	283	280	278	275	270	265
25	377	372	367	364	362	359	357	354	352	349	347	342	337
30	448	443	438	436	433	431	428	426	423	421	418	413	408
35	520	515	510	507	505	502	500	497	495	492	490	485	480
40	591	586	581	578	576	573	571	568	566	563	561	556	551

Table 3-105

Hub Diameter, inch: 8

Leg Distance, inch: 25

SHIM REQUIRED FOR ALIGNMENT

Face - OD Method

FACE TIR MILS	OD TIR, MILS -40	-30	-20	-15	-10	-5	0	5	10	15	20	30	40
-40	-105	-110	-115	-117	-120	-122	-125	-127	-130	-132	-135	-140	-145
-35	-89	-94	-99	-101	-104	-106	-109	-111	-114	-116	-119	-124	-129
-30	-73	-78	-83	-86	-88	-91	-93	-96	-98	-101	-103	-108	-113
-25	-58	-63	-68	-70	-73	-75	-78	-80	-83	-85	-88	-93	-98
-20	-42	-47	-52	-55	-57	-60	-62	-65	-67	-70	-72	-77	-82
-15	-26	-31	-36	-39	-41	-44	-46	-49	-51	-54	-56	-61	-66
-12	-17	-22	-27	-30	-32	-35	-37	-40	-42	-45	-47	-52	-57
-10	-11	-16	-21	-23	-26	-28	-31	-33	-36	-38	-41	-46	-51
-9	-8	-13	-18	-20	-23	-25	-28	-30	-33	-35	-38	-43	-48
-8	-5	-10	-15	-17	-20	-22	-25	-27	-30	-32	-35	-40	-45
-7	-1	-6	-11	-14	-16	-19	-21	-24	-26	-29	-31	-36	-41
-6	1	-3	-8	-11	-13	-16	-18	-21	-23	-26	-28	-33	-38
-5	4	0	-5	-8	-10	-13	-15	-18	-20	-23	-25	-30	-35
-4	7	2	-2	-5	-7	-10	-12	-15	-17	-20	-22	-27	-32
-3	10	5	0	-1	-4	-6	-9	-11	-14	-16	-19	-24	-29
-2	13	8	3	1	-1	-3	-6	-8	-11	-13	-16	-21	-26
-1	16	11	6	4	1	0	-3	-5	-8	-10	-13	-18	-23
0	20	15	10	7	5	2	0	-2	-5	-7	-10	-15	-20
1	23	18	13	10	8	5	3	0	-1	-4	-6	-11	-16
2	26	21	16	13	11	8	6	3	1	-1	-3	-8	-13
3	29	24	19	16	14	11	9	6	4	1	0	-5	-10
4	32	27	22	20	17	15	12	10	7	5	2	-2	-7
5	35	30	25	23	20	18	15	13	10	8	5	0	-4
6	38	33	28	26	23	21	18	16	13	11	8	3	-1
7	41	36	31	29	26	24	21	19	16	14	11	6	1
8	45	40	35	32	30	27	25	22	20	17	15	10	5
9	48	43	38	35	33	30	28	25	23	20	18	13	8
10	51	46	41	38	36	33	31	28	26	23	21	16	11
12	57	52	47	45	42	40	37	35	32	30	27	22	17
15	66	61	56	54	51	49	46	44	41	39	36	31	26
20	82	77	72	70	67	65	62	60	57	55	52	47	42
25	98	93	88	85	83	80	78	75	73	70	68	63	58
30	113	108	103	101	98	96	93	91	88	86	83	78	73
35	129	124	119	116	114	111	109	106	104	101	99	94	89
40	145	140	135	132	130	127	125	122	120	117	115	110	105

Table 3-106

Hub Diameter, inch: 8

Leg Distance, inch: 30

SHIM REQUIRED FOR ALIGNMENT

Face - OD Method

FACE TIR MILS	OD TIR, MILS -40	-30	-20	-15	-10	-5	0	5	10	15	20	30	40
-40	-130	-135	-140	-142	-145	-147	-150	-152	-155	-157	-160	-165	-170
-35	-111	-116	-121	-123	-126	-128	-131	-133	-136	-138	-141	-146	-151
-30	-92	-97	-102	-105	-107	-110	-112	-115	-117	-120	-122	-127	-132
-25	-73	-78	-83	-86	-88	-91	-93	-96	-98	-101	-103	-108	-113
-20	-55	-60	-65	-67	-70	-72	-75	-77	-80	-82	-85	-90	-95
-15	-36	-41	-46	-48	-51	-53	-56	-58	-61	-63	-66	-71	-76
-12	-25	-30	-35	-37	-40	-42	-45	-47	-50	-52	-55	-60	-65
-10	-17	-22	-27	-30	-32	-35	-37	-40	-42	-45	-47	-52	-57
-9	-13	-18	-23	-26	-28	-31	-33	-36	-38	-41	-43	-48	-53
-8	-10	-15	-20	-22	-25	-27	-30	-32	-35	-37	-40	-45	-50
-7	-6	-11	-16	-18	-21	-23	-26	-28	-31	-33	-36	-41	-46
-6	-2	-7	-12	-15	-17	-20	-22	-25	-27	-30	-32	-37	-42
-5	1	-3	-8	-11	-13	-16	-18	-21	-23	-26	-28	-33	-38
-4	5	0	-5	-7	-10	-12	-15	-17	-20	-22	-25	-30	-35
-3	8	3	-1	-3	-6	-8	-11	-13	-16	-18	-21	-26	-31
-2	12	7	2	0	-2	-5	-7	-10	-12	-15	-17	-22	-27
-1	16	11	6	3	1	-1	-3	-6	-8	-11	-13	-18	-23
0	20	15	10	7	5	2	0	-2	-5	-7	-10	-15	-20
1	23	18	13	11	8	6	3	1	-1	-3	-6	-11	-16
2	27	22	17	15	12	10	7	5	2	0	-2	-7	-12
3	31	26	21	18	16	13	11	8	6	3	1	-3	-8
4	35	30	25	22	20	17	15	12	10	7	5	0	-5
5	38	33	28	26	23	21	18	16	13	11	8	3	-1
6	42	37	32	30	27	25	22	20	17	15	12	7	2
7	46	41	36	33	31	28	26	23	21	18	16	11	6
8	50	45	40	37	35	32	30	27	25	22	20	15	10
9	53	48	43	41	38	36	33	31	28	26	23	18	13
10	57	52	47	45	42	40	37	35	32	30	27	22	17
12	65	60	55	52	50	47	45	42	40	37	35	30	25
15	76	71	66	63	61	58	56	53	51	48	46	41	36
20	95	90	85	82	80	77	75	72	70	67	65	60	55
25	113	108	103	101	98	96	93	91	88	86	83	78	73
30	132	127	122	120	117	115	112	110	107	105	102	97	92
35	151	146	141	138	136	133	131	128	126	123	121	116	111
40	170	165	160	157	155	152	150	147	145	142	140	135	130

Table 3-107

Hub Diameter, inch: 8

Leg Distance, inch: 35

SHIM REQUIRED FOR ALIGNMENT

Face – OD Method

FACE TIR MILS	OD TIR, MILS -40	-30	-20	-15	-10	-5	0	5	10	15	20	30	40
-40	-155	-160	-165	-167	-170	-172	-175	-177	-180	-182	-185	-190	-195
-35	-133	-138	-143	-145	-148	-150	-153	-155	-158	-160	-163	-168	-173
-30	-111	-116	-121	-123	-126	-128	-131	-133	-136	-138	-141	-146	-151
-25	-89	-94	-99	-101	-104	-106	-109	-111	-114	-116	-119	-124	-129
-20	-67	-72	-77	-80	-82	-85	-87	-90	-92	-95	-97	-102	-107
-15	-45	-50	-55	-58	-60	-63	-65	-68	-70	-73	-75	-80	-85
-12	-32	-37	-42	-45	-47	-50	-52	-55	-57	-60	-62	-67	-72
-10	-23	-28	-33	-36	-38	-41	-43	-46	-48	-51	-53	-58	-63
-9	-19	-24	-29	-31	-34	-36	-39	-41	-44	-46	-49	-54	-59
-8	-15	-20	-25	-27	-30	-32	-35	-37	-40	-42	-45	-50	-55
-7	-10	-15	-20	-23	-25	-28	-30	-33	-35	-38	-40	-45	-50
-6	-6	-11	-16	-18	-21	-23	-26	-28	-31	-33	-36	-41	-46
-5	-1	-6	-11	-14	-16	-19	-21	-24	-26	-29	-31	-36	-41
-4	2	-2	-7	-10	-12	-15	-17	-20	-22	-25	-27	-32	-37
-3	6	1	-3	-5	-8	-10	-13	-15	-18	-20	-23	-28	-33
-2	11	6	1	-1	-3	-6	-8	-11	-13	-16	-18	-23	-28
-1	15	10	5	3	0	-1	-4	-6	-9	-11	-14	-19	-24
0	20	15	10	7	5	2	0	-2	-5	-7	-10	-15	-20
1	24	19	14	11	9	6	4	1	0	-3	-5	-10	-15
2	28	23	18	16	13	11	8	6	3	1	-1	-6	-11
3	33	28	23	20	18	15	13	10	8	5	3	-1	-6
4	37	32	27	25	22	20	17	15	12	10	7	2	-2
5	41	36	31	29	26	24	21	19	16	14	11	6	1
6	46	41	36	33	31	28	26	23	21	18	16	11	6
7	50	45	40	38	35	33	30	28	25	23	20	15	10
8	55	50	45	42	40	37	35	32	30	27	25	20	15
9	59	54	49	46	44	41	39	36	34	31	29	24	19
10	63	58	53	51	48	46	43	41	38	36	33	28	23
12	72	67	62	60	57	55	52	50	47	45	42	37	32
15	85	80	75	73	70	68	65	63	60	58	55	50	45
20	107	102	97	95	92	90	87	85	82	80	77	72	67
25	129	124	119	116	114	111	109	106	104	101	99	94	89
30	151	146	141	138	136	133	131	128	126	123	121	116	111
35	173	168	163	160	158	155	153	150	148	145	143	138	133
40	195	190	185	182	180	177	175	172	170	167	165	160	155

Table 3-108

Hub Diameter, inch: 8

Leg Distance, inch: 40

SHIM REQUIRED FOR ALIGNMENT

Face - OD Method

FACE TIR MILS	OD TIR, MILS -40	-30	-20	-15	-10	-5	0	5	10	15	20	30	40
-40	-180	-185	-190	-192	-195	-197	-200	-202	-205	-207	-210	-215	-220
-35	-155	-160	-165	-167	-170	-172	-175	-177	-180	-182	-185	-190	-195
-30	-130	-135	-140	-142	-145	-147	-150	-152	-155	-157	-160	-165	-170
-25	-105	-110	-115	-117	-120	-122	-125	-127	-130	-132	-135	-140	-145
-20	-80	-85	-90	-92	-95	-97	-100	-102	-105	-107	-110	-115	-120
-15	-55	-60	-65	-67	-70	-72	-75	-77	-80	-82	-85	-90	-95
-12	-40	-45	-50	-52	-55	-57	-60	-62	-65	-67	-70	-75	-80
-10	-30	-35	-40	-42	-45	-47	-50	-52	-55	-57	-60	-65	-70
-9	-25	-30	-35	-37	-40	-42	-45	-47	-50	-52	-55	-60	-65
-8	-20	-25	-30	-32	-35	-37	-40	-42	-45	-47	-50	-55	-60
-7	-15	-20	-25	-27	-30	-32	-35	-37	-40	-42	-45	-50	-55
-6	-10	-15	-20	-22	-25	-27	-30	-32	-35	-37	-40	-45	-50
-5	-5	-10	-15	-17	-20	-22	-25	-27	-30	-32	-35	-40	-45
-4	0	-5	-10	-12	-15	-17	-20	-22	-25	-27	-30	-35	-40
-3	5	0	-5	-7	-10	-12	-15	-17	-20	-22	-25	-30	-35
-2	10	5	0	-2	-5	-7	-10	-12	-15	-17	-20	-25	-30
-1	15	10	5	2	0	-2	-5	-7	-10	-12	-15	-20	-25
0	20	15	10	7	5	2	0	-2	-5	-7	-10	-15	-20
1	25	20	15	12	10	7	5	2	0	-2	-5	-10	-15
2	30	25	20	17	15	12	10	7	5	2	0	-5	-10
3	35	30	25	22	20	17	15	12	10	7	5	0	-5
4	40	35	30	27	25	22	20	17	15	12	10	5	0
5	45	40	35	32	30	27	25	22	20	17	15	10	5
6	50	45	40	37	35	32	30	27	25	22	20	15	10
7	55	50	45	42	40	37	35	32	30	27	25	20	15
8	60	55	50	47	45	42	40	37	35	32	30	25	20
9	65	60	55	52	50	47	45	42	40	37	35	30	25
10	70	65	60	57	55	52	50	47	45	42	40	35	30
12	80	75	70	67	65	62	60	57	55	52	50	45	40
15	95	90	85	82	80	77	75	72	70	67	65	60	55
20	120	115	110	107	105	102	100	97	95	92	90	85	80
25	145	140	135	132	130	127	125	122	120	117	115	110	105
30	170	165	160	157	155	152	150	147	145	142	140	135	130
35	195	190	185	182	180	177	175	172	170	167	165	160	155
40	220	215	210	207	205	202	200	197	195	192	190	185	180

Table 3-109

Hub Diameter, inch: 8

Leg Distance, inch: 50

SHIM REQUIRED FOR ALIGNMENT

Face - OD Method

FACE TIR MILS	OD TIR, MILS -40	-30	-20	-15	-10	-5	0	5	10	15	20	30	40
-40	-230	-235	-240	-242	-245	-247	-250	-252	-255	-257	-260	-265	-270
-35	-198	-203	-208	-211	-213	-216	-218	-221	-223	-226	-228	-233	-238
-30	-167	-172	-177	-180	-182	-185	-187	-190	-192	-195	-197	-202	-207
-25	-136	-141	-146	-148	-151	-153	-156	-158	-161	-163	-166	-171	-176
-20	-105	-110	-115	-117	-120	-122	-125	-127	-130	-132	-135	-140	-145
-15	-73	-78	-83	-86	-88	-91	-93	-96	-98	-101	-103	-108	-113
-12	-55	-60	-65	-67	-70	-72	-75	-77	-80	-82	-85	-90	-95
-10	-42	-47	-52	-55	-57	-60	-62	-65	-67	-70	-72	-77	-82
-9	-36	-41	-46	-48	-51	-53	-56	-58	-61	-63	-66	-71	-76
-8	-30	-35	-40	-42	-45	-47	-50	-52	-55	-57	-60	-65	-70
-7	-23	-28	-33	-36	-38	-41	-43	-46	-48	-51	-53	-58	-63
-6	-17	-22	-27	-30	-32	-35	-37	-40	-42	-45	-47	-52	-57
-5	-11	-16	-21	-23	-26	-28	-31	-33	-36	-38	-41	-46	-51
-4	-5	-10	-15	-17	-20	-22	-25	-27	-30	-32	-35	-40	-45
-3	1	-3	-8	-11	-13	-16	-18	-21	-23	-26	-28	-33	-38
-2	7	2	-2	-5	-7	-10	-12	-15	-17	-20	-22	-27	-32
-1	13	8	3	1	-1	-3	-6	-8	-11	-13	-16	-21	-26
0	20	15	10	7	5	2	0	-2	-5	-7	-10	-15	-20
1	26	21	16	13	11	8	6	3	1	-1	-3	-8	-13
2	32	27	22	20	17	15	12	10	7	5	2	-2	-7
3	38	33	28	26	23	21	18	16	13	11	8	3	-1
4	45	40	35	32	30	27	25	22	20	17	15	10	5
5	51	46	41	38	36	33	31	28	26	23	21	16	11
6	57	52	47	45	42	40	37	35	32	30	27	22	17
7	63	58	53	51	48	46	43	41	38	36	33	28	23
8	70	65	60	57	55	52	50	47	45	42	40	35	30
9	76	71	66	63	61	58	56	53	51	48	46	41	36
10	82	77	72	70	67	65	62	60	57	55	52	47	42
12	95	90	85	82	80	77	75	72	70	67	65	60	55
15	113	108	103	101	98	96	93	91	88	86	83	78	73
20	145	140	135	132	130	127	125	122	120	117	115	110	105
25	176	171	166	163	161	158	156	153	151	148	146	141	136
30	207	202	197	195	192	190	187	185	182	180	177	172	167
35	238	233	228	226	223	221	218	216	213	211	208	203	198
40	270	265	260	257	255	252	250	247	245	242	240	235	230

Table 3-110

Hub Diameter, inch: 8

Leg Distance, inch: 60

SHIM REQUIRED FOR ALIGNMENT

Face - OD Method

FACE TIR MILS	OD TIR, MILS -40	-30	-20	-15	-10	-5	0	5	10	15	20	30	40
-40	-280	-285	-290	-292	-295	-297	-300	-302	-305	-307	-310	-315	-320
-35	-242	-247	-252	-255	-257	-260	-262	-265	-267	-270	-272	-277	-282
-30	-205	-210	-215	-217	-220	-222	-225	-227	-230	-232	-235	-240	-245
-25	-167	-172	-177	-180	-182	-185	-187	-190	-192	-195	-197	-202	-207
-20	-130	-135	-140	-142	-145	-147	-150	-152	-155	-157	-160	-165	-170
-15	-92	-97	-102	-105	-107	-110	-112	-115	-117	-120	-122	-127	-132
-12	-70	-75	-80	-82	-85	-87	-90	-92	-95	-97	-100	-105	-110
-10	-55	-60	-65	-67	-70	-72	-75	-77	-80	-82	-85	-90	-95
-9	-47	-52	-57	-60	-62	-65	-67	-70	-72	-75	-77	-82	-87
-8	-40	-45	-50	-52	-55	-57	-60	-62	-65	-67	-70	-75	-80
-7	-32	-37	-42	-45	-47	-50	-52	-55	-57	-60	-62	-67	-72
-6	-25	-30	-35	-37	-40	-42	-45	-47	-50	-52	-55	-60	-65
-5	-17	-22	-27	-30	-32	-35	-37	-40	-42	-45	-47	-52	-57
-4	-10	-15	-20	-22	-25	-27	-30	-32	-35	-37	-40	-45	-50
-3	-2	-7	-12	-15	-17	-20	-22	-25	-27	-30	-32	-37	-42
-2	5	0	-5	-7	-10	-12	-15	-17	-20	-22	-25	-30	-35
-1	12	7	2	0	-2	-5	-7	-10	-12	-15	-17	-22	-27
0	20	15	10	7	5	2	0	-2	-5	-7	-10	-15	-20
1	27	22	17	15	12	10	7	5	2	0	-2	-7	-12
2	35	30	25	22	20	17	15	12	10	7	5	0	-5
3	42	37	32	30	27	25	22	20	17	15	12	7	2
4	50	45	40	37	35	32	30	27	25	22	20	15	10
5	57	52	47	45	42	40	37	35	32	30	27	22	17
6	65	60	55	52	50	47	45	42	40	37	35	30	25
7	72	67	62	60	57	55	52	50	47	45	42	37	32
8	80	75	70	67	65	62	60	57	55	52	50	45	40
9	87	82	77	75	72	70	67	65	62	60	57	52	47
10	95	90	85	82	80	77	75	72	70	67	65	60	55
12	110	105	100	97	95	92	90	87	85	82	80	75	70
15	132	127	122	120	117	115	112	110	107	105	102	97	92
20	170	165	160	157	155	152	150	147	145	142	140	135	130
25	207	202	197	195	192	190	187	185	182	180	177	172	167
30	245	240	235	232	230	227	225	222	220	217	215	210	205
35	282	277	272	270	267	265	262	260	257	255	252	247	242
40	320	315	310	307	305	302	300	297	295	292	290	285	280

Table 3-111

Hub Diameter, inch: 8

Leg Distance, inch: 70

SHIM REQUIRED FOR ALIGNMENT

Face - OD Method

FACE TIR MILS	OD TIR, MILS -40	-30	-20	-15	-10	-5	0	5	10	15	20	30	40
-40	-330	-335	-340	-342	-345	-347	-350	-352	-355	-357	-360	-365	-370
-35	-286	-291	-296	-298	-301	-303	-306	-308	-311	-313	-316	-321	-326
-30	-242	-247	-252	-255	-257	-260	-262	-265	-267	-270	-272	-277	-282
-25	-198	-203	-208	-211	-213	-216	-218	-221	-223	-226	-228	-233	-238
-20	-155	-160	-165	-167	-170	-172	-175	-177	-180	-182	-185	-190	-195
-15	-111	-116	-121	-123	-126	-128	-131	-133	-136	-138	-141	-146	-151
-12	-85	-90	-95	-97	-100	-102	-105	-107	-110	-112	-115	-120	-125
-10	-67	-72	-77	-80	-82	-85	-87	-90	-92	-95	-97	-102	-107
-9	-58	-63	-68	-71	-73	-76	-78	-81	-83	-86	-88	-93	-98
-8	-50	-55	-60	-62	-65	-67	-70	-72	-75	-77	-80	-85	-90
-7	-41	-46	-51	-53	-56	-58	-61	-63	-66	-68	-71	-76	-81
-6	-32	-37	-42	-45	-47	-50	-52	-55	-57	-60	-62	-67	-72
-5	-23	-28	-33	-36	-38	-41	-43	-46	-48	-51	-53	-58	-63
-4	-15	-20	-25	-27	-30	-32	-35	-37	-40	-42	-45	-50	-55
-3	-6	-11	-16	-18	-21	-23	-26	-28	-31	-33	-36	-41	-46
-2	2	-2	-7	-10	-12	-15	-17	-20	-22	-25	-27	-32	-37
-1	11	6	1	-1	-3	-6	-8	-11	-13	-16	-18	-23	-28
0	20	15	10	7	5	2	0	-2	-5	-7	-10	-15	-20
1	28	23	18	16	13	11	8	6	3	1	-1	-6	-11
2	37	32	27	25	22	20	17	15	12	10	7	2	-2
3	46	41	36	33	31	28	26	23	21	18	16	11	6
4	55	50	45	42	40	37	35	32	30	27	25	20	15
5	63	58	53	51	48	46	43	41	38	36	33	28	23
6	72	67	62	60	57	55	52	50	47	45	42	37	32
7	81	76	71	68	66	63	61	58	56	53	51	46	41
8	90	85	80	77	75	72	70	67	65	62	60	55	50
9	98	93	88	86	83	81	78	76	73	71	68	63	58
10	107	102	97	95	92	90	87	85	82	80	77	72	67
12	125	120	115	112	110	107	105	102	100	97	95	90	85
15	151	146	141	138	136	133	131	128	126	123	121	116	111
20	195	190	185	182	180	177	175	172	170	167	165	160	155
25	238	233	228	226	223	221	218	216	213	211	208	203	198
30	282	277	272	270	267	265	262	260	257	255	252	247	242
35	326	321	316	313	311	308	306	303	301	298	296	291	286
40	370	365	360	357	355	352	350	347	345	342	340	335	330

Table 3-112

Hub Diameter, inch: 8

Leg Distance, inch: 80

SHIM REQUIRED FOR ALIGNMENT

Face - OD Method

FACE TIR MILS	OD TIR, MILS -40	-30	-20	-15	-10	-5	0	5	10	15	20	30	40
-40	-380	-385	-390	-392	-395	-397	-400	-402	-405	-407	-410	-415	-420
-35	-330	-335	-340	-342	-345	-347	-350	-352	-355	-357	-360	-365	-370
-30	-280	-285	-290	-292	-295	-297	-300	-302	-305	-307	-310	-315	-320
-25	-230	-235	-240	-242	-245	-247	-250	-252	-255	-257	-260	-265	-270
-20	-180	-185	-190	-192	-195	-197	-200	-202	-205	-207	-210	-215	-220
-15	-130	-135	-140	-142	-145	-147	-150	-152	-155	-157	-160	-165	-170
-12	-100	-105	-110	-112	-115	-117	-120	-122	-125	-127	-130	-135	-140
-10	-80	-85	-90	-92	-95	-97	-100	-102	-105	-107	-110	-115	-120
-9	-70	-75	-80	-82	-85	-87	-90	-92	-95	-97	-100	-105	-110
-8	-60	-65	-70	-72	-75	-77	-80	-82	-85	-87	-90	-95	-100
-7	-50	-55	-60	-62	-65	-67	-70	-72	-75	-77	-80	-85	-90
-6	-40	-45	-50	-52	-55	-57	-60	-62	-65	-67	-70	-75	-80
-5	-30	-35	-40	-42	-45	-47	-50	-52	-55	-57	-60	-65	-70
-4	-20	-25	-30	-32	-35	-37	-40	-42	-45	-47	-50	-55	-60
-3	-10	-15	-20	-22	-25	-27	-30	-32	-35	-37	-40	-45	-50
-2	0	-5	-10	-12	-15	-17	-20	-22	-25	-27	-30	-35	-40
-1	10	5	0	-2	-5	-7	-10	-12	-15	-17	-20	-25	-30
0	20	15	10	7	5	2	0	-2	-5	-7	-10	-15	-20
1	30	25	20	17	15	12	10	7	5	2	0	-5	-10
2	40	35	30	27	25	22	20	17	15	12	10	5	0
3	50	45	40	37	35	32	30	27	25	22	20	15	10
4	60	55	50	47	45	42	40	37	35	32	30	25	20
5	70	65	60	57	55	52	50	47	45	42	40	35	30
6	80	75	70	67	65	62	60	57	55	52	50	45	40
7	90	85	80	77	75	72	70	67	65	62	60	55	50
8	100	95	90	87	85	82	80	77	75	72	70	65	60
9	110	105	100	97	95	92	90	87	85	82	80	75	70
10	120	115	110	107	105	102	100	97	95	92	90	85	80
12	140	135	130	127	125	122	120	117	115	112	110	105	100
15	170	165	160	157	155	152	150	147	145	142	140	135	130
20	220	215	210	207	205	202	200	197	195	192	190	185	180
25	270	265	260	257	255	252	250	247	245	242	240	235	230
30	320	315	310	307	305	302	300	297	295	292	290	285	280
35	370	365	360	357	355	352	350	347	345	342	340	335	330
40	420	415	410	407	405	402	400	397	395	392	390	385	380

Table 3-113

Hub Diameter, inch: 8
Leg Distance, inch: 90

SHIM REQUIRED FOR ALIGNMENT

Face - OD Method

FACE TIR MILS	OD TIR, MILS -40	-30	-20	-15	-10	-5	0	5	10	15	20	30	40
-40	-430	-435	-440	-442	-445	-447	-450	-452	-455	-457	-460	-465	-470
-35	-373	-378	-383	-386	-388	-391	-393	-396	-398	-401	-403	-408	-413
-30	-317	-322	-327	-330	-332	-335	-337	-340	-342	-345	-347	-352	-357
-25	-261	-266	-271	-273	-276	-278	-281	-283	-286	-288	-291	-296	-301
-20	-205	-210	-215	-217	-220	-222	-225	-227	-230	-232	-235	-240	-245
-15	-148	-153	-158	-161	-163	-166	-168	-171	-173	-176	-178	-183	-188
-12	-115	-120	-125	-127	-130	-132	-135	-137	-140	-142	-145	-150	-155
-10	-92	-97	-102	-105	-107	-110	-112	-115	-117	-120	-122	-127	-132
-9	-81	-86	-91	-93	-96	-98	-101	-103	-106	-108	-111	-116	-121
-8	-70	-75	-80	-82	-85	-87	-90	-92	-95	-97	-100	-105	-110
-7	-58	-63	-68	-71	-73	-76	-78	-81	-83	-86	-88	-93	-98
-6	-47	-52	-57	-60	-62	-65	-67	-70	-72	-75	-77	-82	-87
-5	-36	-41	-46	-48	-51	-53	-56	-58	-61	-63	-66	-71	-76
-4	-25	-30	-35	-37	-40	-42	-45	-47	-50	-52	-55	-60	-65
-3	-13	-18	-23	-26	-28	-31	-33	-36	-38	-41	-43	-48	-53
-2	-2	-7	-12	-15	-17	-20	-22	-25	-27	-30	-32	-37	-42
-1	8	3	-1	-3	-6	-8	-11	-13	-16	-18	-21	-26	-31
0	20	15	10	7	5	2	0	-2	-5	-7	-10	-15	-20
1	31	26	21	18	16	13	11	8	6	3	1	-3	-8
2	42	37	32	30	27	25	22	20	17	15	12	7	2
3	53	48	43	41	38	36	33	31	28	26	23	18	13
4	65	60	55	52	50	47	45	42	40	37	35	30	25
5	76	71	66	63	61	58	56	53	51	48	46	41	36
6	87	82	77	75	72	70	67	65	62	60	57	52	47
7	98	93	88	86	83	81	78	76	73	71	68	63	58
8	110	105	100	97	95	92	90	87	85	82	80	75	70
9	121	116	111	108	106	103	101	98	96	93	91	86	81
10	132	127	122	120	117	115	112	110	107	105	102	97	92
12	155	150	145	142	140	137	135	132	130	127	125	120	115
15	188	183	178	176	173	171	168	166	163	161	158	153	148
20	245	240	235	232	230	227	225	222	220	217	215	210	205
25	301	296	291	288	286	283	281	278	276	273	271	266	261
30	357	352	347	345	342	340	337	335	332	330	327	322	317
35	413	408	403	401	398	396	393	391	388	386	383	378	373
40	470	465	460	457	455	452	450	447	445	442	440	435	430

Table 3-114

Hub Diameter, inch: 8

Leg Distance, inch: 100

SHIM REQUIRED FOR ALIGNMENT

Face – OD Method

FACE TIR MILS	OD TIR, MILS -40	-30	-20	-15	-10	-5	0	5	10	15	20	30	40
-40	-480	-485	-490	-492	-495	-497	-500	-502	-505	-507	-510	-515	-520
-35	-417	-422	-427	-430	-432	-435	-437	-440	-442	-445	-447	-452	-457
-30	-355	-360	-365	-367	-370	-372	-375	-377	-380	-382	-385	-390	-395
-25	-292	-297	-302	-305	-307	-310	-312	-315	-317	-320	-322	-327	-332
-20	-230	-235	-240	-242	-245	-247	-250	-252	-255	-257	-260	-265	-270
-15	-167	-172	-177	-180	-182	-185	-187	-190	-192	-195	-197	-202	-207
-12	-130	-135	-140	-142	-145	-147	-150	-152	-155	-157	-160	-165	-170
-10	-105	-110	-115	-117	-120	-122	-125	-127	-130	-132	-135	-140	-145
-9	-92	-97	-102	-105	-107	-110	-112	-115	-117	-120	-122	-127	-132
-8	-80	-85	-90	-92	-95	-97	-100	-102	-105	-107	-110	-115	-120
-7	-67	-72	-77	-80	-82	-85	-87	-90	-92	-95	-97	-102	-107
-6	-55	-60	-65	-67	-70	-72	-75	-77	-80	-82	-85	-90	-95
-5	-42	-47	-52	-55	-57	-60	-62	-65	-67	-70	-72	-77	-82
-4	-30	-35	-40	-42	-45	-47	-50	-52	-55	-57	-60	-65	-70
-3	-17	-22	-27	-30	-32	-35	-37	-40	-42	-45	-47	-52	-57
-2	-5	-10	-15	-17	-20	-22	-25	-27	-30	-32	-35	-40	-45
-1	7	2	-2	-5	-7	-10	-12	-15	-17	-20	-22	-27	-32
0	20	15	10	7	5	2	0	-2	-5	-7	-10	-15	-20
1	32	27	22	20	17	15	12	10	7	5	2	-2	-7
2	45	40	35	32	30	27	25	22	20	17	15	10	5
3	57	52	47	45	42	40	37	35	32	30	27	22	17
4	70	65	60	57	55	52	50	47	45	42	40	35	30
5	82	77	72	70	67	65	62	60	57	55	52	47	42
6	95	90	85	82	80	77	75	72	70	67	65	60	55
7	107	102	97	95	92	90	87	85	82	80	77	72	67
8	120	115	110	107	105	102	100	97	95	92	90	85	80
9	132	127	122	120	117	115	112	110	107	105	102	97	92
10	145	140	135	132	130	127	125	122	120	117	115	110	105
12	170	165	160	157	155	152	150	147	145	142	140	135	130
15	207	202	197	195	192	190	187	185	182	180	177	172	167
20	270	265	260	257	255	252	250	247	245	242	240	235	230
25	332	327	322	320	317	315	312	310	307	305	302	297	292
30	395	390	385	382	380	377	375	372	370	367	365	360	355
35	457	452	447	445	442	440	437	435	432	430	427	422	417
40	520	515	510	507	505	502	500	497	495	492	490	485	480

Table 3-115

Hub Diameter, inch: 9

Leg Distance, inch: 25

SHIM REQUIRED FOR ALIGNMENT

Face - OD Method

FACE TIR MILS	OD TIR, MILS -40	-30	-20	-15	-10	-5	0	5	10	15	20	30	40
-40	-91	-96	-101	-103	-106	-108	-111	-113	-116	-118	-121	-126	-131
-35	-77	-82	-87	-89	-92	-94	-97	-99	-102	-104	-107	-112	-117
-30	-63	-68	-73	-75	-78	-80	-83	-85	-88	-90	-93	-98	-103
-25	-49	-54	-59	-61	-64	-66	-69	-71	-74	-76	-79	-84	-89
-20	-35	-40	-45	-48	-50	-53	-55	-58	-60	-63	-65	-70	-75
-15	-21	-26	-31	-34	-36	-39	-41	-44	-46	-49	-51	-56	-61
-12	-13	-18	-23	-25	-28	-30	-33	-35	-38	-40	-43	-48	-53
-10	-7	-12	-17	-20	-22	-25	-27	-30	-32	-35	-37	-42	-47
-9	-5	-10	-15	-17	-20	-22	-25	-27	-30	-32	-35	-40	-45
-8	-2	-7	-12	-14	-17	-19	-22	-24	-27	-29	-32	-37	-42
-7	0	-4	-9	-11	-14	-16	-19	-21	-24	-26	-29	-34	-39
-6	3	-1	-6	-9	-11	-14	-16	-19	-21	-24	-26	-31	-36
-5	6	1	-3	-6	-8	-11	-13	-16	-18	-21	-23	-28	-33
-4	8	3	-1	-3	-6	-8	-11	-13	-16	-18	-21	-26	-31
-3	11	6	1	0	-3	-5	-8	-10	-13	-15	-18	-23	-28
-2	14	9	4	1	0	-3	-5	-8	-10	-13	-15	-20	-25
-1	17	12	7	4	2	0	-2	-5	-7	-10	-12	-17	-22
0	20	15	10	7	5	2	0	-2	-5	-7	-10	-15	-20
1	22	17	12	10	7	5	2	0	-2	-4	-7	-12	-17
2	25	20	15	13	10	8	5	3	0	-1	-4	-9	-14
3	28	23	18	15	13	10	8	5	3	0	-1	-6	-11
4	31	26	21	18	16	13	11	8	6	3	1	-3	-8
5	33	28	23	21	18	16	13	11	8	6	3	-1	-6
6	36	31	26	24	21	19	16	14	11	9	6	1	-3
7	39	34	29	26	24	21	19	16	14	11	9	4	0
8	42	37	32	29	27	24	22	19	17	14	12	7	2
9	45	40	35	32	30	27	25	22	20	17	15	10	5
10	47	42	37	35	32	30	27	25	22	20	17	12	7
12	53	48	43	40	38	35	33	30	28	25	23	18	13
15	61	56	51	49	46	44	41	39	36	34	31	26	21
20	75	70	65	63	60	58	55	53	50	48	45	40	35
25	89	84	79	76	74	71	69	66	64	61	59	54	49
30	103	98	93	90	88	85	83	80	78	75	73	68	63
35	117	112	107	104	102	99	97	94	92	89	87	82	77
40	131	126	121	118	116	113	111	108	106	103	101	96	91

Table 3-116

Hub Diameter, inch: 9

Leg Distance, inch: 30

SHIM REQUIRED FOR ALIGNMENT

Face - OD Method

FACE TIR MILS	OD TIR, MILS -40	-30	-20	-15	-10	-5	0	5	10	15	20	30	40
-40	-113	-118	-123	-125	-128	-130	-133	-135	-138	-140	-143	-148	-153
-35	-96	-101	-106	-109	-111	-114	-116	-119	-121	-124	-126	-131	-136
-30	-80	-85	-90	-92	-95	-97	-100	-102	-105	-107	-110	-115	-120
-25	-63	-68	-73	-75	-78	-80	-83	-85	-88	-90	-93	-98	-103
-20	-46	-51	-56	-59	-61	-64	-66	-69	-71	-74	-76	-81	-86
-15	-30	-35	-40	-42	-45	-47	-50	-52	-55	-57	-60	-65	-70
-12	-20	-25	-30	-32	-35	-37	-40	-42	-45	-47	-50	-55	-60
-10	-13	-18	-23	-25	-28	-30	-33	-35	-38	-40	-43	-48	-53
-9	-10	-15	-20	-22	-25	-27	-30	-32	-35	-37	-40	-45	-50
-8	-6	-11	-16	-19	-21	-24	-26	-29	-31	-34	-36	-41	-46
-7	-3	-8	-13	-15	-18	-20	-23	-25	-28	-30	-33	-38	-43
-6	0	-5	-10	-12	-15	-17	-20	-22	-25	-27	-30	-35	-40
-5	3	-1	-6	-9	-11	-14	-16	-19	-21	-24	-26	-31	-36
-4	6	1	-3	-5	-8	-10	-13	-15	-18	-20	-23	-28	-33
-3	10	5	0	-2	-5	-7	-10	-12	-15	-17	-20	-25	-30
-2	13	8	3	0	-1	-4	-6	-9	-11	-14	-16	-21	-26
-1	16	11	6	4	1	0	-3	-5	-8	-10	-13	-18	-23
0	20	15	10	7	5	2	0	-2	-5	-7	-10	-15	-20
1	23	18	13	10	8	5	3	0	-1	-4	-6	-11	-16
2	26	21	16	14	11	9	6	4	1	0	-3	-8	-13
3	30	25	20	17	15	12	10	7	5	2	0	-5	-10
4	33	28	23	20	18	15	13	10	8	5	3	-1	-6
5	36	31	26	24	21	19	16	14	11	9	6	1	-3
6	40	35	30	27	25	22	20	17	15	12	10	5	0
7	43	38	33	30	28	25	23	20	18	15	13	8	3
8	46	41	36	34	31	29	26	24	21	19	16	11	6
9	50	45	40	37	35	32	30	27	25	22	20	15	10
10	53	48	43	40	38	35	33	30	28	25	23	18	13
12	60	55	50	47	45	42	40	37	35	32	30	25	20
15	70	65	60	57	55	52	50	47	45	42	40	35	30
20	86	81	76	74	71	69	66	64	61	59	56	51	46
25	103	98	93	90	88	85	83	80	78	75	73	68	63
30	120	115	110	107	105	102	100	97	95	92	90	85	80
35	136	131	126	124	121	119	116	114	111	109	106	101	96
40	153	148	143	140	138	135	133	130	128	125	123	118	113

Table 3-117

Hub Diameter, inch: 9

Leg Distance, inch: 35

SHIM REQUIRED FOR ALIGNMENT

Face - OD Method

FACE TIR MILS	OD TIR, MILS -40	-30	-20	-15	-10	-5	0	5	10	15	20	30	40
-40	-135	-140	-145	-148	-150	-153	-155	-158	-160	-163	-165	-170	-175
-35	-116	-121	-126	-128	-131	-133	-136	-138	-141	-143	-146	-151	-156
-30	-96	-101	-106	-109	-111	-114	-116	-119	-121	-124	-126	-131	-136
-25	-77	-82	-87	-89	-92	-94	-97	-99	-102	-104	-107	-112	-117
-20	-57	-62	-67	-70	-72	-75	-77	-80	-82	-85	-87	-92	-97
-15	-38	-43	-48	-50	-53	-55	-58	-60	-63	-65	-68	-73	-78
-12	-26	-31	-36	-39	-41	-44	-46	-49	-51	-54	-56	-61	-66
-10	-18	-23	-28	-31	-33	-36	-38	-41	-43	-46	-48	-53	-58
-9	-15	-20	-25	-27	-30	-32	-35	-37	-40	-42	-45	-50	-55
-8	-11	-16	-21	-23	-26	-28	-31	-33	-36	-38	-41	-46	-51
-7	-7	-12	-17	-19	-22	-24	-27	-29	-32	-34	-37	-42	-47
-6	-3	-8	-13	-15	-18	-20	-23	-25	-28	-30	-33	-38	-43
-5	0	-4	-9	-11	-14	-16	-19	-21	-24	-26	-29	-34	-39
-4	4	0	-5	-8	-10	-13	-15	-18	-20	-23	-25	-30	-35
-3	8	3	-1	-4	-6	-9	-11	-14	-16	-19	-21	-26	-31
-2	12	7	2	0	-2	-5	-7	-10	-12	-15	-17	-22	-27
-1	16	11	6	3	1	-1	-3	-6	-8	-11	-13	-18	-23
0	20	15	10	7	5	2	0	-2	-5	-7	-10	-15	-20
1	23	18	13	11	8	6	3	1	-1	-3	-6	-11	-16
2	27	22	17	15	12	10	7	5	2	0	-2	-7	-12
3	31	26	21	19	16	14	11	9	6	4	1	-3	-8
4	35	30	25	23	20	18	15	13	10	8	5	0	-4
5	39	34	29	26	24	21	19	16	14	11	9	4	0
6	43	38	33	30	28	25	23	20	18	15	13	8	3
7	47	42	37	34	32	29	27	24	22	19	17	12	7
8	51	46	41	38	36	33	31	28	26	23	21	16	11
9	55	50	45	42	40	37	35	32	30	27	25	20	15
10	58	53	48	46	43	41	38	36	33	31	28	23	18
12	66	61	56	54	51	49	46	44	41	39	36	31	26
15	78	73	68	65	63	60	58	55	53	50	48	43	38
20	97	92	87	85	82	80	77	75	72	70	67	62	57
25	117	112	107	104	102	99	97	94	92	89	87	82	77
30	136	131	126	124	121	119	116	114	111	109	106	101	96
35	156	151	146	143	141	138	136	133	131	128	126	121	116
40	175	170	165	163	160	158	155	153	150	148	145	140	135

Table 3-118

Hub Diameter, inch: 9

Leg Distance, inch: 40

SHIM REQUIRED FOR ALIGNMENT

Face - OD Method

FACE TIR MILS	OD TIR, MILS -40	-30	-20	-15	-10	-5	0	5	10	15	20	30	40
-40	-157	-162	-167	-170	-172	-175	-177	-180	-182	-185	-187	-192	-197
-35	-135	-140	-145	-148	-150	-153	-155	-158	-160	-163	-165	-170	-175
-30	-113	-118	-123	-125	-128	-130	-133	-135	-138	-140	-143	-148	-153
-25	-91	-96	-101	-103	-106	-108	-111	-113	-116	-118	-121	-126	-131
-20	-68	-73	-78	-81	-83	-86	-88	-91	-93	-96	-98	-103	-108
-15	-46	-51	-56	-59	-61	-64	-66	-69	-71	-74	-76	-81	-86
-12	-33	-38	-43	-45	-48	-50	-53	-55	-58	-60	-63	-68	-73
-10	-24	-29	-34	-36	-39	-41	-44	-46	-49	-51	-54	-59	-64
-9	-20	-25	-30	-32	-35	-37	-40	-42	-45	-47	-50	-55	-60
-8	-15	-20	-25	-28	-30	-33	-35	-38	-40	-43	-45	-50	-55
-7	-11	-16	-21	-23	-26	-28	-31	-33	-36	-38	-41	-46	-51
-6	-6	-11	-16	-19	-21	-24	-26	-29	-31	-34	-36	-41	-46
-5	-2	-7	-12	-14	-17	-19	-22	-24	-27	-29	-32	-37	-42
-4	2	-2	-7	-10	-12	-15	-17	-20	-22	-25	-27	-32	-37
-3	6	1	-3	-5	-8	-10	-13	-15	-18	-20	-23	-28	-33
-2	11	6	1	-1	-3	-6	-8	-11	-13	-16	-18	-23	-28
-1	15	10	5	3	0	-1	-4	-6	-9	-11	-14	-19	-24
0	20	15	10	7	5	2	0	-2	-5	-7	-10	-15	-20
1	24	19	14	11	9	6	4	1	0	-3	-5	-10	-15
2	28	23	18	16	13	11	8	6	3	1	-1	-6	-11
3	33	28	23	20	18	15	13	10	8	5	3	-1	-6
4	37	32	27	25	22	20	17	15	12	10	7	2	-2
5	42	37	32	29	27	24	22	19	17	14	12	7	2
6	46	41	36	34	31	29	26	24	21	19	16	11	6
7	51	46	41	38	36	33	31	28	26	23	21	16	11
8	55	50	45	43	40	38	35	33	30	28	25	20	15
9	60	55	50	47	45	42	40	37	35	32	30	25	20
10	64	59	54	51	49	46	44	41	39	36	34	29	24
12	73	68	63	60	58	55	53	50	48	45	43	38	33
15	86	81	76	74	71	69	66	64	61	59	56	51	46
20	108	103	98	96	93	91	88	86	83	81	78	73	68
25	131	126	121	118	116	113	111	108	106	103	101	96	91
30	153	148	143	140	138	135	133	130	128	125	123	118	113
35	175	170	165	163	160	158	155	153	150	148	145	140	135
40	197	192	187	185	182	180	177	175	172	170	167	162	157

Table 3-119

Hub Diameter, inch: 9

Leg Distance, inch: 50

SHIM REQUIRED FOR ALIGNMENT

Face - OD Method

FACE TIR MILS	OD TIR, MILS -40	-30	-20	-15	-10	-5	0	5	10	15	20	30	40
-40	-202	-207	-212	-214	-217	-219	-222	-224	-227	-229	-232	-237	-242
-35	-174	-179	-184	-186	-189	-191	-194	-196	-199	-201	-204	-209	-214
-30	-146	-151	-156	-159	-161	-164	-166	-169	-171	-174	-176	-181	-186
-25	-118	-123	-128	-131	-133	-136	-138	-141	-143	-146	-148	-153	-158
-20	-91	-96	-101	-103	-106	-108	-111	-113	-116	-118	-121	-126	-131
-15	-63	-68	-73	-75	-78	-80	-83	-85	-88	-90	-93	-98	-103
-12	-46	-51	-56	-59	-61	-64	-66	-69	-71	-74	-76	-81	-86
-10	-35	-40	-45	-48	-50	-53	-55	-58	-60	-63	-65	-70	-75
-9	-30	-35	-40	-42	-45	-47	-50	-52	-55	-57	-60	-65	-70
-8	-24	-29	-34	-36	-39	-41	-44	-46	-49	-51	-54	-59	-64
-7	-18	-23	-28	-31	-33	-36	-38	-41	-43	-46	-48	-53	-58
-6	-13	-18	-23	-25	-28	-30	-33	-35	-38	-40	-43	-48	-53
-5	-7	-12	-17	-20	-22	-25	-27	-30	-32	-35	-37	-42	-47
-4	-2	-7	-12	-14	-17	-19	-22	-24	-27	-29	-32	-37	-42
-3	3	-1	-6	-9	-11	-14	-16	-19	-21	-24	-26	-31	-36
-2	8	3	-1	-3	-6	-8	-11	-13	-16	-18	-21	-26	-31
-1	14	9	4	1	0	-3	-5	-8	-10	-13	-15	-20	-25
0	20	15	10	7	5	2	0	-2	-5	-7	-10	-15	-20
1	25	20	15	13	10	8	5	3	0	-1	-4	-9	-14
2	31	26	21	18	16	13	11	8	6	3	1	-3	-8
3	36	31	26	24	21	19	16	14	11	9	6	1	-3
4	42	37	32	29	27	24	22	19	17	14	12	7	2
5	47	42	37	35	32	30	27	25	22	20	17	12	7
6	53	48	43	40	38	35	33	30	28	25	23	18	13
7	58	53	48	46	43	41	38	36	33	31	28	23	18
8	64	59	54	51	49	46	44	41	39	36	34	29	24
9	70	65	60	57	55	52	50	47	45	42	40	35	30
10	75	70	65	63	60	58	55	53	50	48	45	40	35
12	86	81	76	74	71	69	66	64	61	59	56	51	46
15	103	98	93	90	88	85	83	80	78	75	73	68	63
20	131	126	121	118	116	113	111	108	106	103	101	96	91
25	158	153	148	146	143	141	138	136	133	131	128	123	118
30	186	181	176	174	171	169	166	164	161	159	156	151	146
35	214	209	204	201	199	196	194	191	189	186	184	179	174
40	242	237	232	229	227	224	222	219	217	214	212	207	202

Table 3-120

Hub Diameter, inch: 9

Leg Distance, inch: 60

SHIM REQUIRED FOR ALIGNMENT

Face - OD Method

FACE TIR MILS	OD TIR, MILS -40	-30	-20	-15	-10	-5	0	5	10	15	20	30	40
-40	-246	-251	-256	-259	-261	-264	-266	-269	-271	-274	-276	-281	-286
-35	-213	-218	-223	-225	-228	-230	-233	-235	-238	-240	-243	-248	-253
-30	-180	-185	-190	-192	-195	-197	-200	-202	-205	-207	-210	-215	-220
-25	-146	-151	-156	-159	-161	-164	-166	-169	-171	-174	-176	-181	-186
-20	-113	-118	-123	-125	-128	-130	-133	-135	-138	-140	-143	-148	-153
-15	-80	-85	-90	-92	-95	-97	-100	-102	-105	-107	-110	-115	-120
-12	-60	-65	-70	-72	-75	-77	-80	-82	-85	-87	-90	-95	-100
-10	-46	-51	-56	-59	-61	-64	-66	-69	-71	-74	-76	-81	-86
-9	-40	-45	-50	-52	-55	-57	-60	-62	-65	-67	-70	-75	-80
-8	-33	-38	-43	-45	-48	-50	-53	-55	-58	-60	-63	-68	-73
-7	-26	-31	-36	-39	-41	-44	-46	-49	-51	-54	-56	-61	-66
-6	-20	-25	-30	-32	-35	-37	-40	-42	-45	-47	-50	-55	-60
-5	-13	-18	-23	-25	-28	-30	-33	-35	-38	-40	-43	-48	-53
-4	-6	-11	-16	-19	-21	-24	-26	-29	-31	-34	-36	-41	-46
-3	0	-5	-10	-12	-15	-17	-20	-22	-25	-27	-30	-35	-40
-2	6	1	-3	-5	-8	-10	-13	-15	-18	-20	-23	-28	-33
-1	13	8	3	0	-1	-4	-6	-9	-11	-14	-16	-21	-26
0	20	15	10	7	5	2	0	-2	-5	-7	-10	-15	-20
1	26	21	16	14	11	9	6	4	1	0	-3	-8	-13
2	33	28	23	20	18	15	13	10	8	5	3	-1	-6
3	40	35	30	27	25	22	20	17	15	12	10	5	0
4	46	41	36	34	31	29	26	24	21	19	16	11	6
5	53	48	43	40	38	35	33	30	28	25	23	18	13
6	60	55	50	47	45	42	40	37	35	32	30	25	20
7	66	61	56	54	51	49	46	44	41	39	36	31	26
8	73	68	63	60	58	55	53	50	48	45	43	38	33
9	80	75	70	67	65	62	60	57	55	52	50	45	40
10	86	81	76	74	71	69	66	64	61	59	56	51	46
12	100	95	90	87	85	82	80	77	75	72	70	65	60
15	120	115	110	107	105	102	100	97	95	92	90	85	80
20	153	148	143	140	138	135	133	130	128	125	123	118	113
25	186	181	176	174	171	169	166	164	161	159	156	151	146
30	220	215	210	207	205	202	200	197	195	192	190	185	180
35	253	248	243	240	238	235	233	230	228	225	223	218	213
40	286	281	276	274	271	269	266	264	261	259	256	251	246

Table 3-121

Hub Diameter, inch: 9

Leg Distance, inch: 70

SHIM REQUIRED FOR ALIGNMENT

Face - OD Method

FACE TIR MILS	OD TIR, MILS -40	-30	-20	-15	-10	-5	0	5	10	15	20	30	40
-40	-291	-296	-301	-303	-306	-308	-311	-313	-316	-318	-321	-326	-331
-35	-252	-257	-262	-264	-267	-269	-272	-274	-277	-279	-282	-287	-292
-30	-213	-218	-223	-225	-228	-230	-233	-235	-238	-240	-243	-248	-253
-25	-174	-179	-184	-186	-189	-191	-194	-196	-199	-201	-204	-209	-214
-20	-135	-140	-145	-148	-150	-153	-155	-158	-160	-163	-165	-170	-175
-15	-96	-101	-106	-109	-111	-114	-116	-119	-121	-124	-126	-131	-136
-12	-73	-78	-83	-85	-88	-90	-93	-95	-98	-100	-103	-108	-113
-10	-57	-62	-67	-70	-72	-75	-77	-80	-82	-85	-87	-92	-97
-9	-50	-55	-60	-62	-65	-67	-70	-72	-75	-77	-80	-85	-90
-8	-42	-47	-52	-54	-57	-59	-62	-64	-67	-69	-72	-77	-82
-7	-34	-39	-44	-46	-49	-51	-54	-56	-59	-61	-64	-69	-74
-6	-26	-31	-36	-39	-41	-44	-46	-49	-51	-54	-56	-61	-66
-5	-18	-23	-28	-31	-33	-36	-38	-41	-43	-46	-48	-53	-58
-4	-11	-16	-21	-23	-26	-28	-31	-33	-36	-38	-41	-46	-51
-3	-3	-8	-13	-15	-18	-20	-23	-25	-28	-30	-33	-38	-43
-2	4	0	-5	-8	-10	-13	-15	-18	-20	-23	-25	-30	-35
-1	12	7	2	0	-2	-5	-7	-10	-12	-15	-17	-22	-27
0	20	15	10	7	5	2	0	-2	-5	-7	-10	-15	-20
1	27	22	17	15	12	10	7	5	2	0	-2	-7	-12
2	35	30	25	23	20	18	15	13	10	8	5	0	-4
3	43	38	33	30	28	25	23	20	18	15	13	8	3
4	51	46	41	38	36	33	31	28	26	23	21	16	11
5	58	53	48	46	43	41	38	36	33	31	28	23	18
6	66	61	56	54	51	49	46	44	41	39	36	31	26
7	74	69	64	61	59	56	54	51	49	46	44	39	34
8	82	77	72	69	67	64	62	59	57	54	52	47	42
9	90	85	80	77	75	72	70	67	65	62	60	55	50
10	97	92	87	85	82	80	77	75	72	70	67	62	57
12	113	108	103	100	98	95	93	90	88	85	83	78	73
15	136	131	126	124	121	119	116	114	111	109	106	101	96
20	175	170	165	163	160	158	155	153	150	148	145	140	135
25	214	209	204	201	199	196	194	191	189	186	184	179	174
30	253	248	243	240	238	235	233	230	228	225	223	218	213
35	292	287	282	279	277	274	272	269	267	264	262	257	252
40	331	326	321	318	316	313	311	308	306	303	301	296	291

Table 3-122

Hub Diameter, inch: 9

Leg Distance, inch: 80

SHIM REQUIRED FOR ALIGNMENT

Face - OD Method

FACE TIR MILS	OD TIR, MILS -40	-30	-20	-15	-10	-5	0	5	10	15	20	30	40
-40	-335	-340	-345	-348	-350	-353	-355	-358	-360	-363	-365	-370	-375
-35	-291	-296	-301	-303	-306	-308	-311	-313	-316	-318	-321	-326	-331
-30	-246	-251	-256	-259	-261	-264	-266	-269	-271	-274	-276	-281	-286
-25	-202	-207	-212	-214	-217	-219	-222	-224	-227	-229	-232	-237	-242
-20	-157	-162	-167	-170	-172	-175	-177	-180	-182	-185	-187	-192	-197
-15	-113	-118	-123	-125	-128	-130	-133	-135	-138	-140	-143	-148	-153
-12	-86	-91	-96	-99	-101	-104	-106	-109	-111	-114	-116	-121	-126
-10	-68	-73	-78	-81	-83	-86	-88	-91	-93	-96	-98	-103	-108
-9	-60	-65	-70	-72	-75	-77	-80	-82	-85	-87	-90	-95	-100
-8	-51	-56	-61	-63	-66	-68	-71	-73	-76	-78	-81	-86	-91
-7	-42	-47	-52	-54	-57	-59	-62	-64	-67	-69	-72	-77	-82
-6	-33	-38	-43	-45	-48	-50	-53	-55	-58	-60	-63	-68	-73
-5	-24	-29	-34	-36	-39	-41	-44	-46	-49	-51	-54	-59	-64
-4	-15	-20	-25	-28	-30	-33	-35	-38	-40	-43	-45	-50	-55
-3	-6	-11	-16	-19	-21	-24	-26	-29	-31	-34	-36	-41	-46
-2	2	-2	-7	-10	-12	-15	-17	-20	-22	-25	-27	-32	-37
-1	11	6	1	-1	-3	-6	-8	-11	-13	-16	-18	-23	-28
0	20	15	10	7	5	2	0	-2	-5	-7	-10	-15	-20
1	28	23	18	16	13	11	8	6	3	1	-1	-6	-11
2	37	32	27	25	22	20	17	15	12	10	7	2	-2
3	46	41	36	34	31	29	26	24	21	19	16	11	6
4	55	50	45	43	40	38	35	33	30	28	25	20	15
5	64	59	54	51	49	46	44	41	39	36	34	29	24
6	73	68	63	60	58	55	53	50	48	45	43	38	33
7	82	77	72	69	67	64	62	59	57	54	52	47	42
8	91	86	81	78	76	73	71	68	66	63	61	56	51
9	100	95	90	87	85	82	80	77	75	72	70	65	60
10	108	103	98	96	93	91	88	86	83	81	78	73	68
12	126	121	116	114	111	109	106	104	101	99	96	91	86
15	153	148	143	140	138	135	133	130	128	125	123	118	113
20	197	192	187	185	182	180	177	175	172	170	167	162	157
25	242	237	232	229	227	224	222	219	217	214	212	207	202
30	286	281	276	274	271	269	266	264	261	259	256	251	246
35	331	326	321	318	316	313	311	308	306	303	301	296	291
40	375	370	365	363	360	358	355	353	350	348	345	340	335

Table 3-123

Hub Diameter, inch: 9

Leg Distance, inch: 90

SHIM REQUIRED FOR ALIGNMENT

Face - OD Method

FACE TIR MILS	OD TIR, MILS -40	-30	-20	-15	-10	-5	0	5	10	15	20	30	40
-40	-380	-385	-390	-392	-395	-397	-400	-402	-405	-407	-410	-415	-420
-35	-330	-335	-340	-342	-345	-347	-350	-352	-355	-357	-360	-365	-370
-30	-280	-285	-290	-292	-295	-297	-300	-302	-305	-307	-310	-315	-320
-25	-230	-235	-240	-242	-245	-247	-250	-252	-255	-257	-260	-265	-270
-20	-180	-185	-190	-192	-195	-197	-200	-202	-205	-207	-210	-215	-220
-15	-130	-135	-140	-142	-145	-147	-150	-152	-155	-157	-160	-165	-170
-12	-100	-105	-110	-112	-115	-117	-120	-122	-125	-127	-130	-135	-140
-10	-80	-85	-90	-92	-95	-97	-100	-102	-105	-107	-110	-115	-120
-9	-70	-75	-80	-82	-85	-87	-90	-92	-95	-97	-100	-105	-110
-8	-60	-65	-70	-72	-75	-77	-80	-82	-85	-87	-90	-95	-100
-7	-50	-55	-60	-62	-65	-67	-70	-72	-75	-77	-80	-85	-90
-6	-40	-45	-50	-52	-55	-57	-60	-62	-65	-67	-70	-75	-80
-5	-30	-35	-40	-42	-45	-47	-50	-52	-55	-57	-60	-65	-70
-4	-20	-25	-30	-32	-35	-37	-40	-42	-45	-47	-50	-55	-60
-3	-10	-15	-20	-22	-25	-27	-30	-32	-35	-37	-40	-45	-50
-2	0	-5	-10	-12	-15	-17	-20	-22	-25	-27	-30	-35	-40
-1	10	5	0	-2	-5	-7	-10	-12	-15	-17	-20	-25	-30
0	20	15	10	7	5	2	0	-2	-5	-7	-10	-15	-20
1	30	25	20	17	15	12	10	7	5	2	0	-5	-10
2	40	35	30	27	25	22	20	17	15	12	10	5	0
3	50	45	40	37	35	32	30	27	25	22	20	15	10
4	60	55	50	47	45	42	40	37	35	32	30	25	20
5	70	65	60	57	55	52	50	47	45	42	40	35	30
6	80	75	70	67	65	62	60	57	55	52	50	45	40
7	90	85	80	77	75	72	70	67	65	62	60	55	50
8	100	95	90	87	85	82	80	77	75	72	70	65	60
9	110	105	100	97	95	92	90	87	85	82	80	75	70
10	120	115	110	107	105	102	100	97	95	92	90	85	80
12	140	135	130	127	125	122	120	117	115	112	110	105	100
15	170	165	160	157	155	152	150	147	145	142	140	135	130
20	220	215	210	207	205	202	200	197	195	192	190	185	180
25	270	265	260	257	255	252	250	247	245	242	240	235	230
30	320	315	310	307	305	302	300	297	295	292	290	285	280
35	370	365	360	357	355	352	350	347	345	342	340	335	330
40	420	415	410	407	405	402	400	397	395	392	390	385	380

Table 3-124

Hub Diameter, inch: 9

Leg Distance, inch: 100

SHIM REQUIRED FOR ALIGNMENT

Face - OD Method

FACE TIR MILS	OD TIR, MILS -40	-30	-20	-15	-10	-5	0	5	10	15	20	30	40
-40	-424	-429	-434	-436	-439	-441	-444	-446	-449	-451	-454	-459	-464
-35	-368	-373	-378	-381	-383	-386	-388	-391	-393	-396	-398	-403	-408
-30	-313	-318	-323	-325	-328	-330	-333	-335	-338	-340	-343	-348	-353
-25	-257	-262	-267	-270	-272	-275	-277	-280	-282	-285	-287	-292	-297
-20	-202	-207	-212	-214	-217	-219	-222	-224	-227	-229	-232	-237	-242
-15	-146	-151	-156	-159	-161	-164	-166	-169	-171	-174	-176	-181	-186
-12	-113	-118	-123	-125	-128	-130	-133	-135	-138	-140	-143	-148	-153
-10	-91	-96	-101	-103	-106	-108	-111	-113	-116	-118	-121	-126	-131
-9	-80	-85	-90	-92	-95	-97	-100	-102	-105	-107	-110	-115	-120
-8	-68	-73	-78	-81	-83	-86	-88	-91	-93	-96	-98	-103	-108
-7	-57	-62	-67	-70	-72	-75	-77	-80	-82	-85	-87	-92	-97
-6	-46	-51	-56	-59	-61	-64	-66	-69	-71	-74	-76	-81	-86
-5	-35	-40	-45	-48	-50	-53	-55	-58	-60	-63	-65	-70	-75
-4	-24	-29	-34	-36	-39	-41	-44	-46	-49	-51	-54	-59	-64
-3	-13	-18	-23	-25	-28	-30	-33	-35	-38	-40	-43	-48	-53
-2	-2	-7	-12	-14	-17	-19	-22	-24	-27	-29	-32	-37	-42
-1	8	3	-1	-3	-6	-8	-11	-13	-16	-18	-21	-26	-31
0	20	15	10	7	5	2	0	-2	-5	-7	-10	-15	-20
1	31	26	21	18	16	13	11	8	6	3	1	-3	-8
2	42	37	32	29	27	24	22	19	17	14	12	7	2
3	53	48	43	40	38	35	33	30	28	25	23	18	13
4	64	59	54	51	49	46	44	41	39	36	34	29	24
5	75	70	65	63	60	58	55	53	50	48	45	40	35
6	86	81	76	74	71	69	66	64	61	59	56	51	46
7	97	92	87	85	82	80	77	75	72	70	67	62	57
8	108	103	98	96	93	91	88	86	83	81	78	73	68
9	120	115	110	107	105	102	100	97	95	92	90	85	80
10	131	126	121	118	116	113	111	108	106	103	101	96	91
12	153	148	143	140	138	135	133	130	128	125	123	118	113
15	186	181	176	174	171	169	166	164	161	159	156	151	146
20	242	237	232	229	227	224	222	219	217	214	212	207	202
25	297	292	287	285	282	280	277	275	272	270	267	262	257
30	353	348	343	340	338	335	333	330	328	325	323	318	313
35	408	403	398	396	393	391	388	386	383	381	378	373	368
40	464	459	454	451	449	446	444	441	439	436	434	429	424

Table 3-125

Hub Diameter, inch: 10

Leg Distance, inch: 25

SHIM REQUIRED FOR ALIGNMENT

Face - OD Method

FACE TIR MILS	OD TIR, MILS -40	-30	-20	-15	-10	-5	0	5	10	15	20	30	40
-40	-80	-85	-90	-92	-95	-97	-100	-102	-105	-107	-110	-115	-120
-35	-67	-72	-77	-80	-82	-85	-87	-90	-92	-95	-97	-102	-107
-30	-55	-60	-65	-67	-70	-72	-75	-77	-80	-82	-85	-90	-95
-25	-42	-47	-52	-55	-57	-60	-62	-65	-67	-70	-72	-77	-82
-20	-30	-35	-40	-42	-45	-47	-50	-52	-55	-57	-60	-65	-70
-15	-17	-22	-27	-30	-32	-35	-37	-40	-42	-45	-47	-52	-57
-12	-10	-15	-20	-22	-25	-27	-30	-32	-35	-37	-40	-45	-50
-10	-5	-10	-15	-17	-20	-22	-25	-27	-30	-32	-35	-40	-45
-9	-2	-7	-12	-15	-17	-20	-22	-25	-27	-30	-32	-37	-42
-8	0	-5	-10	-12	-15	-17	-20	-22	-25	-27	-30	-35	-40
-7	2	-2	-7	-10	-12	-15	-17	-20	-22	-25	-27	-32	-37
-6	5	0	-5	-7	-10	-12	-15	-17	-20	-22	-25	-30	-35
-5	7	2	-2	-5	-7	-10	-12	-15	-17	-20	-22	-27	-32
-4	10	5	0	-2	-5	-7	-10	-12	-15	-17	-20	-25	-30
-3	12	7	2	0	-2	-5	-7	-10	-12	-15	-17	-22	-27
-2	15	10	5	2	0	-2	-5	-7	-10	-12	-15	-20	-25
-1	17	12	7	5	2	0	-2	-5	-7	-10	-12	-17	-22
0	20	15	10	7	5	2	0	-2	-5	-7	-10	-15	-20
1	22	17	12	10	7	5	2	0	-2	-5	-7	-12	-17
2	25	20	15	12	10	7	5	2	0	-2	-5	-10	-15
3	27	22	17	15	12	10	7	5	2	0	-2	-7	-12
4	30	25	20	17	15	12	10	7	5	2	0	-5	-10
5	32	27	22	20	17	15	12	10	7	5	2	-2	-7
6	35	30	25	22	20	17	15	12	10	7	5	0	-5
7	37	32	27	25	22	20	17	15	12	10	7	2	-2
8	40	35	30	27	25	22	20	17	15	12	10	5	0
9	42	37	32	30	27	25	22	20	17	15	12	7	2
10	45	40	35	32	30	27	25	22	20	17	15	10	5
12	50	45	40	37	35	32	30	27	25	22	20	15	10
15	57	52	47	45	42	40	37	35	32	30	27	22	17
20	70	65	60	57	55	52	50	47	45	42	40	35	30
25	82	77	72	70	67	65	62	60	57	55	52	47	42
30	95	90	85	82	80	77	75	72	70	67	65	60	55
35	107	102	97	95	92	90	87	85	82	80	77	72	67
40	120	115	110	107	105	102	100	97	95	92	90	85	80

Table 3-126

Hub Diameter, inch: 10

Leg Distance, inch: 30

SHIM REQUIRED FOR ALIGNMENT

Face – OD Method

FACE TIR MILS	OD TIR, MILS -40	-30	-20	-15	-10	-5	0	5	10	15	20	30	40
-40	-100	-105	-110	-112	-115	-117	-120	-122	-125	-127	-130	-135	-140
-35	-85	-90	-95	-97	-100	-102	-105	-107	-110	-112	-115	-120	-125
-30	-70	-75	-80	-82	-85	-87	-90	-92	-95	-97	-100	-105	-110
-25	-55	-60	-65	-67	-70	-72	-75	-77	-80	-82	-85	-90	-95
-20	-40	-45	-50	-52	-55	-57	-60	-62	-65	-67	-70	-75	-80
-15	-25	-30	-35	-37	-40	-42	-45	-47	-50	-52	-55	-60	-65
-12	-16	-21	-26	-28	-31	-33	-36	-38	-41	-43	-46	-51	-56
-10	-10	-15	-20	-22	-25	-27	-30	-32	-35	-37	-40	-45	-50
-9	-7	-12	-17	-19	-22	-24	-27	-29	-32	-34	-37	-42	-47
-8	-4	-9	-14	-16	-19	-21	-24	-26	-29	-31	-34	-39	-44
-7	-1	-6	-11	-13	-16	-18	-21	-23	-26	-28	-31	-36	-41
-6	2	-3	-8	-10	-13	-15	-18	-20	-23	-25	-28	-33	-38
-5	5	0	-5	-7	-10	-12	-15	-17	-20	-22	-25	-30	-35
-4	8	3	-2	-4	-7	-9	-12	-14	-17	-19	-22	-27	-32
-3	11	6	1	-1	-4	-6	-9	-11	-14	-16	-19	-24	-29
-2	14	9	4	1	-1	-3	-6	-8	-11	-13	-16	-21	-26
-1	17	12	7	4	2	0	-3	-5	-8	-10	-13	-18	-23
0	20	15	10	7	5	2	0	-2	-5	-7	-10	-15	-20
1	23	18	13	10	8	5	3	0	-2	-4	-7	-12	-17
2	26	21	16	13	11	8	6	3	1	-1	-4	-9	-14
3	29	24	19	16	14	11	9	6	4	1	-1	-6	-11
4	32	27	22	19	17	14	12	9	7	4	2	-3	-8
5	35	30	25	22	20	17	15	12	10	7	5	0	-5
6	38	33	28	25	23	20	18	15	13	10	8	3	-2
7	41	36	31	28	26	23	21	18	16	13	11	6	1
8	44	39	34	31	29	26	24	21	19	16	14	9	4
9	47	42	37	34	32	29	27	24	22	19	17	12	7
10	50	45	40	37	35	32	30	27	25	22	20	15	10
12	56	51	46	43	41	38	36	33	31	28	26	21	16
15	65	60	55	52	50	47	45	42	40	37	35	30	25
20	80	75	70	67	65	62	60	57	55	52	50	45	40
25	95	90	85	82	80	77	75	72	70	67	65	60	55
30	110	105	100	97	95	92	90	87	85	82	80	75	70
35	125	120	115	112	110	107	105	102	100	97	95	90	85
40	140	135	130	127	125	122	120	117	115	112	110	105	100

Table 3-127

Hub Diameter, inch: 10

Leg Distance, inch: 35

SHIM REQUIRED FOR ALIGNMENT

Face - OD Method

FACE TIR MILS	OD TIR, MILS -40	-30	-20	-15	-10	-5	0	5	10	15	20	30	40
-40	-120	-125	-130	-132	-135	-137	-140	-142	-145	-147	-150	-155	-160
-35	-102	-107	-112	-115	-117	-120	-122	-125	-127	-130	-132	-137	-142
-30	-85	-90	-95	-97	-100	-102	-105	-107	-110	-112	-115	-120	-125
-25	-67	-72	-77	-80	-82	-85	-87	-90	-92	-95	-97	-102	-107
-20	-50	-55	-60	-62	-65	-67	-70	-72	-75	-77	-80	-85	-90
-15	-32	-37	-42	-45	-47	-50	-52	-55	-57	-60	-62	-67	-72
-12	-22	-27	-32	-34	-37	-39	-42	-44	-47	-49	-52	-57	-62
-10	-15	-20	-25	-27	-30	-32	-35	-37	-40	-42	-45	-50	-55
-9	-11	-16	-21	-24	-26	-29	-31	-34	-36	-39	-41	-46	-51
-8	-8	-13	-18	-20	-23	-25	-28	-30	-33	-35	-38	-43	-48
-7	-4	-9	-14	-17	-19	-22	-24	-27	-29	-32	-34	-39	-44
-6	-1	-6	-11	-13	-16	-18	-21	-23	-26	-28	-31	-36	-41
-5	2	-2	-7	-10	-12	-15	-17	-20	-22	-25	-27	-32	-37
-4	6	1	-4	-6	-9	-11	-14	-16	-19	-21	-24	-29	-34
-3	9	4	0	-3	-5	-8	-10	-13	-15	-18	-20	-25	-30
-2	13	8	3	0	-2	-4	-7	-9	-12	-14	-17	-22	-27
-1	16	11	6	4	1	-1	-3	-6	-8	-11	-13	-18	-23
0	20	15	10	7	5	2	0	-2	-5	-7	-10	-15	-20
1	23	18	13	11	8	6	3	1	-1	-4	-6	-11	-16
2	27	22	17	14	12	9	7	4	2	0	-3	-8	-13
3	30	25	20	18	15	13	10	8	5	3	0	-4	-9
4	34	29	24	21	19	16	14	11	9	6	4	-1	-6
5	37	32	27	25	22	20	17	15	12	10	7	2	-2
6	41	36	31	28	26	23	21	18	16	13	11	6	1
7	44	39	34	32	29	27	24	22	19	17	14	9	4
8	48	43	38	35	33	30	28	25	23	20	18	13	8
9	51	46	41	39	36	34	31	29	26	24	21	16	11
10	55	50	45	42	40	37	35	32	30	27	25	20	15
12	62	57	52	49	47	44	42	39	37	34	32	27	22
15	72	67	62	60	57	55	52	50	47	45	42	37	32
20	90	85	80	77	75	72	70	67	65	62	60	55	50
25	107	102	97	95	92	90	87	85	82	80	77	72	67
30	125	120	115	112	110	107	105	102	100	97	95	90	85
35	142	137	132	130	127	125	122	120	117	115	112	107	102
40	160	155	150	147	145	142	140	137	135	132	130	125	120

Table 3-128

Hub Diameter, inch: 10

Leg Distance, inch: 40

SHIM REQUIRED FOR ALIGNMENT

Face – OD Method

FACE TIR MILS	OD TIR, MILS -40	-30	-20	-15	-10	-5	0	5	10	15	20	30	40
-40	-140	-145	-150	-152	-155	-157	-160	-162	-165	-167	-170	-175	-180
-35	-120	-125	-130	-132	-135	-137	-140	-142	-145	-147	-150	-155	-160
-30	-100	-105	-110	-112	-115	-117	-120	-122	-125	-127	-130	-135	-140
-25	-80	-85	-90	-92	-95	-97	-100	-102	-105	-107	-110	-115	-120
-20	-60	-65	-70	-72	-75	-77	-80	-82	-85	-87	-90	-95	-100
-15	-40	-45	-50	-52	-55	-57	-60	-62	-65	-67	-70	-75	-80
-12	-28	-33	-38	-40	-43	-45	-48	-50	-53	-55	-58	-63	-68
-10	-20	-25	-30	-32	-35	-37	-40	-42	-45	-47	-50	-55	-60
-9	-16	-21	-26	-28	-31	-33	-36	-38	-41	-43	-46	-51	-56
-8	-12	-17	-22	-24	-27	-29	-32	-34	-37	-39	-42	-47	-52
-7	-8	-13	-18	-20	-23	-25	-28	-30	-33	-35	-38	-43	-48
-6	-4	-9	-14	-16	-19	-21	-24	-26	-29	-31	-34	-39	-44
-5	0	-5	-10	-12	-15	-17	-20	-22	-25	-27	-30	-35	-40
-4	4	-1	-6	-8	-11	-13	-16	-18	-21	-23	-26	-31	-36
-3	8	3	-2	-4	-7	-9	-12	-14	-17	-19	-22	-27	-32
-2	12	7	2	0	-3	-5	-8	-10	-13	-15	-18	-23	-28
-1	16	11	6	3	1	-1	-4	-6	-9	-11	-14	-19	-24
0	20	15	10	7	5	2	0	-2	-5	-7	-10	-15	-20
1	24	19	14	11	9	6	4	1	-1	-3	-6	-11	-16
2	28	23	18	15	13	10	8	5	3	0	-2	-7	-12
3	32	27	22	19	17	14	12	9	7	4	2	-3	-8
4	36	31	26	23	21	18	16	13	11	8	6	1	-4
5	40	35	30	27	25	22	20	17	15	12	10	5	0
6	44	39	34	31	29	26	24	21	19	16	14	9	4
7	48	43	38	35	33	30	28	25	23	20	18	13	8
8	52	47	42	39	37	34	32	29	27	24	22	17	12
9	56	51	46	43	41	38	36	33	31	28	26	21	16
10	60	55	50	47	45	42	40	37	35	32	30	25	20
12	68	63	58	55	53	50	48	45	43	40	38	33	28
15	80	75	70	67	65	62	60	57	55	52	50	45	40
20	100	95	90	87	85	82	80	77	75	72	70	65	60
25	120	115	110	107	105	102	100	97	95	92	90	85	80
30	140	135	130	127	125	122	120	117	115	112	110	105	100
35	160	155	150	147	145	142	140	137	135	132	130	125	120
40	180	175	170	167	165	162	160	157	155	152	150	145	140

Table 3-129

Hub Diameter, inch: 10

Leg Distance, inch: 50

SHIM REQUIRED FOR ALIGNMENT

Face - OD Method

FACE TIR MILS	OD TIR, MILS -40	-30	-20	-15	-10	-5	0	5	10	15	20	30	40
-40	-180	-185	-190	-192	-195	-197	-200	-202	-205	-207	-210	-215	-220
-35	-155	-160	-165	-167	-170	-172	-175	-177	-180	-182	-185	-190	-195
-30	-130	-135	-140	-142	-145	-147	-150	-152	-155	-157	-160	-165	-170
-25	-105	-110	-115	-117	-120	-122	-125	-127	-130	-132	-135	-140	-145
-20	-80	-85	-90	-92	-95	-97	-100	-102	-105	-107	-110	-115	-120
-15	-55	-60	-65	-67	-70	-72	-75	-77	-80	-82	-85	-90	-95
-12	-40	-45	-50	-52	-55	-57	-60	-62	-65	-67	-70	-75	-80
-10	-30	-35	-40	-42	-45	-47	-50	-52	-55	-57	-60	-65	-70
-9	-25	-30	-35	-37	-40	-42	-45	-47	-50	-52	-55	-60	-65
-8	-20	-25	-30	-32	-35	-37	-40	-42	-45	-47	-50	-55	-60
-7	-15	-20	-25	-27	-30	-32	-35	-37	-40	-42	-45	-50	-55
-6	-10	-15	-20	-22	-25	-27	-30	-32	-35	-37	-40	-45	-50
-5	-5	-10	-15	-17	-20	-22	-25	-27	-30	-32	-35	-40	-45
-4	0	-5	-10	-12	-15	-17	-20	-22	-25	-27	-30	-35	-40
-3	5	0	-5	-7	-10	-12	-15	-17	-20	-22	-25	-30	-35
-2	10	5	0	-2	-5	-7	-10	-12	-15	-17	-20	-25	-30
-1	15	10	5	2	0	-2	-5	-7	-10	-12	-15	-20	-25
0	20	15	10	7	5	2	0	-2	-5	-7	-10	-15	-20
1	25	20	15	12	10	7	5	2	0	-2	-5	-10	-15
2	30	25	20	17	15	12	10	7	5	2	0	-5	-10
3	35	30	25	22	20	17	15	12	10	7	5	0	-5
4	40	35	30	27	25	22	20	17	15	12	10	5	0
5	45	40	35	32	30	27	25	22	20	17	15	10	5
6	50	45	40	37	35	32	30	27	25	22	20	15	10
7	55	50	45	42	40	37	35	32	30	27	25	20	15
8	60	55	50	47	45	42	40	37	35	32	30	25	20
9	65	60	55	52	50	47	45	42	40	37	35	30	25
10	70	65	60	57	55	52	50	47	45	42	40	35	30
12	80	75	70	67	65	62	60	57	55	52	50	45	40
15	95	90	85	82	80	77	75	72	70	67	65	60	55
20	120	115	110	107	105	102	100	97	95	92	90	85	80
25	145	140	135	132	130	127	125	122	120	117	115	110	105
30	170	165	160	157	155	152	150	147	145	142	140	135	130
35	195	190	185	182	180	177	175	172	170	167	165	160	155
40	220	215	210	207	205	202	200	197	195	192	190	185	180

Table 3-130

Hub Diameter, inch: 10

Leg Distance, inch: 60

SHIM REQUIRED FOR ALIGNMENT

Face - OD Method

FACE TIR MILS	OD TIR, MILS -40	-30	-20	-15	-10	-5	0	5	10	15	20	30	40
-40	-220	-225	-230	-232	-235	-237	-240	-242	-245	-247	-250	-255	-260
-35	-190	-195	-200	-202	-205	-207	-210	-212	-215	-217	-220	-225	-230
-30	-160	-165	-170	-172	-175	-177	-180	-182	-185	-187	-190	-195	-200
-25	-130	-135	-140	-142	-145	-147	-150	-152	-155	-157	-160	-165	-170
-20	-100	-105	-110	-112	-115	-117	-120	-122	-125	-127	-130	-135	-140
-15	-70	-75	-80	-82	-85	-87	-90	-92	-95	-97	-100	-105	-110
-12	-52	-57	-62	-64	-67	-69	-72	-74	-77	-79	-82	-87	-92
-10	-40	-45	-50	-52	-55	-57	-60	-62	-65	-67	-70	-75	-80
-9	-34	-39	-44	-46	-49	-51	-54	-56	-59	-61	-64	-69	-74
-8	-28	-33	-38	-40	-43	-45	-48	-50	-53	-55	-58	-63	-68
-7	-22	-27	-32	-34	-37	-39	-42	-44	-47	-49	-52	-57	-62
-6	-16	-21	-26	-28	-31	-33	-36	-38	-41	-43	-46	-51	-56
-5	-10	-15	-20	-22	-25	-27	-30	-32	-35	-37	-40	-45	-50
-4	-4	-9	-14	-16	-19	-21	-24	-26	-29	-31	-34	-39	-44
-3	2	-3	-8	-10	-13	-15	-18	-20	-23	-25	-28	-33	-38
-2	8	3	-2	-4	-7	-9	-12	-14	-17	-19	-22	-27	-32
-1	14	9	4	1	-1	-3	-6	-8	-11	-13	-16	-21	-26
0	20	15	10	7	5	2	0	-2	-5	-7	-10	-15	-20
1	26	21	16	13	11	8	6	3	1	-1	-4	-9	-14
2	32	27	22	19	17	14	12	9	7	4	2	-3	-8
3	38	33	28	25	23	20	18	15	13	10	8	3	-2
4	44	39	34	31	29	26	24	21	19	16	14	9	4
5	50	45	40	37	35	32	30	27	25	22	20	15	10
6	56	51	46	43	41	38	36	33	31	28	26	21	16
7	62	57	52	49	47	44	42	39	37	34	32	27	22
8	68	63	58	55	53	50	48	45	43	40	38	33	28
9	74	69	64	61	59	56	54	51	49	46	44	39	34
10	80	75	70	67	65	62	60	57	55	52	50	45	40
12	92	87	82	79	77	74	72	69	67	64	62	57	52
15	110	105	100	97	95	92	90	87	85	82	80	75	70
20	140	135	130	127	125	122	120	117	115	112	110	105	100
25	170	165	160	157	155	152	150	147	145	142	140	135	130
30	200	195	190	187	185	182	180	177	175	172	170	165	160
35	230	225	220	217	215	212	210	207	205	202	200	195	190
40	260	255	250	247	245	242	240	237	235	232	230	225	220

Table 3-131

Hub Diameter, inch: 10

Leg Distance, inch: 70

SHIM REQUIRED FOR ALIGNMENT

Face - OD Method

FACE TIR MILS	OD TIR, MILS -40	-30	-20	-15	-10	-5	0	5	10	15	20	30	40
-40	-260	-265	-270	-272	-275	-277	-280	-282	-285	-287	-290	-295	-300
-35	-225	-230	-235	-237	-240	-242	-245	-247	-250	-252	-255	-260	-265
-30	-190	-195	-200	-202	-205	-207	-210	-212	-215	-217	-220	-225	-230
-25	-155	-160	-165	-167	-170	-172	-175	-177	-180	-182	-185	-190	-195
-20	-120	-125	-130	-132	-135	-137	-140	-142	-145	-147	-150	-155	-160
-15	-85	-90	-95	-97	-100	-102	-105	-107	-110	-112	-115	-120	-125
-12	-64	-69	-74	-76	-79	-81	-84	-86	-89	-91	-94	-99	-104
-10	-50	-55	-60	-62	-65	-67	-70	-72	-75	-77	-80	-85	-90
-9	-43	-48	-53	-55	-58	-60	-63	-65	-68	-70	-73	-78	-83
-8	-36	-41	-46	-48	-51	-53	-56	-58	-61	-63	-66	-71	-76
-7	-29	-34	-39	-41	-44	-46	-49	-51	-54	-56	-59	-64	-69
-6	-22	-27	-32	-34	-37	-39	-42	-44	-47	-49	-52	-57	-62
-5	-15	-20	-25	-27	-30	-32	-35	-37	-40	-42	-45	-50	-55
-4	-8	-13	-18	-20	-23	-25	-28	-30	-33	-35	-38	-43	-48
-3	-1	-6	-11	-13	-16	-18	-21	-23	-26	-28	-31	-36	-41
-2	6	1	-4	-6	-9	-11	-14	-16	-19	-21	-24	-29	-34
-1	13	8	3	0	-2	-4	-7	-9	-12	-14	-17	-22	-27
0	20	15	10	7	5	2	0	-2	-5	-7	-10	-15	-20
1	27	22	17	14	12	9	7	4	2	0	-3	-8	-13
2	34	29	24	21	19	16	14	11	9	6	4	-1	-6
3	41	36	31	28	26	23	21	18	16	13	11	6	1
4	48	43	38	35	33	30	28	25	23	20	18	13	8
5	55	50	45	42	40	37	35	32	30	27	25	20	15
6	62	57	52	49	47	44	42	39	37	34	32	27	22
7	69	64	59	56	54	51	49	46	44	41	39	34	29
8	76	71	66	63	61	58	56	53	51	48	46	41	36
9	83	78	73	70	68	65	63	60	58	55	53	48	43
10	90	85	80	77	75	72	70	67	65	62	60	55	50
12	104	99	94	91	89	86	84	81	79	76	74	69	64
15	125	120	115	112	110	107	105	102	100	97	95	90	85
20	160	155	150	147	145	142	140	137	135	132	130	125	120
25	195	190	185	182	180	177	175	172	170	167	165	160	155
30	230	225	220	217	215	212	210	207	205	202	200	195	190
35	265	260	255	252	250	247	245	242	240	237	235	230	225
40	300	295	290	287	285	282	280	277	275	272	270	265	260

Table 3-132

Hub Diameter, inch: 10

Leg Distance, inch: 80

SHIM REQUIRED FOR ALIGNMENT

Face - OD Method

FACE TIR MILS	OD TIR, MILS -40	-30	-20	-15	-10	-5	0	5	10	15	20	30	40
-40	-300	-305	-310	-312	-315	-317	-320	-322	-325	-327	-330	-335	-340
-35	-260	-265	-270	-272	-275	-277	-280	-282	-285	-287	-290	-295	-300
-30	-220	-225	-230	-232	-235	-237	-240	-242	-245	-247	-250	-255	-260
-25	-180	-185	-190	-192	-195	-197	-200	-202	-205	-207	-210	-215	-220
-20	-140	-145	-150	-152	-155	-157	-160	-162	-165	-167	-170	-175	-180
-15	-100	-105	-110	-112	-115	-117	-120	-122	-125	-127	-130	-135	-140
-12	-76	-81	-86	-88	-91	-93	-96	-98	-101	-103	-106	-111	-116
-10	-60	-65	-70	-72	-75	-77	-80	-82	-85	-87	-90	-95	-100
-9	-52	-57	-62	-64	-67	-69	-72	-74	-77	-79	-82	-87	-92
-8	-44	-49	-54	-56	-59	-61	-64	-66	-69	-71	-74	-79	-84
-7	-36	-41	-46	-48	-51	-53	-56	-58	-61	-63	-66	-71	-76
-6	-28	-33	-38	-40	-43	-45	-48	-50	-53	-55	-58	-63	-68
-5	-20	-25	-30	-32	-35	-37	-40	-42	-45	-47	-50	-55	-60
-4	-12	-17	-22	-24	-27	-29	-32	-34	-37	-39	-42	-47	-52
-3	-4	-9	-14	-16	-19	-21	-24	-26	-29	-31	-34	-39	-44
-2	4	-1	-6	-8	-11	-13	-16	-18	-21	-23	-26	-31	-36
-1	12	7	2	0	-3	-5	-8	-10	-13	-15	-18	-23	-28
0	20	15	10	7	5	2	0	-2	-5	-7	-10	-15	-20
1	28	23	18	15	13	10	8	5	3	0	-2	-7	-12
2	36	31	26	23	21	18	16	13	11	8	6	1	-4
3	44	39	34	31	29	26	24	21	19	16	14	9	4
4	52	47	42	39	37	34	32	29	27	24	22	17	12
5	60	55	50	47	45	42	40	37	35	32	30	25	20
6	68	63	58	55	53	50	48	45	43	40	38	33	28
7	76	71	66	63	61	58	56	53	51	48	46	41	36
8	84	79	74	71	69	66	64	61	59	56	54	49	44
9	92	87	82	79	77	74	72	69	67	64	62	57	52
10	100	95	90	87	85	82	80	77	75	72	70	65	60
12	116	111	106	103	101	98	96	93	91	88	86	81	76
15	140	135	130	127	125	122	120	117	115	112	110	105	100
20	180	175	170	167	165	162	160	157	155	152	150	145	140
25	220	215	210	207	205	202	200	197	195	192	190	185	180
30	260	255	250	247	245	242	240	237	235	232	230	225	220
35	300	295	290	287	285	282	280	277	275	272	270	265	260
40	340	335	330	327	325	322	320	317	315	312	310	305	300

Table 3-133

Hub Diameter, inch: 10

Leg Distance, inch: 90

SHIM REQUIRED FOR ALIGNMENT

Face - OD Method

FACE TIR MILS	OD TIR, MILS -40	-30	-20	-15	-10	-5	0	5	10	15	20	30	40
-40	-340	-345	-350	-352	-355	-357	-360	-362	-365	-367	-370	-375	-380
-35	-295	-300	-305	-307	-310	-312	-315	-317	-320	-322	-325	-330	-335
-30	-250	-255	-260	-262	-265	-267	-270	-272	-275	-277	-280	-285	-290
-25	-205	-210	-215	-217	-220	-222	-225	-227	-230	-232	-235	-240	-245
-20	-160	-165	-170	-172	-175	-177	-180	-182	-185	-187	-190	-195	-200
-15	-115	-120	-125	-127	-130	-132	-135	-137	-140	-142	-145	-150	-155
-12	-88	-93	-98	-100	-103	-105	-108	-110	-113	-115	-118	-123	-128
-10	-70	-75	-80	-82	-85	-87	-90	-92	-95	-97	-100	-105	-110
-9	-61	-66	-71	-73	-76	-78	-81	-83	-86	-88	-91	-96	-101
-8	-52	-57	-62	-64	-67	-69	-72	-74	-77	-79	-82	-87	-92
-7	-43	-48	-53	-55	-58	-60	-63	-65	-68	-70	-73	-78	-83
-6	-34	-39	-44	-46	-49	-51	-54	-56	-59	-61	-64	-69	-74
-5	-25	-30	-35	-37	-40	-42	-45	-47	-50	-52	-55	-60	-65
-4	-16	-21	-26	-28	-31	-33	-36	-38	-41	-43	-46	-51	-56
-3	-7	-12	-17	-19	-22	-24	-27	-29	-32	-34	-37	-42	-47
-2	2	-3	-8	-10	-13	-15	-18	-20	-23	-25	-28	-33	-38
-1	11	6	1	-1	-4	-6	-9	-11	-14	-16	-19	-24	-29
0	20	15	10	7	5	2	0	-2	-5	-7	-10	-15	-20
1	29	24	19	16	14	11	9	6	4	1	-1	-6	-11
2	38	33	28	25	23	20	18	15	13	10	8	3	-2
3	47	42	37	34	32	29	27	24	22	19	17	12	7
4	56	51	46	43	41	38	36	33	31	28	26	21	16
5	65	60	55	52	50	47	45	42	40	37	35	30	25
6	74	69	64	61	59	56	54	51	49	46	44	39	34
7	83	78	73	70	68	65	63	60	58	55	53	48	43
8	92	87	82	79	77	74	72	69	67	64	62	57	52
9	101	96	91	88	86	83	81	78	76	73	71	66	61
10	110	105	100	97	95	92	90	87	85	82	80	75	70
12	128	123	118	115	113	110	108	105	103	100	98	93	88
15	155	150	145	142	140	137	135	132	130	127	125	120	115
20	200	195	190	187	185	182	180	177	175	172	170	165	160
25	245	240	235	232	230	227	225	222	220	217	215	210	205
30	290	285	280	277	275	272	270	267	265	262	260	255	250
35	335	330	325	322	320	317	315	312	310	307	305	300	295
40	380	375	370	367	365	362	360	357	355	352	350	345	340

Table 3-134

Hub Diameter, inch: 10

Leg Distance, inch: 100

SHIM REQUIRED FOR ALIGNMENT

Face - OD Method

FACE TIR MILS	OD TIR, MILS -40	-30	-20	-15	-10	-5	0	5	10	15	20	30	40
-40	-380	-385	-390	-392	-395	-397	-400	-402	-405	-407	-410	-415	-420
-35	-330	-335	-340	-342	-345	-347	-350	-352	-355	-357	-360	-365	-370
-30	-280	-285	-290	-292	-295	-297	-300	-302	-305	-307	-310	-315	-320
-25	-230	-235	-240	-242	-245	-247	-250	-252	-255	-257	-260	-265	-270
-20	-180	-185	-190	-192	-195	-197	-200	-202	-205	-207	-210	-215	-220
-15	-130	-135	-140	-142	-145	-147	-150	-152	-155	-157	-160	-165	-170
-12	-100	-105	-110	-112	-115	-117	-120	-122	-125	-127	-130	-135	-140
-10	-80	-85	-90	-92	-95	-97	-100	-102	-105	-107	-110	-115	-120
-9	-70	-75	-80	-82	-85	-87	-90	-92	-95	-97	-100	-105	-110
-8	-60	-65	-70	-72	-75	-77	-80	-82	-85	-87	-90	-95	-100
-7	-50	-55	-60	-62	-65	-67	-70	-72	-75	-77	-80	-85	-90
-6	-40	-45	-50	-52	-55	-57	-60	-62	-65	-67	-70	-75	-80
-5	-30	-35	-40	-42	-45	-47	-50	-52	-55	-57	-60	-65	-70
-4	-20	-25	-30	-32	-35	-37	-40	-42	-45	-47	-50	-55	-60
-3	-10	-15	-20	-22	-25	-27	-30	-32	-35	-37	-40	-45	-50
-2	0	-5	-10	-12	-15	-17	-20	-22	-25	-27	-30	-35	-40
-1	10	5	0	-2	-5	-7	-10	-12	-15	-17	-20	-25	-30
0	20	15	10	7	5	2	0	-2	-5	-7	-10	-15	-20
1	30	25	20	17	15	12	10	7	5	2	0	-5	-10
2	40	35	30	27	25	22	20	17	15	12	10	5	0
3	50	45	40	37	35	32	30	27	25	22	20	15	10
4	60	55	50	47	45	42	40	37	35	32	30	25	20
5	70	65	60	57	55	52	50	47	45	42	40	35	30
6	80	75	70	67	65	62	60	57	55	52	50	45	40
7	90	85	80	77	75	72	70	67	65	62	60	55	50
8	100	95	90	87	85	82	80	77	75	72	70	65	60
9	110	105	100	97	95	92	90	87	85	82	80	75	70
10	120	115	110	107	105	102	100	97	95	92	90	85	80
12	140	135	130	127	125	122	120	117	115	112	110	105	100
15	170	165	160	157	155	152	150	147	145	142	140	135	130
20	220	215	210	207	205	202	200	197	195	192	190	185	180
25	270	265	260	257	255	252	250	247	245	242	240	235	230
30	320	315	310	307	305	302	300	297	295	292	290	285	280
35	370	365	360	357	355	352	350	347	345	342	340	335	330
40	420	415	410	407	405	402	400	397	395	392	390	385	380

Chapter 4

Reverse Indicator Method

The Reverse Indicator Method of taking alignment data is becoming increasingly popular. It is the method most often selected for alignment of large, high-speed machines. Although it does not appear as straightforward as the Face-OD Method, the concept of the Reverse Indicator Method is also simple.

In the Reverse Indicator Method, a dial indicator is mounted on one shaft and rotated so that it indicates on the other shaft. It is next mounted on the second shaft and rotated so that it indicates on the first shaft. Hence the name, "Reverse Indicator Method."

In actual practice, the coupling hubs are mounted on the shafts and the readings are taken on the coupling hubs. Normally, two indicators are used to gather the data quickly.

Setting Up

The equations, forms, and tables described for the Reverse Indicator Method are based on the following assumptions:

1. The indicators rotate with the shaft and read the other shaft (or coupling).

2. The indicators are "zeroed" at the top of the shaft (or coupling).
3. There is no indicator sag, or the readings have been corrected for sag before using the tables. (See "Indicator Sag" later in this chapter.)

A common problem is to align a centrifugal pump with a motor driver. The pump is connected to piping and baseplate properly and will not be moved. The coupling spacer is installed so that the shafts may be rotated together. The dial indicator is mounted on the motor coupling hub and extended to read the pump coupling hub. The indicator is adjusted to read zero at the top of the pump coupling hub OD. Readings are taken at horizontal and vertical centerlines. The indicator is then reversed to read the motor coupling hub. Again, the indicator is zeroed at the top and readings are taken at horizontal and vertical centerlines. Shims are added (or removed) from the motor driver feet.

Many people prefer to use an alignment jig that allows both shafts to rotate together. This can speed up gathering data and improve accuracy. The tables are also valid for this procedure. However, for purposes of explanation, the shafts will be considered to be separated with one rotating and one stationary.

A typical setup is shown in Figure 4-1.

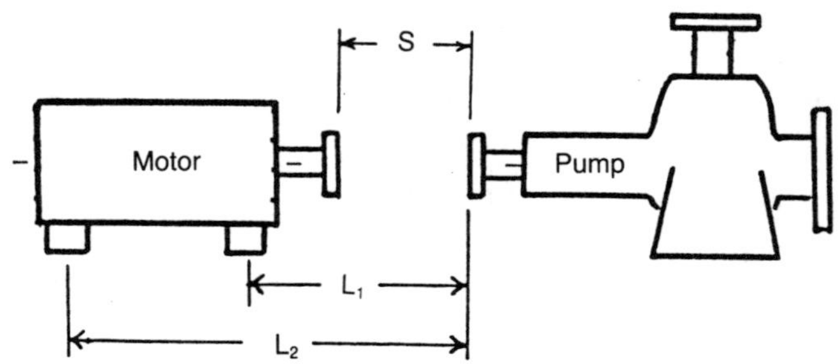

Figure 4-1. A typical pump—motor arrangement showing separation (S) and leg distances (L_1 and L_2) used in the Reverse Indicator Method.

Data Forms

Figure 4-2 is a useful form for recording alignment data. It is good practice to record both preliminary and final readings as well as operating (hot) readings. This can help speed up future re-alignments.

REVERSE INDICATOR ALIGNMENT RECORD

Area ____________________ Date ____________________
DriveN ____________________ Type ____________________
DriveR ____________________ Type ____________________

Coupling ____
Size ____________ Distance between shaft ends ____________

Ambient temperature, °F ____ Operating temperature, °F ____
Expected growth of DriveR leg, mils: IB ________ OB ________
DriveN coupling face to DriveR leg, in.: IB ______ OB ______
Indicator sag, mils TIR ____________

Mount indicators on the DriveR and read the DriveN hub:

OD Readings, TIR at DriveR Hub — OD Readings, TIR at DriveN Hub

The vertical and horizontal sums of the readings should be equal.
Sum of DriveR Readings: Vertical ______ Horizontal ______
Sum of DriveN Readings: Vertical ______ Horizontal ______

Data Used in Tables
TIR (corrected for sag), mils: Face __________ OD __________

Shim required for IB leg, mils ________ (Actual used ________)
Shim required for OB leg, mils ________ (Actual used ________)

Remarks

____
____
Data taken by____________________

Figure 4-2. This form is helpful in recording Reverse Indicator alignment data.

Equations and Calculations

Once the reverse readings have been taken, the calculation of the required shim is straightforward:

$$\text{Shim} = \frac{1}{2}\left[(\text{DriveN} + \text{DriveR})\left(\frac{\text{Leg}}{\text{Spacer}}\right) - \text{DriveN}\right]$$

DriveR = Total Indicator Reading, mils at DriveR hub

DriveN = Total Indicator Reading, mils at DriveN hub

Spacer = Distance between the two indicators

Leg = Distance to leg which is to be shimmed, in.

Shim = Required shim, mils (Add shim for +, remove for −)

Be sure to use the + or − sign of the reading in the calculation.

The alignment tables in this book were constructed using this equation. They are much handier to use than the equation and reduce the chances of error in calculation.

To use the tables, first find the section for your spacer length. Then go to the page for the leg you need to shim. Find your DriveR and DriveN readings and read the shim required in the proper column.

The "spacer length" is actually the distance between the dial indicators. This will be approximately the coupling spacer length. However, you may be able to set up for a particular length by adjusting the alignment jig.

If the table does not have your exact readings, you can determine the required shim by looking in the nearby rows and columns and estimating. In some cases, you may find that apparently large changes in readings require very little change in shim.

If you wish to be more accurate than an estimate, you can break up your readings into parts, get the shim for each partial reading and combine the partial shim to get the total. For example, suppose your DriveR reading is −7 mils TIR and the DriveN reading is 0. This is not in the tables. However, you can look up the shim for −5 mils and −2 mils. The shim required for −7 mils is the sum of these two partial readings.

If one reading is not zero, you can still break up your data into two or more equivalent partial sets. For example, turn to Table 5-90 for a 10-in. spacer and 40-in. leg. Suppose your data showed −15 on the DriveR and +12 on the DriveN. The shim would be −12 mils. Now break up the data into two sets. Let the first set be −10 on the DriveR and +8 on the DriveN. This would require −8 mils of shim. The second set is −5 on the DriveR and +4 on the DriveN. This would require −4 mils of shim. The total shim is still −12 mils, just as before.

For large misalignments, you can divide the DriveR and DriveN readings by 10, enter the proper table, find the shim, and multiply it by 10 to get the correct value.

Examples

Example 1: The Reverse Indicator Method is used to align a motor to a pump. The indicators are zeroed at the top. The DriveR reading is 10 mils TIR. The DriveN reading is 5 mils TIR. The spacer length is 5 in. The distance from the pump coupling face to the near motor leg is 25 in.; to the far leg 50 in. Indicator sag is zero. What shim is required?

Use the tables for a 5-in. spacer. Reading directly from Table 5-42 for a 25-in. leg shows to add 35 mils to this leg. Reading directly from Table 5-46 for a 50-in. leg shows to add 72 mils to this leg.

Example 2: The Reverse Indicator Method is used to align a steam turbine to a pump. The indicators are zeroed at the top. The DriveR readings is −2 mils TIR. The DriveN reading is 6 mils TIR. The spacer length is 7 in. The turbine legs are 31 in. and 67 in. from the pump coupling face. The indicator jig has a sag of 5 mils TIR. From past experience, you want to set the front legs 5 mils low and the back legs 15 mils low. What shim is needed?

First, correct the readings for sag. The DriveR reading should be $-2 + 5 = +3$ mils TIR. The DriveN reading should be $6 + 5 = 11$ mils TIR.

Now, turn to Table 5-61 for a 7-in. spacer and 30-in. leg. The table does not have +3 mils for the DriveR and +11 mils for the DriveN. You will have to estimate for these values. For 2 mils DriveR and 10 mils DriveN the shim is +20. For 2 mils DriveR and 12 mils DriveN the shim is +24. This means that for 2 mils DriveR and 11 mils Drive N the shim would be 22 mils. Now, notice that for 12 mils on the DriveN that going from 2 to 5 mils on the DriveR adds 6 mils to the shim. Therefore, going from 2 to 3 mils on the DriveR would add 2 mil. So, the shim required for the 30-in leg would be 24 mils; however, the actual leg distance is 31 in.

Next, turn to Table 5-62 for a 7-in. spacer and 35-in. leg. You should estimate about 29 mils of shim for 3 mils on the DriveR, and 11 mils on the DriveN. Since you would add 24 mils for a 30-in. leg and 29 mils for a 35-in. leg, you would add about 25 mils for the 31-in. leg. (The calculated exact value is 25.5 mils.)

Remember that you want to set this leg to be 5 mils low. Therefore, you actually add only 20 mils of shim.

In a similar way, you use the 60-in. leg and 70-in. leg tables to determine a "cold" shim of 62 mils for the 67-in. leg. To set this leg 15 mils low, you actually add only 47 mils of shim.

Alignment Tolerances

You will not be able to align the shafts perfectly. If you ever think you have achieved that perfect alignment, re-check it the next day, at night, or during a rain. Alignment is a dynamic condition; it is constantly changing. Your objective is to obtain an alignment that is within the operating range of the coupling and drive train system. Since you do not know the direction or amount of change, your goal should be to align to within the tolerances of your ability to measure and shim.

A common reason for not obtaining close alignment is that the required shim is not properly calculated. This means that the alignment procedure is actually a trial and error method. The alignment crew members become discouraged after a few tries and rationalize that the alignment is acceptable.

By using the alignment tables in this book, it should be easier to get "perfect" cold alignment rather than previous "acceptable" alignment with trial and error routines. Since the alignment tables should make complete correction possible the first time (and certainly the second!), there should be no reason to settle for less. However, things being the way they are, here are some criteria currently in use:

1. References 2, 6, and 9 recommend 3 mils TIR at DriveR and DriveN.
2. Reference 4 recommends .5 mil TIR for each inch of spacer length.

A number of flexible disc coupling manufacturers give the maximum misalignment capability of their couplings as 1/2 degree of angularity. This is based on misalignment before fatiguing the discs and should not be used as an operating alignment criteria. In practice, metal disc failures have been observed with disc packs at less than 1/4 degree.

With the tables, it should not be difficult to obtain alignment well within the 3 mil TIR criteria. After all, this would still mean about 15 mils of shim are needed for a 7 inch spacer and 40-in. leg distance.

Errors in Taking Data

There are ample opportunities to err while taking alignment data. Some of the more common ones include:

1. Poor clamps or clamps not tight
2. Broken dial indicators
3. Indicator bag sag
4. Bent shaft
5. Coupling hub not fitted properly to shaft
6. Read dial indicator wrong (especially + and − readings)

Fortunately, the machines serve as a final check. That is, if the calculated correction does not make a significant improvement in alignment, a mistake has been made.

There is a simple way of checking your readings before adding or removing shims. The horizontal readings and the vertical readings should have the same sum. This applies to both DriveR and DriveN values. For example, consider the following readings:

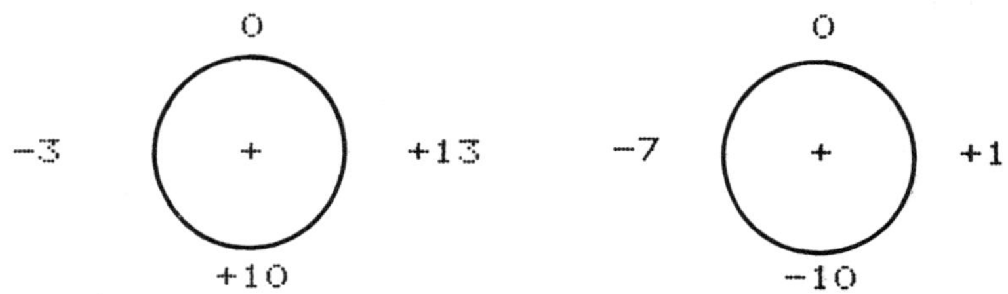

DriveR DriveN

For the DriveR, the vertical readings total 0 + 10 = +10. The horizontal readings total +13 − 3 = +10. Therefore, this is probably good data. However, the DriveN readings total −10 vertically and −6 horizontally. Since the vertical and horizontal sums are different, the DriveN readings are suspect and should be checked. Since the cause of the error may also have affected the DriveR reading, the DriveR readings should also be checked.

If the accuracy of the data cannot be improved, check the significance of the error by determining the shim required for either case.

Indicator Sag

Indicator sag is caused by gravity and will always be present to some degree. You can minimize the amount of sag by using light indicators with stiff clamps and extensions. Table 4-1 gives some idea of the amount of sag to expect and the effect of weight and extension length. The table assumes a rigid clamping arrangement. To know the actual sag, you should test the alignment jig by setting up on a shaft that is between centers in a lathe. Arrange the indicator to read the OD of the same shaft it is mounted on. Zero the indicator at the top and rotate the shaft 180 degrees. The indicator now shows TIR sag for the OD readings.

Table 4-1
Typical Indicator Sag
(OD TIR, mils, for 1/2″ steel rod)

Indicator Weight (oz)	Length of Extension Rod (in.)				
	5	10	15	20	25
0	−.1	−2	−8	−24	−59
3	−.3	−3	−12	−35	−80
6	−.4	−4	−17	−46	−101
9	−.6	−6	−21	−57	−122

If you test the jig as described, you will find that the indicator shows a negative reading when the shaft is rotated. In other words, a situation expected to read zero TIR actually reads as a minus TIR. For example, the reading might be −11 mils on the sag test shaft. This is why typical sag values are also shown as negative values.

Now, suppose you take that same jig and use it to align equipment. When the indicator shows −11 mils TIR, you know that the true reading is 0. If the indicator shows −20 mils, the true reading would be −9 mils (−20 + 11). If the indicator shows a +7 mils, the true reading would be 18 mils. Mathematically, this is:

$$\text{Corrected Reading} = \text{Reading} - \text{Sag}$$

Remember to take the negative sign of the sag into account.

Hot Alignment

So far, the techniques presented have been for alignment of "cold" (that is, not operating) equipment. Good cold alignment is the basis for establishing the proper hot or operating alignment. If you know the true effects of piping and temperature, the cold alignment can be the final alignment.

A word of caution: Take manufacturers' suggested cold and hot alignment with a lump of salt. They do not know the effect of your piping, foundation and operations as well as you do.

With the alignment tables, offsetting for thermal growth is simple. For example, suppose you have read +5 mils TIR on the DriveR of a 7-in. spacer and +7 mils TIR on the DriveN. The leg distance is 50 in. to a leg that you think will "grow" by 30 mils when at operating temperature. By referring to the alignment tables, you see that the required cold correction is to add 39 mils of shim to that leg. Since you think the leg will grow 30 mils, you actually only add 9 mils of shim. If you check the alignment after adding the 9 mils of shim, you probably will get readings of, say, 3 mils TIR on the DriveR and 7 mils TIR on the DriveN. For these readings, the alignment table will indicate an additional 30 mils is required. This is the amount that you allowed for thermal growth of the leg.

In the same example, suppose you take hot alignment readings after adding the 9 mils of shim and the DriveR is now −2; the DriveN is −1. According to the alignment table, you should take off 10 mils of shim. This means that the leg actually grew by 40 mils instead of 30 mils. You should record that the leg grows by 40 mils and use this correction in the future.

Chapter 5

Reverse Indicator Alignment Tables

Table 5-1

Separation, inch: 2

Leg Distance, inch: 5

SHIM REQUIRED FOR ALIGNMENT

Reverse Indicator Method

riveN IR ILS | DriveR TIR, MILS

riveN IR ILS	-40	-30	-20	-15	-10	-5	-2	0	2	5	10	15	20	30	40
-40	-80	-67	-55	-48	-42	-36	-32	-30	-27	-23	-17	-11	-5	7	20
-30	-72	-60	-47	-41	-35	-28	-25	-22	-20	-16	-10	-3	2	15	27
-25	-68	-56	-43	-37	-31	-25	-21	-18	-16	-12	-6	0	6	18	31
-20	-65	-52	-40	-33	-27	-21	-17	-15	-12	-8	-2	3	10	22	35
-18	-63	-51	-38	-32	-26	-19	-16	-13	-11	-7	-1	5	11	24	36
-16	-62	-49	-37	-30	-24	-18	-14	-12	-9	-5	0	6	13	25	38
-14	-60	-48	-35	-29	-23	-16	-13	-10	-8	-4	2	8	14	27	39
-12	-59	-46	-34	-27	-21	-15	-11	-9	-6	-2	3	9	16	28	41
-10	-57	-45	-32	-26	-20	-13	-10	-7	-5	-1	5	11	17	30	42
-8	-56	-43	-31	-24	-18	-12	-8	-6	-3	0	6	12	19	31	44
-7	-55	-42	-30	-24	-17	-11	-7	-5	-2	1	7	13	19	32	44
-6	-54	-42	-29	-23	-17	-10	-7	-4	-2	1	8	14	20	33	45
-5	-53	-41	-28	-22	-16	-10	-6	-3	-1	2	8	15	21	33	46
-4	-53	-40	-28	-21	-15	-9	-5	-3	0	3	9	15	22	34	47
-3	-52	-39	-27	-21	-14	-8	-4	-2	0	4	10	16	22	35	47
-2	-51	-39	-26	-20	-14	-7	-4	-1	1	4	11	17	23	36	48
-1	-50	-38	-25	-19	-13	-7	-3	0	1	5	11	18	24	36	49
0	-50	-37	-25	-18	-12	-6	-2	0	2	6	12	18	25	37	50
1	-49	-36	-24	-18	-11	-5	-1	0	3	7	13	19	25	38	50
2	-48	-36	-23	-17	-11	-4	-1	1	4	7	14	20	26	39	51
3	-47	-35	-22	-16	-10	-4	0	2	4	8	14	21	27	39	52
4	-47	-34	-22	-15	-9	-3	0	3	5	9	15	21	28	40	53
5	-46	-33	-21	-15	-8	-2	1	3	6	10	16	22	28	41	53
6	-45	-33	-20	-14	-8	-1	2	4	7	10	17	23	29	42	54
7	-44	-32	-19	-13	-7	-1	2	5	7	11	17	24	30	42	55
8	-44	-31	-19	-12	-6	0	3	6	8	12	18	24	31	43	56
10	-42	-30	-17	-11	-5	1	5	7	10	13	20	26	32	45	57
12	-41	-28	-16	-9	-3	2	6	9	11	15	21	27	34	46	59
14	-39	-27	-14	-8	-2	4	8	10	13	16	23	29	35	48	60
16	-38	-25	-13	-6	0	5	9	12	14	18	24	30	37	49	62
18	-36	-24	-11	-5	1	7	11	13	16	19	26	32	38	51	63
20	-35	-22	-10	-3	2	8	12	15	17	21	27	33	40	52	65
25	-31	-18	-6	0	6	12	16	18	21	25	31	37	43	56	68
30	-27	-15	-2	3	10	16	20	22	25	28	35	41	47	60	72
40	-20	-7	5	11	17	23	27	30	32	36	42	48	55	67	80

Table 5-2

Separation, inch: 2

Leg Distance, inch: 10

SHIM REQUIRED FOR ALIGNMENT

Reverse Indicator Method

DriveN TIR MILS	DriveR TIR, MILS -40	-30	-20	-15	-10	-5	-2	0	2	5	10	15	20	30	40
-40	-180	-155	-130	-117	-105	-92	-85	-80	-75	-67	-55	-42	-30	-5	20
-30	-160	-135	-110	-97	-85	-72	-65	-60	-55	-47	-35	-22	-10	15	40
-25	-150	-125	-100	-87	-75	-62	-55	-50	-45	-37	-25	-12	0	25	50
-20	-140	-115	-90	-77	-65	-52	-45	-40	-35	-27	-15	-2	10	35	60
-18	-136	-111	-86	-73	-61	-48	-41	-36	-31	-23	-11	1	14	39	64
-16	-132	-107	-82	-69	-57	-44	-37	-32	-27	-19	-7	5	18	43	68
-14	-128	-103	-78	-65	-53	-40	-33	-28	-23	-15	-3	9	22	47	72
-12	-124	-99	-74	-61	-49	-36	-29	-24	-19	-11	1	13	26	51	76
-10	-120	-95	-70	-57	-45	-32	-25	-20	-15	-7	5	17	30	55	80
-8	-116	-91	-66	-53	-41	-28	-21	-16	-11	-3	9	21	34	59	84
-7	-114	-89	-64	-51	-39	-26	-19	-14	-9	-1	11	23	36	61	86
-6	-112	-87	-62	-49	-37	-24	-17	-12	-7	0	13	25	38	63	88
-5	-110	-85	-60	-47	-35	-22	-15	-10	-5	2	15	27	40	65	90
-4	-108	-83	-58	-45	-33	-20	-13	-8	-3	4	17	29	42	67	92
-3	-106	-81	-56	-43	-31	-18	-11	-6	-1	6	19	31	44	69	94
-2	-104	-79	-54	-41	-29	-16	-9	-4	1	8	21	33	46	71	96
-1	-102	-77	-52	-39	-27	-14	-7	-2	3	10	23	35	48	73	98
0	-100	-75	-50	-37	-25	-12	-5	0	5	12	25	37	50	75	100
1	-98	-73	-48	-35	-23	-10	-3	2	7	14	27	39	52	77	102
2	-96	-71	-46	-33	-21	-8	-1	4	9	16	29	41	54	79	104
3	-94	-69	-44	-31	-19	-6	1	6	11	18	31	43	56	81	106
4	-92	-67	-42	-29	-17	-4	3	8	13	20	33	45	58	83	108
5	-90	-65	-40	-27	-15	-2	5	10	15	22	35	47	60	85	110
6	-88	-63	-38	-25	-13	0	7	12	17	24	37	49	62	87	112
7	-86	-61	-36	-23	-11	1	9	14	19	26	39	51	64	89	114
8	-84	-59	-34	-21	-9	3	11	16	21	28	41	53	66	91	116
10	-80	-55	-30	-17	-5	7	15	20	25	32	45	57	70	95	120
12	-76	-51	-26	-13	-1	11	19	24	29	36	49	61	74	99	124
14	-72	-47	-22	-9	3	15	23	28	33	40	53	65	78	103	128
16	-68	-43	-18	-5	7	19	27	32	37	44	57	69	82	107	132
18	-64	-39	-14	-1	11	23	31	36	41	48	61	73	86	111	136
20	-60	-35	-10	2	15	27	35	40	45	52	65	77	90	115	140
25	-50	-25	0	12	25	37	45	50	55	62	75	87	100	125	150
30	-40	-15	10	22	35	47	55	60	65	72	85	97	110	135	160
40	-20	5	30	42	55	67	75	80	85	92	105	117	130	155	180

Table 5-3

Separation, inch: 2

Leg Distance, inch: 15

SHIM REQUIRED FOR ALIGNMENT

Reverse Indicator Method

-iveN IR (LS	DriveR TIR, MILS														
	-40	-30	-20	-15	-10	-5	-2	0	2	5	10	15	20	30	40
-40	-280	-242	-205	-186	-167	-148	-137	-130	-122	-111	-92	-73	-55	-17	20
-30	-247	-210	-172	-153	-135	-116	-105	-97	-90	-78	-60	-41	-22	15	52
-25	-231	-193	-156	-137	-118	-100	-88	-81	-73	-62	-43	-25	-6	31	68
-20	-215	-177	-140	-121	-102	-83	-72	-65	-57	-46	-27	-8	10	47	85
-18	-208	-171	-133	-114	-96	-77	-66	-58	-51	-39	-21	-2	16	54	91
-16	-202	-164	-127	-108	-89	-70	-59	-52	-44	-33	-14	4	23	60	98
-14	-195	-158	-120	-101	-83	-64	-53	-45	-38	-26	-8	10	29	67	104
-12	-189	-151	-114	-95	-76	-57	-46	-39	-31	-20	-1	17	36	73	111
-10	-182	-145	-107	-88	-70	-51	-40	-32	-25	-13	5	23	42	80	117
-8	-176	-138	-101	-82	-63	-44	-33	-26	-18	-7	11	30	49	86	124
-7	-172	-135	-97	-79	-60	-41	-30	-22	-15	-4	14	33	52	89	127
-6	-169	-132	-94	-75	-57	-38	-27	-19	-12	0	18	36	55	93	130
-5	-166	-128	-91	-72	-53	-35	-23	-16	-8	2	21	40	58	96	133
-4	-163	-125	-88	-69	-50	-31	-20	-13	-5	5	24	43	62	99	137
-3	-159	-122	-84	-66	-47	-28	-17	-9	-2	9	27	46	65	102	140
-2	-156	-119	-81	-62	-44	-25	-14	-6	1	12	31	49	68	106	143
-1	-153	-115	-78	-59	-40	-22	-10	-3	4	15	34	53	71	109	146
0	-150	-112	-75	-56	-37	-18	-7	0	7	18	37	56	75	112	150
1	-146	-109	-71	-53	-34	-15	-4	3	10	22	40	59	78	115	153
2	-143	-106	-68	-49	-31	-12	-1	6	14	25	44	62	81	119	156
3	-140	-102	-65	-46	-27	-9	2	9	17	28	47	66	84	122	159
4	-137	-99	-62	-43	-24	-5	5	13	20	31	50	69	88	125	163
5	-133	-96	-58	-40	-21	-2	8	16	23	35	53	72	91	128	166
6	-130	-93	-55	-36	-18	0	12	19	27	38	57	75	94	132	169
7	-127	-89	-52	-33	-14	4	15	22	30	41	60	79	97	135	172
8	-124	-86	-49	-30	-11	7	18	26	33	44	63	82	101	138	176
10	-117	-80	-42	-23	-5	13	25	32	40	51	70	88	107	145	182
12	-111	-73	-36	-17	1	20	31	39	46	57	76	95	114	151	189
14	-104	-67	-29	-10	8	26	38	45	53	64	83	101	120	158	195
16	-98	-60	-23	-4	14	33	44	52	59	70	89	108	127	164	202
18	-91	-54	-16	2	21	39	51	58	66	77	96	114	133	171	208
20	-85	-47	-10	8	27	46	57	65	72	83	102	121	140	177	215
25	-68	-31	6	25	43	62	73	81	88	100	118	137	156	193	231
30	-52	-15	22	41	60	78	90	97	105	116	135	153	172	210	247
40	-20	17	55	73	92	111	122	130	137	148	167	186	205	242	280

Table 5-4

Separation, inch: 2

Leg Distance, inch: 20

SHIM REQUIRED FOR ALIGNMENT

Reverse Indicator Method

DriveN TIR MILS	DriveR TIR, MILS -40	-30	-20	-15	-10	-5	-2	0	2	5	10	15	20	30	40
-40	-380	-330	-280	-255	-230	-205	-190	-180	-170	-155	-130	-105	-80	-30	20
-30	-335	-285	-235	-210	-185	-160	-145	-135	-125	-110	-85	-60	-35	15	65
-25	-312	-262	-212	-187	-162	-137	-122	-112	-102	-87	-62	-37	-12	37	87
-20	-290	-240	-190	-165	-140	-115	-100	-90	-80	-65	-40	-15	10	60	110
-18	-281	-231	-181	-156	-131	-106	-91	-81	-71	-56	-31	-6	19	69	119
-16	-272	-222	-172	-147	-122	-97	-82	-72	-62	-47	-22	3	28	78	128
-14	-263	-213	-163	-138	-113	-88	-73	-63	-53	-38	-13	12	37	87	137
-12	-254	-204	-154	-129	-104	-79	-64	-54	-44	-29	-4	21	46	96	146
-10	-245	-195	-145	-120	-95	-70	-55	-45	-35	-20	5	30	55	105	155
-8	-236	-186	-136	-111	-86	-61	-46	-36	-26	-11	14	39	64	114	164
-7	-231	-181	-131	-106	-81	-56	-41	-31	-21	-6	18	43	68	118	168
-6	-227	-177	-127	-102	-77	-52	-37	-27	-17	-2	23	48	73	123	173
-5	-222	-172	-122	-97	-72	-47	-32	-22	-12	2	27	52	77	127	177
-4	-218	-168	-118	-93	-68	-43	-28	-18	-8	7	32	57	82	132	182
-3	-213	-163	-113	-88	-63	-38	-23	-13	-3	11	36	61	86	136	186
-2	-209	-159	-109	-84	-59	-34	-19	-9	1	16	41	66	91	141	191
-1	-204	-154	-104	-79	-54	-29	-14	-4	5	20	45	70	95	145	195
0	-200	-150	-100	-75	-50	-25	-10	0	10	25	50	75	100	150	200
1	-195	-145	-95	-70	-45	-20	-5	4	14	29	54	79	104	154	204
2	-191	-141	-91	-66	-41	-16	-1	9	19	34	59	84	109	159	209
3	-186	-136	-86	-61	-36	-11	3	13	23	38	63	88	113	163	213
4	-182	-132	-82	-57	-32	-7	8	18	28	43	68	93	118	168	218
5	-177	-127	-77	-52	-27	-2	12	22	32	47	72	97	122	172	222
6	-173	-123	-73	-48	-23	2	17	27	37	52	77	102	127	177	227
7	-168	-118	-68	-43	-18	6	21	31	41	56	81	106	131	181	231
8	-164	-114	-64	-39	-14	11	26	36	46	61	86	111	136	186	236
10	-155	-105	-55	-30	-5	20	35	45	55	70	95	120	145	195	245
12	-146	-96	-46	-21	4	29	44	54	64	79	104	129	154	204	254
14	-137	-87	-37	-12	13	38	53	63	73	88	113	138	163	213	263
16	-128	-78	-28	-3	22	47	62	72	82	97	122	147	172	222	272
18	-119	-69	-19	6	31	56	71	81	91	106	131	156	181	231	281
20	-110	-60	-10	15	40	65	80	90	100	115	140	165	190	240	290
25	-87	-37	12	37	62	87	102	112	122	137	162	187	212	262	312
30	-65	-15	35	60	85	110	125	135	145	160	185	210	235	285	335
40	-20	30	80	105	130	155	170	180	190	205	230	255	280	330	380

Table 5-5

Separation, inch: 2
Leg Distance, inch: 25

SHIM REQUIRED FOR ALIGNMENT

Reverse Indicator Method

DriveR TIR, MILS

riveN IR ILS	-40	-30	-20	-15	-10	-5	-2	0	2	5	10	15	20	30	40
-40	-480	-417	-355	-323	-292	-261	-242	-230	-217	-198	-167	-136	-105	-42	20
-30	-422	-360	-297	-266	-235	-203	-185	-172	-160	-141	-110	-78	-47	15	77
-25	-393	-331	-268	-237	-206	-175	-156	-143	-131	-112	-81	-50	-18	43	106
-20	-365	-302	-240	-208	-177	-146	-127	-115	-102	-83	-52	-21	10	72	135
-18	-353	-291	-228	-197	-166	-134	-116	-103	-91	-72	-41	-9	21	84	146
-16	-342	-279	-217	-185	-154	-123	-104	-92	-79	-60	-29	1	33	95	158
-14	-330	-268	-205	-174	-143	-111	-93	-80	-68	-49	-18	13	44	107	169
-12	-319	-256	-194	-162	-131	-100	-81	-69	-56	-37	-6	24	56	118	181
-10	-307	-245	-182	-151	-120	-88	-70	-57	-45	-26	5	36	67	130	192
-8	-296	-233	-171	-139	-108	-77	-58	-46	-33	-14	16	47	79	141	204
-7	-290	-227	-165	-134	-102	-71	-52	-40	-27	-9	22	53	84	147	209
-6	-284	-222	-159	-128	-97	-65	-47	-34	-22	-3	28	59	90	153	215
-5	-278	-216	-153	-122	-91	-60	-41	-28	-16	2	33	65	96	158	221
-4	-273	-210	-148	-116	-85	-54	-35	-23	-10	8	39	70	102	164	227
-3	-267	-204	-142	-111	-79	-48	-29	-17	-4	14	45	76	107	170	232
-2	-261	-199	-136	-105	-74	-42	-24	-11	1	19	51	82	113	176	238
-1	-255	-193	-130	-99	-68	-37	-18	-5	6	25	56	88	119	181	244
0	-250	-187	-125	-93	-62	-31	-12	0	12	31	62	93	125	187	250
1	-244	-181	-119	-88	-56	-25	-6	5	18	37	68	99	130	193	255
2	-238	-176	-113	-82	-51	-19	-1	11	24	42	74	105	136	199	261
3	-232	-170	-107	-76	-45	-14	4	17	29	48	79	111	142	204	267
4	-227	-164	-102	-70	-39	-8	10	23	35	54	85	116	148	210	273
5	-221	-158	-96	-65	-33	-2	16	28	41	60	91	122	153	216	278
6	-215	-153	-90	-59	-28	3	22	34	47	65	97	128	159	222	284
7	-209	-147	-84	-53	-22	9	27	40	52	71	102	134	165	227	290
8	-204	-141	-79	-47	-16	14	33	46	58	77	108	139	171	233	296
10	-192	-130	-67	-36	-5	26	45	57	70	88	120	151	182	245	307
12	-181	-118	-56	-24	6	37	56	69	81	100	131	162	194	256	319
14	-169	-107	-44	-13	18	49	68	80	93	111	143	174	205	268	330
16	-158	-95	-33	-1	29	60	79	92	104	123	154	185	217	279	342
18	-146	-84	-21	9	41	72	91	103	116	134	166	197	228	291	353
20	-135	-72	-10	21	52	83	102	115	127	146	177	208	240	302	365
25	-106	-43	18	50	81	112	131	143	156	175	206	237	268	331	393
30	-77	-15	47	78	110	141	160	172	185	203	235	266	297	360	422
40	-20	42	105	136	167	198	217	230	242	261	292	323	355	417	480

Table 5-6

Separation, inch: 2

Leg Distance, inch: 30

SHIM REQUIRED FOR ALIGNMENT

Reverse Indicator Method

DriveN TIR MILS	DriveR TIR, MILS -40	-30	-20	-15	-10	-5	-2	0	2	5	10	15	20	30	40
-40	-580	-505	-430	-392	-355	-317	-295	-280	-265	-242	-205	-167	-130	-55	20
-30	-510	-435	-360	-322	-285	-247	-225	-210	-195	-172	-135	-97	-60	15	90
-25	-475	-400	-325	-287	-250	-212	-190	-175	-160	-137	-100	-62	-25	50	125
-20	-440	-365	-290	-252	-215	-177	-155	-140	-125	-102	-65	-27	10	85	160
-18	-426	-351	-276	-238	-201	-163	-141	-126	-111	-88	-51	-13	24	99	174
-16	-412	-337	-262	-224	-187	-149	-127	-112	-97	-74	-37	0	38	113	188
-14	-398	-323	-248	-210	-173	-135	-113	-98	-83	-60	-23	14	52	127	202
-12	-384	-309	-234	-196	-159	-121	-99	-84	-69	-46	-9	28	66	141	216
-10	-370	-295	-220	-182	-145	-107	-85	-70	-55	-32	5	42	80	155	230
-8	-356	-281	-206	-168	-131	-93	-71	-56	-41	-18	19	56	94	169	244
-7	-349	-274	-199	-161	-124	-86	-64	-49	-34	-11	26	63	101	176	251
-6	-342	-267	-192	-154	-117	-79	-57	-42	-27	-4	33	70	108	183	258
-5	-335	-260	-185	-147	-110	-72	-50	-35	-20	2	40	77	115	190	265
-4	-328	-253	-178	-140	-103	-65	-43	-28	-13	9	47	84	122	197	272
-3	-321	-246	-171	-133	-96	-58	-36	-21	-6	16	54	91	129	204	279
-2	-314	-239	-164	-126	-89	-51	-29	-14	1	23	61	98	136	211	286
-1	-307	-232	-157	-119	-82	-44	-22	-7	8	30	68	105	143	218	293
0	-300	-225	-150	-112	-75	-37	-15	0	15	37	75	112	150	225	300
1	-293	-218	-143	-105	-68	-30	-8	7	22	44	82	119	157	232	307
2	-286	-211	-136	-98	-61	-23	-1	14	29	51	89	126	164	239	314
3	-279	-204	-129	-91	-54	-16	6	21	36	58	96	133	171	246	321
4	-272	-197	-122	-84	-47	-9	13	28	43	65	103	140	178	253	328
5	-265	-190	-115	-77	-40	-2	20	35	50	72	110	147	185	260	335
6	-258	-183	-108	-70	-33	4	27	42	57	79	117	154	192	267	342
7	-251	-176	-101	-63	-26	11	34	49	64	86	124	161	199	274	349
8	-244	-169	-94	-56	-19	18	41	56	71	93	131	168	206	281	356
10	-230	-155	-80	-42	-5	32	55	70	85	107	145	182	220	295	370
12	-216	-141	-66	-28	9	46	69	84	99	121	159	196	234	309	384
14	-202	-127	-52	-14	23	60	83	98	113	135	173	210	248	323	398
16	-188	-113	-38	0	37	74	97	112	127	149	187	224	262	337	412
18	-174	-99	-24	13	51	88	111	126	141	163	201	238	276	351	426
20	-160	-85	-10	27	65	102	125	140	155	177	215	252	290	365	440
25	-125	-50	25	62	100	137	160	175	190	212	250	287	325	400	475
30	-90	-15	60	97	135	172	195	210	225	247	285	322	360	435	510
40	-20	55	130	167	205	242	265	280	295	317	355	392	430	505	580

Table 5-7

Separation, inch: 2.5

Leg Distance, inch: 5

SHIM REQUIRED FOR ALIGNMENT

Reverse Indicator Method

DriveN TIR MILS	DriveR TIR, MILS -40	-30	-20	-15	-10	-5	-2	0	2	5	10	15	20	30	40
-40	-60	-50	-40	-35	-30	-25	-22	-20	-18	-15	-10	-5	0	10	20
-30	-55	-45	-35	-30	-25	-20	-17	-15	-13	-10	-5	0	5	15	25
-25	-52	-42	-32	-27	-22	-17	-14	-12	-10	-7	-2	2	7	17	27
-20	-50	-40	-30	-25	-20	-15	-12	-10	-8	-5	0	5	10	20	30
-18	-49	-39	-29	-24	-19	-14	-11	-9	-7	-4	1	6	11	21	31
-16	-48	-38	-28	-23	-18	-13	-10	-8	-6	-3	2	7	12	22	32
-14	-47	-37	-27	-22	-17	-12	-9	-7	-5	-2	3	8	13	23	33
-12	-46	-36	-26	-21	-16	-11	-8	-6	-4	-1	4	9	14	24	34
-10	-45	-35	-25	-20	-15	-10	-7	-5	-3	0	5	10	15	25	35
-8	-44	-34	-24	-19	-14	-9	-6	-4	-2	1	6	11	16	26	36
-7	-43	-33	-23	-18	-13	-8	-5	-3	-1	1	6	11	16	26	36
-6	-43	-33	-23	-18	-13	-8	-5	-3	-1	2	7	12	17	27	37
-5	-42	-32	-22	-17	-12	-7	-4	-2	0	2	7	12	17	27	37
-4	-42	-32	-22	-17	-12	-7	-4	-2	0	3	8	13	18	28	38
-3	-41	-31	-21	-16	-11	-6	-3	-1	0	3	8	13	18	28	38
-2	-41	-31	-21	-16	-11	-6	-3	-1	1	4	9	14	19	29	39
-1	-40	-30	-20	-15	-10	-5	-2	0	1	4	9	14	19	29	39
0	-40	-30	-20	-15	-10	-5	-2	0	2	5	10	15	20	30	40
1	-39	-29	-19	-14	-9	-4	-1	0	2	5	10	15	20	30	40
2	-39	-29	-19	-14	-9	-4	-1	1	3	6	11	16	21	31	41
3	-38	-28	-18	-13	-8	-3	0	1	3	6	11	16	21	31	41
4	-38	-28	-18	-13	-8	-3	0	2	4	7	12	17	22	32	42
5	-37	-27	-17	-12	-7	-2	0	2	4	7	12	17	22	32	42
6	-37	-27	-17	-12	-7	-2	1	3	5	8	13	18	23	33	43
7	-36	-26	-16	-11	-6	-1	1	3	5	8	13	18	23	33	43
8	-36	-26	-16	-11	-6	-1	2	4	6	9	14	19	24	34	44
10	-35	-25	-15	-10	-5	0	3	5	7	10	15	20	25	35	45
12	-34	-24	-14	-9	-4	1	4	6	8	11	16	21	26	36	46
14	-33	-23	-13	-8	-3	2	5	7	9	12	17	22	27	37	47
16	-32	-22	-12	-7	-2	3	6	8	10	13	18	23	28	38	48
18	-31	-21	-11	-6	-1	4	7	9	11	14	19	24	29	39	49
20	-30	-20	-10	-5	0	5	8	10	12	15	20	25	30	40	50
25	-27	-17	-7	-2	2	7	10	12	14	17	22	27	32	42	52
30	-25	-15	-5	0	5	10	13	15	17	20	25	30	35	45	55
40	-20	-10	0	5	10	15	18	20	22	25	30	35	40	50	60

Table 5-8

Separation, inch: 2.5

Leg Distance, inch: 10

SHIM REQUIRED FOR ALIGNMENT

Reverse Indicator Method

DriveN TIR MILS	DriveR TIR, MILS -40	-30	-20	-15	-10	-5	-2	0	2	5	10	15	20	30	40
-40	-140	-120	-100	-90	-80	-70	-64	-60	-56	-50	-40	-30	-20	0	20
-30	-125	-105	-85	-75	-65	-55	-49	-45	-41	-35	-25	-15	-5	15	35
-25	-117	-97	-77	-67	-57	-47	-41	-37	-33	-27	-17	-7	2	22	42
-20	-110	-90	-70	-60	-50	-40	-34	-30	-26	-20	-10	0	10	30	50
-18	-107	-87	-67	-57	-47	-37	-31	-27	-23	-17	-7	3	13	33	53
-16	-104	-84	-64	-54	-44	-34	-28	-24	-20	-14	-4	6	16	36	56
-14	-101	-81	-61	-51	-41	-31	-25	-21	-17	-11	-1	9	19	39	59
-12	-98	-78	-58	-48	-38	-28	-22	-18	-14	-8	2	12	22	42	62
-10	-95	-75	-55	-45	-35	-25	-19	-15	-11	-5	5	15	25	45	65
-8	-92	-72	-52	-42	-32	-22	-16	-12	-8	-2	8	18	28	48	68
-7	-90	-70	-50	-40	-30	-20	-14	-10	-6	0	9	19	29	49	69
-6	-89	-69	-49	-39	-29	-19	-13	-9	-5	1	11	21	31	51	71
-5	-87	-67	-47	-37	-27	-17	-11	-7	-3	2	12	22	32	52	72
-4	-86	-66	-46	-36	-26	-16	-10	-6	-2	4	14	24	34	54	74
-3	-84	-64	-44	-34	-24	-14	-8	-4	0	5	15	25	35	55	75
-2	-83	-63	-43	-33	-23	-13	-7	-3	1	7	17	27	37	57	77
-1	-81	-61	-41	-31	-21	-11	-5	-1	2	8	18	28	38	58	78
0	-80	-60	-40	-30	-20	-10	-4	0	4	10	20	30	40	60	80
1	-78	-58	-38	-28	-18	-8	-2	1	5	11	21	31	41	61	81
2	-77	-57	-37	-27	-17	-7	-1	3	7	13	23	33	43	63	83
3	-75	-55	-35	-25	-15	-5	0	4	8	14	24	34	44	64	84
4	-74	-54	-34	-24	-14	-4	2	6	10	16	26	36	46	66	86
5	-72	-52	-32	-22	-12	-2	3	7	11	17	27	37	47	67	87
6	-71	-51	-31	-21	-11	-1	5	9	13	19	29	39	49	69	89
7	-69	-49	-29	-19	-9	0	6	10	14	20	30	40	50	70	90
8	-68	-48	-28	-18	-8	2	8	12	16	22	32	42	52	72	92
10	-65	-45	-25	-15	-5	5	11	15	19	25	35	45	55	75	95
12	-62	-42	-22	-12	-2	8	14	18	22	28	38	48	58	78	98
14	-59	-39	-19	-9	1	11	17	21	25	31	41	51	61	81	101
16	-56	-36	-16	-6	4	14	20	24	28	34	44	54	64	84	104
18	-53	-33	-13	-3	7	17	23	27	31	37	47	57	67	87	107
20	-50	-30	-10	0	10	20	26	30	34	40	50	60	70	90	110
25	-42	-22	-2	7	17	27	33	37	41	47	57	67	77	97	117
30	-35	-15	5	15	25	35	41	45	49	55	65	75	85	105	125
40	-20	0	20	30	40	50	56	60	64	70	80	90	100	120	140

Table 5-9

Separation, inch: 2.5

Leg Distance, inch: 15

SHIM REQUIRED FOR ALIGNMENT

Reverse Indicator Method

riveN IR ILS	DriveR TIR, MILS														
	-40	-30	-20	-15	-10	-5	-2	0	2	5	10	15	20	30	40
-40	-220	-190	-160	-145	-130	-115	-106	-100	-94	-85	-70	-55	-40	-10	20
-30	-195	-165	-135	-120	-105	-90	-81	-75	-69	-60	-45	-30	-15	15	45
-25	-182	-152	-122	-107	-92	-77	-68	-62	-56	-47	-32	-17	-2	27	57
-20	-170	-140	-110	-95	-80	-65	-56	-50	-44	-35	-20	-5	10	40	70
-18	-165	-135	-105	-90	-75	-60	-51	-45	-39	-30	-15	0	15	45	75
-16	-160	-130	-100	-85	-70	-55	-46	-40	-34	-25	-10	5	20	50	80
-14	-155	-125	-95	-80	-65	-50	-41	-35	-29	-20	-5	10	25	55	85
-12	-150	-120	-90	-75	-60	-45	-36	-30	-24	-15	0	15	30	60	90
-10	-145	-115	-85	-70	-55	-40	-31	-25	-19	-10	5	20	35	65	95
-8	-140	-110	-80	-65	-50	-35	-26	-20	-14	-5	10	25	40	70	100
-7	-137	-107	-77	-62	-47	-32	-23	-17	-11	-2	12	27	42	72	102
-6	-135	-105	-75	-60	-45	-30	-21	-15	-9	0	15	30	45	75	105
-5	-132	-102	-72	-57	-42	-27	-18	-12	-6	2	17	32	47	77	107
-4	-130	-100	-70	-55	-40	-25	-16	-10	-4	5	20	35	50	80	110
-3	-127	-97	-67	-52	-37	-22	-13	-7	-1	7	22	37	52	82	112
-2	-125	-95	-65	-50	-35	-20	-11	-5	1	10	25	40	55	85	115
-1	-122	-92	-62	-47	-32	-17	-8	-2	3	12	27	42	57	87	117
0	-120	-90	-60	-45	-30	-15	-6	0	6	15	30	45	60	90	120
1	-117	-87	-57	-42	-27	-12	-3	2	8	17	32	47	62	92	122
2	-115	-85	-55	-40	-25	-10	-1	5	11	20	35	50	65	95	125
3	-112	-82	-52	-37	-22	-7	1	7	13	22	37	52	67	97	127
4	-110	-80	-50	-35	-20	-5	4	10	16	25	40	55	70	100	130
5	-107	-77	-47	-32	-17	-2	6	12	18	27	42	57	72	102	132
6	-105	-75	-45	-30	-15	0	9	15	21	30	45	60	75	105	135
7	-102	-72	-42	-27	-12	2	11	17	23	32	47	62	77	107	137
8	-100	-70	-40	-25	-10	5	14	20	26	35	50	65	80	110	140
10	-95	-65	-35	-20	-5	10	19	25	31	40	55	70	85	115	145
12	-90	-60	-30	-15	0	15	24	30	36	45	60	75	90	120	150
14	-85	-55	-25	-10	5	20	29	35	41	50	65	80	95	125	155
16	-80	-50	-20	-5	10	25	34	40	46	55	70	85	100	130	160
18	-75	-45	-15	0	15	30	39	45	51	60	75	90	105	135	165
20	-70	-40	-10	5	20	35	44	50	56	65	80	95	110	140	170
25	-57	-27	2	17	32	47	56	62	68	77	92	107	122	152	182
30	-45	-15	15	30	45	60	69	75	81	90	105	120	135	165	195
40	-20	10	40	55	70	85	94	100	106	115	130	145	160	190	220

Table 5-10

Separation, inch: 2.5

Leg Distance, inch: 20

SHIM REQUIRED FOR ALIGNMENT

Reverse Indicator Method

DriveN TIR MILS	DriveR TIR, MILS -40	-30	-20	-15	-10	-5	-2	0	2	5	10	15	20	30	40
-40	-300	-260	-220	-200	-180	-160	-148	-140	-132	-120	-100	-80	-60	-20	20
-30	-265	-225	-185	-165	-145	-125	-113	-105	-97	-85	-65	-45	-25	15	55
-25	-247	-207	-167	-147	-127	-107	-95	-87	-79	-67	-47	-27	-7	32	72
-20	-230	-190	-150	-130	-110	-90	-78	-70	-62	-50	-30	-10	10	50	90
-18	-223	-183	-143	-123	-103	-83	-71	-63	-55	-43	-23	-3	17	57	97
-16	-216	-176	-136	-116	-96	-76	-64	-56	-48	-36	-16	4	24	64	104
-14	-209	-169	-129	-109	-89	-69	-57	-49	-41	-29	-9	11	31	71	111
-12	-202	-162	-122	-102	-82	-62	-50	-42	-34	-22	-2	18	38	78	118
-10	-195	-155	-115	-95	-75	-55	-43	-35	-27	-15	5	25	45	85	125
-8	-188	-148	-108	-88	-68	-48	-36	-28	-20	-8	12	32	52	92	132
-7	-184	-144	-104	-84	-64	-44	-32	-24	-16	-4	15	35	55	95	135
-6	-181	-141	-101	-81	-61	-41	-29	-21	-13	-1	19	39	59	99	139
-5	-177	-137	-97	-77	-57	-37	-25	-17	-9	2	22	42	62	102	142
-4	-174	-134	-94	-74	-54	-34	-22	-14	-6	6	26	46	66	106	146
-3	-170	-130	-90	-70	-50	-30	-18	-10	-2	9	29	49	69	109	149
-2	-167	-127	-87	-67	-47	-27	-15	-7	1	13	33	53	73	113	153
-1	-163	-123	-83	-63	-43	-23	-11	-3	4	16	36	56	76	116	156
0	-160	-120	-80	-60	-40	-20	-8	0	8	20	40	60	80	120	160
1	-156	-116	-76	-56	-36	-16	-4	3	11	23	43	63	83	123	163
2	-153	-113	-73	-53	-33	-13	-1	7	15	27	47	67	87	127	167
3	-149	-109	-69	-49	-29	-9	2	10	18	30	50	70	90	130	170
4	-146	-106	-66	-46	-26	-6	6	14	22	34	54	74	94	134	174
5	-142	-102	-62	-42	-22	-2	9	17	25	37	57	77	97	137	177
6	-139	-99	-59	-39	-19	1	13	21	29	41	61	81	101	141	181
7	-135	-95	-55	-35	-15	4	16	24	32	44	64	84	104	144	184
8	-132	-92	-52	-32	-12	8	20	28	36	48	68	88	108	148	188
10	-125	-85	-45	-25	-5	15	27	35	43	55	75	95	115	155	195
12	-118	-78	-38	-18	2	22	34	42	50	62	82	102	122	162	202
14	-111	-71	-31	-11	9	29	41	49	57	69	89	109	129	169	209
16	-104	-64	-24	-4	16	36	48	56	64	76	96	116	136	176	216
18	-97	-57	-17	3	23	43	55	63	71	83	103	123	143	183	223
20	-90	-50	-10	10	30	50	62	70	78	90	110	130	150	190	230
25	-72	-32	7	27	47	67	79	87	95	107	127	147	167	207	247
30	-55	-15	25	45	65	85	97	105	113	125	145	165	185	225	265
40	-20	20	60	80	100	120	132	140	148	160	180	200	220	260	300

Table 5-11

Separation, inch: 2.5

Leg Distance, inch: 25

SHIM REQUIRED FOR ALIGNMENT

Reverse Indicator Method

riveN IR ILS	DriveR TIR, MILS -40	-30	-20	-15	-10	-5	-2	0	2	5	10	15	20	30	40
-40	-380	-330	-280	-255	-230	-205	-190	-180	-170	-155	-130	-105	-80	-30	20
-30	-335	-285	-235	-210	-185	-160	-145	-135	-125	-110	-85	-60	-35	15	65
-25	-312	-262	-212	-187	-162	-137	-122	-112	-102	-87	-62	-37	-12	37	87
-20	-290	-240	-190	-165	-140	-115	-100	-90	-80	-65	-40	-15	10	60	110
-18	-281	-231	-181	-156	-131	-106	-91	-81	-71	-56	-31	-6	19	69	119
-16	-272	-222	-172	-147	-122	-97	-82	-72	-62	-47	-22	3	28	78	128
-14	-263	-213	-163	-138	-113	-88	-73	-63	-53	-38	-13	12	37	87	137
-12	-254	-204	-154	-129	-104	-79	-64	-54	-44	-29	-4	21	46	96	146
-10	-245	-195	-145	-120	-95	-70	-55	-45	-35	-20	5	30	55	105	155
-8	-236	-186	-136	-111	-86	-61	-46	-36	-26	-11	14	39	64	114	164
-7	-231	-181	-131	-106	-81	-56	-41	-31	-21	-6	18	43	68	118	168
-6	-227	-177	-127	-102	-77	-52	-37	-27	-17	-2	23	48	73	123	173
-5	-222	-172	-122	-97	-72	-47	-32	-22	-12	2	27	52	77	127	177
-4	-218	-168	-118	-93	-68	-43	-28	-18	-8	7	32	57	82	132	182
-3	-213	-163	-113	-88	-63	-38	-23	-13	-3	11	36	61	86	136	186
-2	-209	-159	-109	-84	-59	-34	-19	-9	1	16	41	66	91	141	191
-1	-204	-154	-104	-79	-54	-29	-14	-4	5	20	45	70	95	145	195
0	-200	-150	-100	-75	-50	-25	-10	0	10	25	50	75	100	150	200
1	-195	-145	-95	-70	-45	-20	-5	4	14	29	54	79	104	154	204
2	-191	-141	-91	-66	-41	-16	-1	9	19	34	59	84	109	159	209
3	-186	-136	-86	-61	-36	-11	3	13	23	38	63	88	113	163	213
4	-182	-132	-82	-57	-32	-7	8	18	28	43	68	93	118	168	218
5	-177	-127	-77	-52	-27	-2	12	22	32	47	72	97	122	172	222
6	-173	-123	-73	-48	-23	2	17	27	37	52	77	102	127	177	227
7	-168	-118	-68	-43	-18	6	21	31	41	56	81	106	131	181	231
8	-164	-114	-64	-39	-14	11	26	36	46	61	86	111	136	186	236
10	-155	-105	-55	-30	-5	20	35	45	55	70	95	120	145	195	245
12	-146	-96	-46	-21	4	29	44	54	64	79	104	129	154	204	254
14	-137	-87	-37	-12	13	38	53	63	73	88	113	138	163	213	263
16	-128	-78	-28	-3	22	47	62	72	82	97	122	147	172	222	272
18	-119	-69	-19	6	31	56	71	81	91	106	131	156	181	231	281
20	-110	-60	-10	15	40	65	80	90	100	115	140	165	190	240	290
25	-87	-37	12	37	62	87	102	112	122	137	162	187	212	262	312
30	-65	-15	35	60	85	110	125	135	145	160	185	210	235	285	335
40	-20	30	80	105	130	155	170	180	190	205	230	255	280	330	380

Table 5-12

Separation, inch: 2.5

Leg Distance, inch: 30

SHIM REQUIRED FOR ALIGNMENT

Reverse Indicator Method

DriveN TIR MILS	DriveR TIR, MILS -40	-30	-20	-15	-10	-5	-2	0	2	5	10	15	20	30	40
-40	-460	-400	-340	-310	-280	-250	-232	-220	-208	-190	-160	-130	-100	-40	20
-30	-405	-345	-285	-255	-225	-195	-177	-165	-153	-135	-105	-75	-45	15	75
-25	-377	-317	-257	-227	-197	-167	-149	-137	-125	-107	-77	-47	-17	42	102
-20	-350	-290	-230	-200	-170	-140	-122	-110	-98	-80	-50	-20	10	70	130
-18	-339	-279	-219	-189	-159	-129	-111	-99	-87	-69	-39	-9	21	81	141
-16	-328	-268	-208	-178	-148	-118	-100	-88	-76	-58	-28	2	32	92	152
-14	-317	-257	-197	-167	-137	-107	-89	-77	-65	-47	-17	13	43	103	163
-12	-306	-246	-186	-156	-126	-96	-78	-66	-54	-36	-6	24	54	114	174
-10	-295	-235	-175	-145	-115	-85	-67	-55	-43	-25	5	35	65	125	185
-8	-284	-224	-164	-134	-104	-74	-56	-44	-32	-14	16	46	76	136	196
-7	-278	-218	-158	-128	-98	-68	-50	-38	-26	-8	21	51	81	141	201
-6	-273	-213	-153	-123	-93	-63	-45	-33	-21	-3	27	57	87	147	207
-5	-267	-207	-147	-117	-87	-57	-39	-27	-15	2	32	62	92	152	212
-4	-262	-202	-142	-112	-82	-52	-34	-22	-10	8	38	68	98	158	218
-3	-256	-196	-136	-106	-76	-46	-28	-16	-4	13	43	73	103	163	223
-2	-251	-191	-131	-101	-71	-41	-23	-11	1	19	49	79	109	169	229
-1	-245	-185	-125	-95	-65	-35	-17	-5	6	24	54	84	114	174	234
0	-240	-180	-120	-90	-60	-30	-12	0	12	30	60	90	120	180	240
1	-234	-174	-114	-84	-54	-24	-6	5	17	35	65	95	125	185	245
2	-229	-169	-109	-79	-49	-19	-1	11	23	41	71	101	131	191	251
3	-223	-163	-103	-73	-43	-13	4	16	28	46	76	106	136	196	256
4	-218	-158	-98	-68	-38	-8	10	22	34	52	82	112	142	202	262
5	-212	-152	-92	-62	-32	-2	15	27	39	57	87	117	147	207	267
6	-207	-147	-87	-57	-27	3	21	33	45	63	93	123	153	213	273
7	-201	-141	-81	-51	-21	8	26	38	50	68	98	128	158	218	278
8	-196	-136	-76	-46	-16	14	32	44	56	74	104	134	164	224	284
10	-185	-125	-65	-35	-5	25	43	55	67	85	115	145	175	235	295
12	-174	-114	-54	-24	6	36	54	66	78	96	126	156	186	246	306
14	-163	-103	-43	-13	17	47	65	77	89	107	137	167	197	257	317
16	-152	-92	-32	-2	28	58	76	88	100	118	148	178	208	268	328
18	-141	-81	-21	9	39	69	87	99	111	129	159	189	219	279	339
20	-130	-70	-10	20	50	80	98	110	122	140	170	200	230	290	350
25	-102	-42	17	47	77	107	125	137	149	167	197	227	257	317	377
30	-75	-15	45	75	105	135	153	165	177	195	225	255	285	345	405
40	-20	40	100	130	160	190	208	220	232	250	280	310	340	400	460

Table 5-13

Separation, inch: 3

Leg Distance, inch: 10

SHIM REQUIRED FOR ALIGNMENT

Reverse Indicator Method

iveN IR ILS	DriveR TIR, MILS -40	-30	-20	-15	-10	-5	-2	0	2	5	10	15	20	30	40
-40	-113	-96	-80	-71	-63	-55	-50	-46	-43	-38	-30	-21	-13	3	20
-30	-101	-85	-68	-60	-51	-43	-38	-35	-31	-26	-18	-10	-1	15	31
-25	-95	-79	-62	-54	-45	-37	-32	-29	-25	-20	-12	-4	4	20	37
-20	-90	-73	-56	-48	-40	-31	-26	-23	-20	-15	-6	1	10	26	43
-18	-87	-71	-54	-46	-37	-29	-24	-21	-17	-12	-4	4	12	29	45
-16	-85	-68	-52	-43	-35	-27	-22	-18	-15	-10	-2	6	14	31	48
-14	-83	-66	-49	-41	-33	-24	-19	-16	-13	-8	0	8	17	33	50
-12	-80	-64	-47	-39	-30	-22	-17	-14	-10	-5	2	11	19	36	52
-10	-78	-61	-45	-36	-28	-20	-15	-11	-8	-3	5	13	21	38	55
-8	-76	-59	-42	-34	-26	-17	-12	-9	-6	-1	7	15	24	40	57
-7	-74	-58	-41	-33	-24	-16	-11	-8	-4	0	8	16	25	41	58
-6	-73	-57	-40	-32	-23	-15	-10	-7	-3	1	9	18	26	43	59
-5	-72	-55	-39	-30	-22	-14	-9	-5	-2	2	10	19	27	44	60
-4	-71	-54	-38	-29	-21	-13	-8	-4	-1	3	12	20	28	45	62
-3	-70	-53	-36	-28	-20	-11	-6	-3	0	4	13	21	29	46	63
-2	-69	-52	-35	-27	-19	-10	-5	-2	1	6	14	22	31	47	64
-1	-67	-51	-34	-26	-17	-9	-4	-1	2	7	15	23	32	48	65
0	-66	-50	-33	-25	-16	-8	-3	0	3	8	16	25	33	50	66
1	-65	-48	-32	-23	-15	-7	-2	1	4	9	17	26	34	51	67
2	-64	-47	-31	-22	-14	-6	-1	2	5	10	19	27	35	52	69
3	-63	-46	-29	-21	-13	-4	0	3	6	11	20	28	36	53	70
4	-62	-45	-28	-20	-12	-3	1	4	8	13	21	29	38	54	71
5	-60	-44	-27	-19	-10	-2	2	5	9	14	22	30	39	55	72
6	-59	-43	-26	-18	-9	-1	3	7	10	15	23	32	40	57	73
7	-58	-41	-25	-16	-8	0	4	8	11	16	24	33	41	58	74
8	-57	-40	-24	-15	-7	1	6	9	12	17	26	34	42	59	76
10	-55	-38	-21	-13	-5	3	8	11	15	20	28	36	45	61	78
12	-52	-36	-19	-11	-2	5	10	14	17	22	30	39	47	64	80
14	-50	-33	-17	-8	0	8	13	16	19	24	33	41	49	66	83
16	-48	-31	-14	-6	2	10	15	18	22	27	35	43	52	68	85
18	-45	-29	-12	-4	4	12	17	21	24	29	37	46	54	71	87
20	-43	-26	-10	-1	6	15	20	23	26	31	40	48	56	73	90
25	-37	-20	-4	4	12	20	25	29	32	37	45	54	62	79	95
30	-31	-15	1	10	18	26	31	35	38	43	51	60	68	85	101
40	-20	-3	13	21	30	38	43	46	50	55	63	71	80	96	113

Table 5-14

Separation, inch: 3

Leg Distance, inch: 15

SHIM REQUIRED FOR ALIGNMENT

Reverse Indicator Method

DriveN TIR MILS	DriveR TIR, MILS -40	-30	-20	-15	-10	-5	-2	0	2	5	10	15	20	30	40
-40	-180	-155	-130	-117	-105	-92	-85	-80	-75	-67	-55	-42	-30	-5	20
-30	-160	-135	-110	-97	-85	-72	-65	-60	-55	-47	-35	-22	-10	15	40
-25	-150	-125	-100	-87	-75	-62	-55	-50	-45	-37	-25	-12	0	25	50
-20	-140	-115	-90	-77	-65	-52	-45	-40	-35	-27	-15	-2	10	35	60
-18	-136	-111	-86	-73	-61	-48	-41	-36	-31	-23	-11	1	14	39	64
-16	-132	-107	-82	-69	-57	-44	-37	-32	-27	-19	-7	5	18	43	68
-14	-128	-103	-78	-65	-53	-40	-33	-28	-23	-15	-3	9	22	47	72
-12	-124	-99	-74	-61	-49	-36	-29	-24	-19	-11	1	13	26	51	76
-10	-120	-95	-70	-57	-45	-32	-25	-20	-15	-7	5	17	30	55	80
-8	-116	-91	-66	-53	-41	-28	-21	-16	-11	-3	9	21	34	59	84
-7	-114	-89	-64	-51	-39	-26	-19	-14	-9	-1	11	23	36	61	86
-6	-112	-87	-62	-49	-37	-24	-17	-12	-7	0	13	25	38	63	88
-5	-110	-85	-60	-47	-35	-22	-15	-10	-5	2	15	27	40	65	90
-4	-108	-83	-58	-45	-33	-20	-13	-8	-3	4	17	29	42	67	92
-3	-106	-81	-56	-43	-31	-18	-11	-6	-1	6	19	31	44	69	94
-2	-104	-79	-54	-41	-29	-16	-9	-4	1	8	21	33	46	71	96
-1	-102	-77	-52	-39	-27	-14	-7	-2	3	10	23	35	48	73	98
0	-100	-75	-50	-37	-25	-12	-5	0	5	12	25	37	50	75	100
1	-98	-73	-48	-35	-23	-10	-3	2	7	14	27	39	52	77	102
2	-96	-71	-46	-33	-21	-8	-1	4	9	16	29	41	54	79	104
3	-94	-69	-44	-31	-19	-6	1	6	11	18	31	43	56	81	106
4	-92	-67	-42	-29	-17	-4	3	8	13	20	33	45	58	83	108
5	-90	-65	-40	-27	-15	-2	5	10	15	22	35	47	60	85	110
6	-88	-63	-38	-25	-13	0	7	12	17	24	37	49	62	87	112
7	-86	-61	-36	-23	-11	1	9	14	19	26	39	51	64	89	114
8	-84	-59	-34	-21	-9	3	11	16	21	28	41	53	66	91	116
10	-80	-55	-30	-17	-5	7	15	20	25	32	45	57	70	95	120
12	-76	-51	-26	-13	-1	11	19	24	29	36	49	61	74	99	124
14	-72	-47	-22	-9	3	15	23	28	33	40	53	65	78	103	128
16	-68	-43	-18	-5	7	19	27	32	37	44	57	69	82	107	132
18	-64	-39	-14	-1	11	23	31	36	41	48	61	73	86	111	136
20	-60	-35	-10	2	15	27	35	40	45	52	65	77	90	115	140
25	-50	-25	0	12	25	37	45	50	55	62	75	87	100	125	150
30	-40	-15	10	22	35	47	55	60	65	72	85	97	110	135	160
40	-20	5	30	42	55	67	75	80	85	92	105	117	130	155	180

Table 5-15

Separation, inch: 3

Leg Distance, inch: 20

SHIM REQUIRED FOR ALIGNMENT

Reverse Indicator Method

-iveN R ILS	DriveR TIR, MILS -40	-30	-20	-15	-10	-5	-2	0	2	5	10	15	20	30	40
-40	-246	-213	-180	-163	-146	-130	-120	-113	-106	-96	-80	-63	-46	-13	20
-30	-218	-185	-151	-135	-118	-101	-91	-85	-78	-68	-51	-35	-18	15	48
-25	-204	-170	-137	-120	-104	-87	-77	-70	-64	-54	-37	-20	-4	29	62
-20	-190	-156	-123	-106	-90	-73	-63	-56	-50	-40	-23	-6	10	43	76
-18	-184	-151	-117	-101	-84	-67	-57	-51	-44	-34	-17	-1	15	49	82
-16	-178	-145	-112	-95	-78	-62	-52	-45	-38	-28	-12	4	21	54	88
-14	-173	-139	-106	-89	-73	-56	-46	-39	-33	-23	-6	10	27	60	93
-12	-167	-134	-100	-84	-67	-50	-40	-34	-27	-17	0	16	32	66	99
-10	-161	-128	-95	-78	-61	-45	-35	-28	-21	-11	5	21	38	71	105
-8	-156	-122	-89	-72	-56	-39	-29	-22	-16	-6	10	27	44	77	110
-7	-153	-119	-86	-69	-53	-36	-26	-19	-13	-3	13	30	46	80	113
-6	-150	-117	-83	-67	-50	-33	-23	-17	-10	0	16	33	49	83	116
-5	-147	-114	-80	-64	-47	-30	-20	-14	-7	2	19	35	52	85	119
-4	-144	-111	-78	-61	-44	-28	-18	-11	-4	5	22	38	55	88	122
-3	-141	-108	-75	-58	-41	-25	-15	-8	-1	8	24	41	58	91	124
-2	-139	-105	-72	-55	-39	-22	-12	-5	1	11	27	44	61	94	127
-1	-136	-102	-69	-52	-36	-19	-9	-2	3	13	30	47	63	97	130
0	-133	-100	-66	-50	-33	-16	-6	0	6	16	33	50	66	100	133
1	-130	-97	-63	-47	-30	-13	-3	2	9	19	36	52	69	102	136
2	-127	-94	-61	-44	-27	-11	-1	5	12	22	39	55	72	105	139
3	-124	-91	-58	-41	-24	-8	1	8	15	25	41	58	75	108	141
4	-122	-88	-55	-38	-22	-5	4	11	18	28	44	61	78	111	144
5	-119	-85	-52	-35	-19	-2	7	14	20	30	47	64	80	114	147
6	-116	-83	-49	-33	-16	0	10	17	23	33	50	67	83	117	150
7	-113	-80	-46	-30	-13	3	13	19	26	36	53	69	86	119	153
8	-110	-77	-44	-27	-10	6	16	22	29	39	56	72	89	122	156
10	-105	-71	-38	-21	-5	11	21	28	35	45	61	78	95	128	161
12	-99	-66	-32	-16	0	17	27	34	40	50	67	84	100	134	167
14	-93	-60	-27	-10	6	23	33	39	46	56	73	89	106	139	173
16	-88	-54	-21	-4	12	28	38	45	52	62	78	95	112	145	178
18	-82	-49	-15	1	17	34	44	51	57	67	84	101	117	151	184
20	-76	-43	-10	6	23	40	50	56	63	73	90	106	123	156	190
25	-62	-29	4	20	37	54	64	70	77	87	104	120	137	170	204
30	-48	-15	18	35	51	68	78	85	91	101	118	135	151	185	218
40	-20	13	46	63	80	96	106	113	120	130	146	163	180	213	246

Table 5-16

Separation, inch: 3

Leg Distance, inch: 25

SHIM REQUIRED FOR ALIGNMENT

Reverse Indicator Method

DriveN TIR MILS	DriveR TIR, MILS -40	-30	-20	-15	-10	-5	-2	0	2	5	10	15	20	30	40
-40	-313	-271	-230	-209	-188	-167	-155	-146	-138	-125	-105	-84	-63	-21	20
-30	-276	-235	-193	-172	-151	-130	-118	-110	-101	-89	-68	-47	-26	15	56
-25	-258	-216	-175	-154	-133	-112	-100	-91	-83	-70	-50	-29	-8	33	75
-20	-240	-198	-156	-135	-115	-94	-81	-73	-65	-52	-31	-10	10	51	93
-18	-232	-191	-149	-128	-107	-86	-74	-66	-57	-45	-24	-3	17	59	100
-16	-225	-183	-142	-121	-100	-79	-67	-58	-50	-37	-17	3	24	66	108
-14	-218	-176	-134	-113	-93	-72	-59	-51	-43	-30	-9	11	32	73	115
-12	-210	-169	-127	-106	-85	-64	-52	-44	-35	-23	-2	18	39	81	122
-10	-203	-161	-120	-99	-78	-57	-45	-36	-28	-15	5	25	46	88	130
-8	-196	-154	-112	-91	-71	-50	-37	-29	-21	-8	12	33	54	95	137
-7	-192	-150	-109	-88	-67	-46	-34	-25	-17	-4	16	36	57	99	141
-6	-188	-147	-105	-84	-63	-42	-30	-22	-13	-1	19	40	61	103	144
-5	-185	-143	-101	-80	-60	-39	-26	-18	-10	2	23	44	65	106	148
-4	-181	-139	-98	-77	-56	-35	-23	-14	-6	6	27	47	68	110	152
-3	-177	-136	-94	-73	-52	-31	-19	-11	-2	9	30	51	72	114	155
-2	-174	-132	-90	-69	-49	-28	-15	-7	1	13	34	55	76	117	159
-1	-170	-128	-87	-66	-45	-24	-12	-3	4	17	38	58	79	121	163
0	-166	-125	-83	-62	-41	-20	-8	0	8	20	41	62	83	125	166
1	-163	-121	-79	-58	-38	-17	-4	3	12	24	45	66	87	128	170
2	-159	-117	-76	-55	-34	-13	-1	7	15	28	49	69	90	132	174
3	-155	-114	-72	-51	-30	-9	2	11	19	31	52	73	94	136	177
4	-152	-110	-68	-47	-27	-6	6	14	23	35	56	77	98	139	181
5	-148	-106	-65	-44	-23	-2	10	18	26	39	60	80	101	143	185
6	-144	-103	-61	-40	-19	1	13	22	30	42	63	84	105	147	188
7	-141	-99	-57	-36	-16	4	17	25	34	46	67	88	109	150	192
8	-137	-95	-54	-33	-12	8	21	29	37	50	71	91	112	154	196
10	-130	-88	-46	-25	-5	15	28	36	45	57	78	99	120	161	203
12	-122	-81	-39	-18	2	23	35	44	52	64	85	106	127	169	210
14	-115	-73	-32	-11	9	30	43	51	59	72	93	113	134	176	218
16	-108	-66	-24	-3	17	37	50	58	67	79	100	121	142	183	225
18	-100	-59	-17	3	24	45	57	66	74	86	107	128	149	191	232
20	-93	-51	-10	10	31	52	65	73	81	94	115	135	156	198	240
25	-75	-33	8	29	50	70	83	91	100	112	133	154	175	216	258
30	-56	-15	26	47	68	89	101	110	118	130	151	172	193	235	276
40	-20	21	63	84	105	125	138	146	155	167	188	209	230	271	313

Table 5-17

Separation, inch: 3

Leg Distance, inch: 30

SHIM REQUIRED FOR ALIGNMENT

Reverse Indicator Method

iveN IR LS	DriveR TIR, MILS -40	-30	-20	-15	-10	-5	-2	0	2	5	10	15	20	30	40
-40	-380	-330	-280	-255	-230	-205	-190	-180	-170	-155	-130	-105	-80	-30	20
-30	-335	-285	-235	-210	-185	-160	-145	-135	-125	-110	-85	-60	-35	15	65
-25	-312	-262	-212	-187	-162	-137	-122	-112	-102	-87	-62	-37	-12	37	87
-20	-290	-240	-190	-165	-140	-115	-100	-90	-80	-65	-40	-15	10	60	110
-18	-281	-231	-181	-156	-131	-106	-91	-61	-71	-56	-31	-6	19	69	119
-16	-272	-222	-172	-147	-122	-97	-82	-72	-62	-47	-22	3	28	78	128
-14	-263	-213	-163	-138	-113	-88	-73	-63	-53	-38	-13	12	37	87	137
-12	-254	-204	-154	-129	-104	-79	-64	-54	-44	-29	-4	21	46	96	146
-10	-245	-195	-145	-120	-95	-70	-55	-45	-35	-20	5	30	55	105	155
-8	-236	-186	-136	-111	-86	-61	-46	-36	-26	-11	14	39	64	114	164
-7	-231	-181	-131	-106	-81	-56	-41	-31	-21	-6	18	43	68	118	168
-6	-227	-177	-127	-102	-77	-52	-37	-27	-17	-2	23	48	73	123	173
-5	-222	-172	-122	-97	-72	-47	-32	-22	-12	2	27	52	77	127	177
-4	-218	-168	-118	-93	-68	-43	-28	-18	-8	7	32	57	82	132	182
-3	-213	-163	-113	-88	-63	-38	-23	-13	-3	11	36	61	86	136	186
-2	-209	-159	-109	-84	-59	-34	-19	-9	1	16	41	66	91	141	191
-1	-204	-154	-104	-79	-54	-29	-14	-4	5	20	45	70	95	145	195
0	-200	-150	-100	-75	-50	-25	-10	0	10	25	50	75	100	150	200
1	-195	-145	-95	-70	-45	-20	-5	4	14	29	54	79	104	154	204
2	-191	-141	-91	-66	-41	-16	-1	9	19	34	59	84	109	159	209
3	-186	-136	-86	-61	-36	-11	3	13	23	38	63	88	113	163	213
4	-182	-132	-82	-57	-32	-7	8	18	28	43	68	93	118	168	218
5	-177	-127	-77	-52	-27	-2	12	22	32	47	72	97	122	172	222
6	-173	-123	-73	-48	-23	2	17	27	37	52	77	102	127	177	227
7	-168	-118	-68	-43	-18	6	21	31	41	56	81	106	131	181	231
8	-164	-114	-64	-39	-14	11	26	36	46	61	86	111	136	186	236
10	-155	-105	-55	-30	-5	20	35	45	55	70	95	120	145	195	245
12	-146	-96	-46	-21	4	29	44	54	64	79	104	129	154	204	254
14	-137	-87	-37	-12	13	38	53	63	73	88	113	138	163	213	263
16	-128	-78	-28	-3	22	47	62	72	82	97	122	147	172	222	272
18	-119	-69	-19	6	31	56	71	81	91	106	131	156	181	231	281
20	-110	-60	-10	15	40	65	80	90	100	115	140	165	190	240	290
25	-87	-37	12	37	62	87	102	112	122	137	162	187	212	262	312
30	-65	-15	35	60	85	110	125	135	145	160	185	210	235	285	335
40	-20	30	80	105	130	155	170	180	190	205	230	255	280	330	380

Table 5-18

Separation, inch: 3

Leg Distance, inch: 35

SHIM REQUIRED FOR ALIGNMENT

Reverse Indicator Method

DriveN TIR MILS	DriveR TIR, MILS -40	-30	-20	-15	-10	-5	-2	0	2	5	10	15	20	30	40
-40	-446	-388	-330	-300	-271	-242	-225	-213	-201	-184	-155	-125	-96	-38	20
-30	-393	-335	-276	-247	-218	-189	-171	-160	-148	-130	-101	-72	-43	15	73
-25	-366	-308	-250	-220	-191	-162	-145	-133	-121	-104	-75	-45	-16	41	100
-20	-340	-281	-223	-194	-165	-135	-118	-106	-95	-77	-48	-19	10	68	126
-18	-329	-271	-212	-183	-154	-125	-107	-96	-84	-66	-37	-8	20	79	137
-16	-318	-260	-202	-172	-143	-114	-97	-85	-73	-56	-27	2	31	89	148
-14	-308	-249	-191	-162	-133	-103	-86	-74	-63	-45	-16	12	42	100	158
-12	-297	-239	-180	-151	-122	-93	-75	-64	-52	-34	-5	23	52	111	169
-10	-286	-228	-170	-140	-111	-82	-65	-53	-41	-24	5	34	63	121	180
-8	-276	-217	-159	-130	-101	-71	-54	-42	-31	-13	15	44	74	132	190
-7	-270	-212	-154	-124	-95	-66	-49	-37	-25	-8	21	50	79	137	196
-6	-265	-207	-148	-119	-90	-61	-43	-32	-20	-2	26	55	84	143	201
-5	-260	-201	-143	-114	-85	-55	-38	-26	-15	2	31	60	90	148	206
-4	-254	-196	-138	-108	-79	-50	-33	-21	-9	7	37	66	95	153	212
-3	-249	-191	-132	-103	-74	-45	-27	-16	-4	13	42	71	100	159	217
-2	-244	-185	-127	-98	-69	-39	-22	-10	1	18	47	76	106	164	222
-1	-238	-180	-122	-92	-63	-34	-17	-5	6	23	53	82	111	169	228
0	-233	-175	-116	-87	-58	-29	-11	0	11	29	58	87	116	175	233
1	-228	-169	-111	-82	-53	-23	-6	5	17	34	63	92	122	180	238
2	-222	-164	-106	-76	-47	-18	-1	10	22	39	69	98	127	185	244
3	-217	-159	-100	-71	-42	-13	4	16	27	45	74	103	132	191	249
4	-212	-153	-95	-66	-37	-7	9	21	33	50	79	108	138	196	254
5	-206	-148	-90	-60	-31	-2	15	26	38	55	85	114	143	201	260
6	-201	-143	-84	-55	-26	2	20	32	43	61	90	119	148	207	265
7	-196	-137	-79	-50	-21	8	25	37	49	66	95	124	154	212	270
8	-190	-132	-74	-44	-15	13	31	42	54	71	101	130	159	217	276
10	-180	-121	-63	-34	-5	24	41	53	65	82	111	140	170	228	286
12	-169	-111	-52	-23	5	34	52	64	75	93	122	151	180	239	297
14	-158	-100	-42	-12	16	45	63	74	86	103	133	162	191	249	308
16	-148	-89	-31	-2	27	56	73	85	97	114	143	172	202	260	318
18	-137	-79	-20	8	37	66	84	96	107	125	154	183	212	271	329
20	-126	-68	-10	19	48	77	95	106	118	135	165	194	223	281	340
25	-100	-41	16	45	75	104	121	133	145	162	191	220	250	308	366
30	-73	-15	43	72	101	130	148	160	171	189	218	247	276	335	393
40	-20	38	96	125	155	184	201	213	225	242	271	300	330	388	446

Table 5-19

Separation, inch: 3

Leg Distance, inch: 40

SHIM REQUIRED FOR ALIGNMENT

Reverse Indicator Method

riveN IR ILS	DriveR TIR, MILS -40	-30	-20	-15	-10	-5	-2	0	2	5	10	15	20	30	40
-40	-513	-446	-380	-346	-313	-280	-260	-246	-233	-213	-180	-146	-113	-46	20
-30	-451	-385	-318	-285	-251	-218	-198	-185	-171	-151	-118	-85	-51	15	81
-25	-420	-354	-287	-254	-220	-187	-167	-154	-140	-120	-87	-54	-20	45	112
-20	-390	-323	-256	-223	-190	-156	-136	-123	-110	-90	-56	-23	10	76	143
-18	-377	-311	-244	-211	-177	-144	-124	-111	-97	-77	-44	-11	22	89	155
-16	-365	-298	-232	-198	-165	-132	-112	-98	-85	-65	-32	1	34	101	168
-14	-353	-286	-219	-186	-153	-119	-99	-86	-73	-53	-19	13	47	113	180
-12	-340	-274	-207	-174	-140	-107	-87	-74	-60	-40	-7	26	59	126	192
-10	-328	-261	-195	-161	-128	-95	-75	-61	-48	-28	5	38	71	138	205
-8	-316	-249	-182	-149	-116	-82	-62	-49	-36	-16	17	50	84	150	217
-7	-309	-243	-176	-143	-109	-76	-56	-43	-29	-9	23	56	90	156	223
-6	-303	-237	-170	-137	-103	-70	-50	-37	-23	-3	29	63	96	163	229
-5	-297	-230	-164	-130	-97	-64	-44	-30	-17	2	35	69	102	169	235
-4	-291	-224	-158	-124	-91	-58	-38	-24	-11	8	42	75	108	175	242
-3	-285	-218	-151	-118	-85	-51	-31	-18	-5	14	48	81	114	181	248
-2	-279	-212	-145	-112	-79	-45	-25	-12	1	21	54	87	121	187	254
-1	-272	-206	-139	-106	-72	-39	-19	-6	7	27	60	93	127	193	260
0	-266	-200	-133	-100	-66	-33	-13	0	13	33	66	100	133	200	266
1	-260	-193	-127	-93	-60	-27	-7	6	19	39	72	106	139	206	272
2	-254	-187	-121	-87	-54	-21	-1	12	25	45	79	112	145	212	279
3	-248	-181	-114	-81	-48	-14	5	18	31	51	85	118	151	218	285
4	-242	-175	-108	-75	-42	-8	11	24	38	58	91	124	158	224	291
5	-235	-169	-102	-69	-35	-2	17	30	44	64	97	130	164	230	297
6	-229	-163	-96	-63	-29	3	23	37	50	70	103	137	170	237	303
7	-223	-156	-90	-56	-23	9	29	43	56	76	109	143	176	243	309
8	-217	-150	-84	-50	-17	16	36	49	62	82	116	149	182	249	316
10	-205	-138	-71	-38	-5	28	48	61	75	95	128	161	195	261	328
12	-192	-126	-59	-26	7	40	60	74	87	107	140	174	207	274	340
14	-180	-113	-47	-13	19	53	73	86	99	119	153	186	219	286	353
16	-168	-101	-34	-1	32	65	85	98	112	132	165	198	232	298	365
18	-155	-89	-22	11	44	77	97	111	124	144	177	211	244	311	377
20	-143	-76	-10	23	56	90	110	123	136	156	190	223	256	323	390
25	-112	-45	20	54	87	120	140	154	167	187	220	254	287	354	420
30	-81	-15	51	85	118	151	171	185	198	218	251	285	318	385	451
40	-20	46	113	146	180	213	233	246	260	280	313	346	380	446	513

Table 5-20

Separation, inch: 3

Leg Distance, inch: 50

SHIM REQUIRED FOR ALIGNMENT

Reverse Indicator Method

DriveN TIR MILS	DriveR TIR, MILS -40	-30	-20	-15	-10	-5	-2	0	2	5	10	15	20	30	40
-40	-646	-563	-480	-438	-396	-355	-330	-313	-296	-271	-230	-188	-146	-63	20
-30	-568	-485	-401	-360	-318	-276	-251	-235	-218	-193	-151	-110	-68	15	98
-25	-529	-445	-362	-320	-279	-237	-212	-195	-179	-154	-112	-70	-29	54	137
-20	-490	-406	-323	-281	-240	-198	-173	-156	-140	-115	-73	-31	10	93	176
-18	-474	-391	-307	-266	-224	-182	-157	-141	-124	-99	-57	-16	25	109	192
-16	-458	-375	-292	-250	-208	-167	-142	-125	-108	-83	-42	0	41	124	208
-14	-443	-359	-276	-234	-193	-151	-126	-109	-93	-68	-26	15	57	140	223
-12	-427	-344	-260	-219	-177	-135	-110	-94	-77	-52	-10	31	72	156	239
-10	-411	-328	-245	-203	-161	-120	-95	-78	-61	-36	5	46	88	171	255
-8	-396	-312	-229	-187	-146	-104	-79	-62	-46	-21	20	62	104	187	270
-7	-388	-304	-221	-179	-138	-96	-71	-54	-38	-13	28	70	111	195	278
-6	-380	-297	-213	-172	-130	-88	-63	-47	-30	-5	36	78	119	203	286
-5	-372	-289	-205	-164	-122	-80	-55	-39	-22	2	44	85	127	210	294
-4	-364	-281	-198	-156	-114	-73	-48	-31	-14	10	52	93	135	218	302
-3	-356	-273	-190	-148	-106	-65	-40	-23	-6	18	59	101	143	226	309
-2	-349	-265	-182	-140	-99	-57	-32	-15	1	26	67	109	151	234	317
-1	-341	-257	-174	-132	-91	-49	-24	-7	8	33	75	117	158	242	325
0	-333	-250	-166	-125	-83	-41	-16	0	16	41	83	125	166	250	333
1	-325	-242	-158	-117	-75	-33	-8	7	24	49	91	132	174	257	341
2	-317	-234	-151	-109	-67	-26	-1	15	32	57	99	140	182	265	349
3	-309	-226	-143	-101	-59	-18	6	23	40	65	106	148	190	273	356
4	-302	-218	-135	-93	-52	-10	14	31	48	73	114	156	198	281	364
5	-294	-210	-127	-85	-44	-2	22	39	55	80	122	164	205	289	372
6	-286	-203	-119	-78	-36	5	30	47	63	88	130	172	213	297	380
7	-278	-195	-111	-70	-28	13	38	54	71	96	138	179	221	304	388
8	-270	-187	-104	-62	-20	21	46	62	79	104	146	187	229	312	396
10	-255	-171	-88	-46	-5	36	61	78	95	120	161	203	245	328	411
12	-239	-156	-72	-31	10	52	77	94	110	135	177	219	260	344	427
14	-223	-140	-57	-15	26	68	93	109	126	151	193	234	276	359	443
16	-208	-124	-41	0	42	83	108	125	142	167	208	250	292	375	458
18	-192	-109	-25	16	57	99	124	141	157	182	224	266	307	391	474
20	-176	-93	-10	31	73	115	140	156	173	198	240	281	323	406	490
25	-137	-54	29	70	112	154	179	195	212	237	279	320	362	445	529
30	-98	-15	68	110	151	193	218	235	251	276	318	360	401	485	568
40	-20	63	146	188	230	271	296	313	330	355	396	438	480	563	646

Table 5-21

Separation, inch: 3

Leg Distance, inch: 60

SHIM REQUIRED FOR ALIGNMENT

Reverse Indicator Method

DriveN TIR MILS	DriveR TIR, MILS -40	-30	-20	-15	-10	-5	-2	0	2	5	10	15	20	30	40
-40	-780	-680	-580	-530	-480	-430	-400	-380	-360	-330	-280	-230	-180	-80	20
-30	-685	-585	-485	-435	-385	-335	-305	-285	-265	-235	-185	-135	-85	15	115
-25	-637	-537	-437	-387	-337	-287	-257	-237	-217	-187	-137	-87	-37	62	162
-20	-590	-490	-390	-340	-290	-240	-210	-190	-170	-140	-90	-40	10	110	210
-18	-571	-471	-371	-321	-271	-221	-191	-171	-151	-121	-71	-21	29	129	229
-16	-552	-452	-352	-302	-252	-202	-172	-152	-132	-102	-52	-2	48	148	248
-14	-533	-433	-333	-283	-233	-183	-153	-133	-113	-83	-33	17	67	167	267
-12	-514	-414	-314	-264	-214	-164	-134	-114	-94	-64	-14	36	86	186	286
-10	-495	-395	-295	-245	-195	-145	-115	-95	-75	-45	5	55	105	205	305
-8	-476	-376	-276	-226	-176	-126	-96	-76	-56	-26	24	74	124	224	324
-7	-466	-366	-266	-216	-166	-116	-86	-66	-46	-16	33	83	133	233	333
-6	-457	-357	-257	-207	-157	-107	-77	-57	-37	-7	43	93	143	243	343
-5	-447	-347	-247	-197	-147	-97	-67	-47	-27	2	52	102	152	252	352
-4	-438	-338	-238	-188	-138	-88	-58	-38	-18	12	62	112	162	262	362
-3	-428	-328	-228	-178	-128	-78	-48	-28	-8	21	71	121	171	271	371
-2	-419	-319	-219	-169	-119	-69	-39	-19	1	31	81	131	181	281	381
-1	-409	-309	-209	-159	-109	-59	-29	-9	10	40	90	140	190	290	390
0	-400	-300	-200	-150	-100	-50	-20	0	20	50	100	150	200	300	400
1	-390	-290	-190	-140	-90	-40	-10	9	29	59	109	159	209	309	409
2	-381	-281	-181	-131	-81	-31	-1	19	39	69	119	169	219	319	419
3	-371	-271	-171	-121	-71	-21	8	28	48	78	128	178	228	328	428
4	-362	-262	-162	-112	-62	-12	18	38	58	88	138	188	238	338	438
5	-352	-252	-152	-102	-52	-2	27	47	67	97	147	197	247	347	447
6	-343	-243	-143	-93	-43	7	37	57	77	107	157	207	257	357	457
7	-333	-233	-133	-83	-33	16	46	66	86	116	166	216	266	366	466
8	-324	-224	-124	-74	-24	26	56	76	96	126	176	226	276	376	476
10	-305	-205	-105	-55	-5	45	75	95	115	145	195	245	295	395	495
12	-286	-186	-86	-36	14	64	94	114	134	164	214	264	314	414	514
14	-267	-167	-67	-17	33	83	113	133	153	183	233	283	333	433	533
16	-248	-148	-48	2	52	102	132	152	172	202	252	302	352	452	552
18	-229	-129	-29	21	71	121	151	171	191	221	271	321	371	471	571
20	-210	-110	-10	40	90	140	170	190	210	240	290	340	390	490	590
25	-162	-62	37	87	137	187	217	237	257	287	337	387	437	537	637
30	-115	-15	85	135	185	235	265	285	305	335	385	435	485	585	685
40	-20	80	180	230	280	330	360	380	400	430	480	530	580	680	780

Table 5-22

Separation, inch: 3.5

Leg Distance, inch: 10

SHIM REQUIRED FOR ALIGNMENT

Reverse Indicator Method

DriveN TIR MILS	DriveR TIR, MILS -40	-30	-20	-15	-10	-5	-2	0	2	5	10	15	20	30	40
-40	-94	-80	-65	-58	-51	-44	-40	-37	-34	-30	-22	-15	-8	5	20
-30	-85	-70	-56	-49	-42	-35	-30	-27	-25	-20	-13	-6	0	15	29
-25	-80	-66	-51	-44	-37	-30	-26	-23	-20	-16	-8	-1	5	19	33
-20	-75	-61	-47	-40	-32	-25	-21	-18	-15	-11	-4	2	10	24	38
-18	-73	-59	-45	-38	-31	-23	-19	-16	-13	-9	-2	4	11	26	40
-16	-72	-57	-43	-36	-29	-22	-17	-14	-12	-7	0	6	13	28	42
-14	-70	-55	-41	-34	-27	-20	-15	-13	-10	-5	1	8	15	29	44
-12	-68	-54	-39	-32	-25	-18	-14	-11	-8	-4	3	10	17	31	46
-10	-66	-52	-37	-30	-23	-16	-12	-9	-6	-2	5	12	19	33	47
-8	-64	-50	-36	-28	-21	-14	-10	-7	-4	0	6	14	21	35	49
-7	-63	-49	-35	-27	-20	-13	-9	-6	-3	0	7	14	22	36	50
-6	-62	-48	-34	-27	-19	-12	-8	-5	-2	1	8	15	23	37	51
-5	-61	-47	-33	-26	-18	-11	-7	-4	-1	2	9	16	23	38	52
-4	-60	-46	-32	-25	-18	-10	-6	-3	0	3	10	17	24	39	53
-3	-59	-45	-31	-24	-17	-9	-5	-2	0	4	11	18	25	40	54
-2	-59	-44	-30	-23	-16	-9	-4	-1	1	5	12	19	26	41	55
-1	-58	-43	-29	-22	-15	-8	-3	0	1	6	13	20	27	41	56
0	-57	-42	-28	-21	-14	-7	-2	0	2	7	14	21	28	42	57
1	-56	-41	-27	-20	-13	-6	-1	0	3	8	15	22	29	43	58
2	-55	-41	-26	-19	-12	-5	-1	1	4	9	16	23	30	44	59
3	-54	-40	-25	-18	-11	-4	0	2	5	9	17	24	31	45	59
4	-53	-39	-24	-17	-10	-3	0	3	6	10	18	25	32	46	60
5	-52	-38	-23	-16	-9	-2	1	4	7	11	18	26	33	47	61
6	-51	-37	-23	-15	-8	-1	2	5	8	12	19	27	34	48	62
7	-50	-36	-22	-14	-7	0	3	6	9	13	20	27	35	49	63
8	-49	-35	-21	-14	-6	0	4	7	10	14	21	28	36	50	64
10	-47	-33	-19	-12	-5	2	6	9	12	16	23	30	37	52	66
12	-46	-31	-17	-10	-3	4	8	11	14	18	25	32	39	54	68
14	-44	-29	-15	-8	-1	5	10	13	15	20	27	34	41	55	70
16	-42	-28	-13	-6	0	7	12	14	17	22	29	36	43	57	72
18	-40	-26	-11	-4	2	9	13	16	19	23	31	38	45	59	73
20	-38	-24	-10	-2	4	11	15	18	21	25	32	40	47	61	75
25	-33	-19	-5	1	8	16	20	23	26	30	37	44	51	66	80
30	-29	-15	0	6	13	20	25	27	30	35	42	49	56	70	85
40	-20	-5	8	15	22	30	34	37	40	44	51	58	65	80	94

Table 5-23

Separation, inch: 3.5

Leg Distance, inch: 15

SHIM REQUIRED FOR ALIGNMENT

Reverse Indicator Method

DriveN TIR MILS	DriveR TIR, MILS -40	-30	-20	-15	-10	-5	-2	0	2	5	10	15	20	30	40
-40	-151	-130	-108	-97	-87	-76	-70	-65	-61	-55	-44	-33	-22	-1	20
-30	-135	-113	-92	-81	-70	-60	-53	-49	-45	-38	-27	-17	-6	15	36
-25	-126	-105	-83	-73	-62	-51	-45	-41	-36	-30	-19	-8	1	23	44
-20	-118	-97	-75	-65	-54	-43	-37	-32	-28	-22	-11	0	10	31	52
-18	-115	-93	-72	-61	-51	-40	-33	-29	-25	-18	-8	2	13	34	56
-16	-112	-90	-69	-58	-47	-37	-30	-26	-22	-15	-4	5	16	38	59
-14	-108	-87	-65	-55	-44	-33	-27	-23	-18	-12	-1	9	19	41	62
-12	-105	-84	-62	-51	-41	-30	-24	-19	-15	-9	1	12	23	44	66
-10	-102	-80	-59	-48	-37	-27	-20	-16	-12	-5	5	15	26	47	69
-8	-98	-77	-56	-45	-34	-23	-17	-13	-8	-2	8	19	29	51	72
-7	-97	-75	-54	-43	-32	-22	-15	-11	-7	0	9	20	31	52	74
-6	-95	-74	-52	-42	-31	-20	-14	-9	-5	0	11	22	33	54	75
-5	-93	-72	-51	-40	-29	-18	-12	-8	-3	2	13	23	34	56	77
-4	-92	-70	-49	-38	-28	-17	-10	-6	-2	4	14	25	36	57	79
-3	-90	-69	-47	-37	-26	-15	-9	-4	0	5	16	27	37	59	80
-2	-89	-67	-46	-35	-24	-14	-7	-3	1	7	18	28	39	61	82
-1	-87	-65	-44	-33	-23	-12	-5	-1	2	9	19	30	41	62	84
0	-85	-64	-42	-32	-21	-10	-4	0	4	10	21	32	42	64	85
1	-84	-62	-41	-30	-19	-9	-2	1	5	12	23	33	44	65	87
2	-82	-61	-39	-28	-18	-7	-1	3	7	14	24	35	46	67	89
3	-80	-59	-37	-27	-16	-5	0	4	9	15	26	37	47	69	90
4	-79	-57	-36	-25	-14	-4	2	6	10	17	28	38	49	70	92
5	-77	-56	-34	-23	-13	-2	3	8	12	18	29	40	51	72	93
6	-75	-54	-33	-22	-11	0	5	9	14	20	31	42	52	74	95
7	-74	-52	-31	-20	-9	0	7	11	15	22	32	43	54	75	97
8	-72	-51	-29	-19	-8	2	8	13	17	23	34	45	56	77	98
10	-69	-47	-26	-15	-5	5	12	16	20	27	37	48	59	80	102
12	-66	-44	-23	-12	-1	9	15	19	24	30	41	51	62	84	105
14	-62	-41	-19	-9	1	12	18	23	27	33	44	55	65	87	108
16	-59	-38	-16	-5	4	15	22	26	30	37	47	58	69	90	112
18	-56	-34	-13	-2	8	18	25	29	33	40	51	61	72	93	115
20	-52	-31	-10	0	11	22	28	32	37	43	54	65	75	97	118
25	-44	-23	-1	8	19	30	36	41	45	51	62	73	83	105	126
30	-36	-15	6	17	27	38	45	49	53	60	70	81	92	113	135
40	-20	1	22	33	44	55	61	65	70	76	87	97	108	130	151

Table 5-24

Separation, inch: 3.5

Leg Distance, inch: 20

SHIM REQUIRED FOR ALIGNMENT

Reverse Indicator Method

DriveN TIR MILS	DriveR TIR, MILS -40	-30	-20	-15	-10	-5	-2	0	2	5	10	15	20	30	40
-40	-208	-180	-151	-137	-122	-108	-100	-94	-88	-80	-65	-51	-37	-8	20
-30	-185	-156	-127	-113	-99	-85	-76	-70	-65	-56	-42	-27	-13	15	43
-25	-173	-144	-116	-101	-87	-73	-64	-58	-53	-44	-30	-16	-1	26	55
-20	-161	-132	-104	-90	-75	-61	-52	-47	-41	-32	-18	-4	10	38	67
-18	-156	-128	-99	-85	-71	-56	-48	-42	-36	-28	-13	0	14	43	71
-16	-152	-123	-94	-80	-66	-52	-43	-37	-32	-23	-9	5	19	48	76
-14	-147	-118	-90	-75	-61	-47	-38	-33	-27	-18	-4	9	24	52	81
-12	-142	-114	-85	-71	-56	-42	-34	-28	-22	-14	0	14	28	57	86
-10	-137	-109	-80	-66	-52	-37	-29	-23	-17	-9	5	19	33	62	90
-8	-133	-104	-76	-61	-47	-33	-24	-18	-13	-4	9	24	38	66	95
-7	-130	-102	-73	-59	-45	-30	-22	-16	-10	-2	12	26	40	69	97
-6	-128	-99	-71	-57	-42	-28	-19	-14	-8	0	14	28	43	71	100
-5	-126	-97	-68	-54	-40	-26	-17	-11	-6	2	16	31	45	73	102
-4	-123	-95	-66	-52	-38	-23	-15	-9	-3	4	19	33	47	76	104
-3	-121	-92	-64	-49	-35	-21	-12	-7	-1	7	21	35	50	78	107
-2	-119	-90	-61	-47	-33	-19	-10	-4	1	9	23	38	52	81	109
-1	-116	-88	-59	-45	-30	-16	-8	-2	3	11	26	40	54	83	111
0	-114	-85	-57	-42	-28	-14	-5	0	5	14	28	42	57	85	114
1	-111	-83	-54	-40	-26	-11	-3	2	8	16	30	45	59	88	116
2	-109	-81	-52	-38	-23	-9	-1	4	10	19	33	47	61	90	119
3	-107	-78	-50	-35	-21	-7	1	7	12	21	35	49	64	92	121
4	-104	-76	-47	-33	-19	-4	3	9	15	23	38	52	66	95	123
5	-102	-73	-45	-31	-16	-2	6	11	17	26	40	54	68	97	126
6	-100	-71	-43	-28	-14	0	8	14	19	28	42	57	71	99	128
7	-97	-69	-40	-26	-12	2	10	16	22	30	45	59	73	102	130
8	-95	-66	-38	-24	-9	4	13	18	24	33	47	61	76	104	133
10	-90	-62	-33	-19	-5	9	17	23	29	37	52	66	80	109	137
12	-86	-57	-28	-14	0	14	22	28	34	42	56	71	85	114	142
14	-81	-52	-24	-9	4	18	27	33	38	47	61	75	90	118	147
16	-76	-48	-19	-5	9	23	32	37	43	52	66	80	94	123	152
18	-71	-43	-14	0	13	28	36	42	48	56	71	85	99	128	156
20	-67	-38	-10	4	18	32	41	47	52	61	75	90	104	132	161
25	-55	-26	1	16	30	44	53	58	64	73	87	101	116	144	173
30	-43	-15	13	27	42	56	65	70	76	85	99	113	127	156	185
40	-20	8	37	51	65	80	88	94	100	108	122	137	151	180	208

Table 5-25

Separation, inch: 3.5

Leg Distance, inch: 25

SHIM REQUIRED FOR ALIGNMENT

Reverse Indicator Method

DriveN TIR MILS	DriveR TIR, MILS -40	-30	-20	-15	-10	-5	-2	0	2	5	10	15	20	30	40
-40	-265	-230	-194	-176	-158	-140	-130	-122	-115	-105	-87	-69	-51	-15	20
-30	-235	-199	-163	-145	-127	-110	-99	-92	-85	-74	-56	-38	-20	15	50
-25	-219	-183	-148	-130	-112	-94	-83	-76	-69	-58	-41	-23	-5	30	66
-20	-204	-168	-132	-115	-97	-79	-68	-61	-54	-43	-25	-7	10	45	81
-18	-198	-162	-126	-108	-91	-73	-62	-55	-48	-37	-19	-1	16	51	87
-16	-192	-156	-120	-102	-84	-67	-56	-49	-42	-31	-13	4	22	58	93
-14	-185	-150	-114	-96	-78	-60	-50	-43	-35	-25	-7	10	28	64	99
-12	-179	-144	-108	-90	-72	-54	-44	-36	-29	-19	-1	16	34	70	106
-10	-173	-137	-102	-84	-66	-48	-37	-30	-23	-12	5	22	40	76	112
-8	-167	-131	-96	-78	-60	-42	-31	-24	-17	-6	11	29	46	82	118
-7	-164	-128	-92	-75	-57	-39	-28	-21	-14	-3	14	32	49	85	121
-6	-161	-125	-89	-72	-54	-36	-25	-18	-11	0	17	35	53	88	124
-5	-158	-122	-86	-68	-51	-33	-22	-15	-8	2	20	38	56	91	127
-4	-155	-119	-83	-65	-48	-30	-19	-12	-5	5	23	41	59	94	130
-3	-152	-116	-80	-62	-44	-27	-16	-9	-2	8	26	44	62	97	133
-2	-149	-113	-77	-59	-41	-24	-13	-6	1	11	29	47	65	101	136
-1	-145	-110	-74	-56	-38	-20	-10	-3	4	14	32	50	68	104	139
0	-142	-107	-71	-53	-35	-17	-7	0	7	17	35	53	71	107	142
1	-139	-104	-68	-50	-32	-14	-4	3	10	20	38	56	74	110	145
2	-136	-101	-65	-47	-29	-11	-1	6	13	24	41	59	77	113	149
3	-133	-97	-62	-44	-26	-8	2	9	16	27	44	62	80	116	152
4	-130	-94	-59	-41	-23	-5	5	12	19	30	48	65	83	119	155
5	-127	-91	-56	-38	-20	-2	8	15	22	33	51	68	86	122	158
6	-124	-88	-53	-35	-17	0	11	18	25	36	54	72	89	125	161
7	-121	-85	-49	-32	-14	3	14	21	28	39	57	75	92	128	164
8	-118	-82	-46	-29	-11	6	17	24	31	42	60	78	96	131	167
10	-112	-76	-40	-22	-5	12	23	30	37	48	66	84	102	137	173
12	-106	-70	-34	-16	1	19	29	36	44	54	72	90	108	144	179
14	-99	-64	-28	-10	7	25	35	43	50	60	78	96	114	150	185
16	-93	-58	-22	-4	13	31	42	49	56	67	84	102	120	156	192
18	-87	-51	-16	1	19	37	48	55	62	73	91	108	126	162	198
20	-81	-45	-10	7	25	43	54	61	68	79	97	115	132	168	204
25	-66	-30	5	23	41	58	69	76	83	94	112	130	148	183	219
30	-50	-15	20	38	56	74	85	92	99	110	127	145	163	199	235
40	-20	15	51	69	87	105	115	122	130	140	158	176	194	230	265

Table 5-26

Separation, inch: 3.5

Leg Distance, inch: 30

SHIM REQUIRED FOR ALIGNMENT

Reverse Indicator Method

DriveN TIR MILS	DriveR TIR, MILS -40	-30	-20	-15	-10	-5	-2	0	2	5	10	15	20	30	40
-40	-322	-280	-237	-215	-194	-172	-160	-151	-142	-130	-108	-87	-65	-22	20
-30	-285	-242	-199	-177	-156	-135	-122	-113	-105	-92	-70	-49	-27	15	57
-25	-266	-223	-180	-158	-137	-116	-103	-94	-86	-73	-51	-30	-8	33	76
-20	-247	-204	-161	-140	-118	-97	-84	-75	-67	-54	-32	-11	10	52	95
-18	-239	-196	-153	-132	-111	-89	-76	-68	-59	-46	-25	-3	17	60	103
-16	-232	-189	-146	-124	-103	-82	-69	-60	-52	-39	-17	3	25	68	110
-14	-224	-181	-138	-117	-95	-74	-61	-53	-44	-31	-10	11	32	75	118
-12	-216	-174	-131	-109	-88	-66	-54	-45	-36	-24	-2	18	40	83	126
-10	-209	-166	-123	-102	-80	-59	-46	-37	-29	-16	5	26	47	90	133
-8	-201	-158	-116	-94	-73	-51	-38	-30	-21	-8	12	34	55	98	141
-7	-197	-155	-112	-90	-69	-47	-35	-26	-17	-5	16	37	59	102	144
-6	-194	-151	-108	-87	-65	-44	-31	-22	-14	-1	20	41	63	105	148
-5	-190	-147	-104	-83	-61	-40	-27	-18	-10	2	23	45	66	109	152
-4	-186	-143	-100	-79	-58	-36	-23	-15	-6	6	27	49	70	113	156
-3	-182	-139	-97	-75	-54	-32	-19	-11	-2	10	31	52	74	117	160
-2	-179	-136	-93	-71	-50	-29	-16	-7	1	13	35	56	78	121	163
-1	-175	-132	-89	-68	-46	-25	-12	-3	4	17	39	60	81	124	167
0	-171	-128	-85	-64	-42	-21	-8	0	8	21	42	64	85	128	171
1	-167	-124	-81	-60	-39	-17	-4	3	12	25	46	68	89	132	175
2	-163	-121	-78	-56	-35	-13	-1	7	16	29	50	71	93	136	179
3	-160	-117	-74	-52	-31	-10	2	11	19	32	54	75	97	139	182
4	-156	-113	-70	-49	-27	-6	6	15	23	36	58	79	100	143	186
5	-152	-109	-66	-45	-23	-2	10	18	27	40	61	83	104	147	190
6	-148	-105	-63	-41	-20	1	14	22	31	44	65	87	108	151	194
7	-144	-102	-59	-37	-16	5	17	26	35	47	69	90	112	155	197
8	-141	-98	-55	-34	-12	8	21	30	38	51	73	94	116	158	201
10	-133	-90	-47	-26	-5	16	29	37	46	59	80	102	123	166	209
12	-126	-83	-40	-18	2	24	36	45	54	66	88	109	131	174	216
14	-118	-75	-32	-11	10	31	44	53	61	74	95	117	138	181	224
16	-110	-68	-25	-3	17	39	52	60	69	82	103	124	146	189	232
18	-103	-60	-17	3	25	46	59	68	76	89	111	132	153	196	239
20	-95	-52	-10	11	32	54	67	75	84	97	118	140	161	204	247
25	-76	-33	8	30	51	73	86	94	103	116	137	158	180	223	266
30	-57	-15	27	49	70	92	105	113	122	135	156	177	199	242	285
40	-20	22	65	87	108	130	142	151	160	172	194	215	237	280	322

Table 5-27

Separation, inch: 3.5

Leg Distance, inch: 35

SHIM REQUIRED FOR ALIGNMENT

Reverse Indicator Method

riveN IR ILS	DriveR TIR, MILS -40	-30	-20	-15	-10	-5	-2	0	2	5	10	15	20	30	40
-40	-380	-330	-280	-255	-230	-205	-190	-180	-170	-155	-130	-105	-80	-30	20
-30	-335	-285	-235	-210	-185	-160	-145	-135	-125	-110	-85	-60	-35	15	65
-25	-312	-262	-212	-187	-162	-137	-122	-112	-102	-87	-62	-37	-12	37	87
-20	-290	-240	-190	-165	-140	-115	-100	-90	-80	-65	-40	-15	10	60	110
-18	-281	-231	-181	-156	-131	-106	-91	-81	-71	-56	-31	-6	19	69	119
-16	-272	-222	-172	-147	-122	-97	-82	-72	-62	-47	-22	3	28	78	128
-14	-263	-213	-163	-138	-113	-88	-73	-63	-53	-38	-13	12	37	87	137
-12	-254	-204	-154	-129	-104	-79	-64	-54	-44	-29	-4	21	46	96	146
-10	-245	-195	-145	-120	-95	-70	-55	-45	-35	-20	5	30	55	105	155
-8	-236	-186	-136	-111	-86	-61	-46	-36	-26	-11	14	39	64	114	164
-7	-231	-181	-131	-106	-81	-56	-41	-31	-21	-6	18	43	68	118	168
-6	-227	-177	-127	-102	-77	-52	-37	-27	-17	-2	23	48	73	123	173
-5	-222	-172	-122	-97	-72	-47	-32	-22	-12	2	27	52	77	127	177
-4	-218	-168	-118	-93	-68	-43	-28	-18	-8	7	32	57	82	132	182
-3	-213	-163	-113	-88	-63	-38	-23	-13	-3	11	36	61	86	136	186
-2	-209	-159	-109	-84	-59	-34	-19	-9	1	16	41	66	91	141	191
-1	-204	-154	-104	-79	-54	-29	-14	-4	5	20	45	70	95	145	195
0	-200	-150	-100	-75	-50	-25	-10	0	10	25	50	75	100	150	200
1	-195	-145	-95	-70	-45	-20	-5	4	14	29	54	79	104	154	204
2	-191	-141	-91	-66	-41	-16	-1	9	19	34	59	84	109	159	209
3	-186	-136	-86	-61	-36	-11	3	13	23	38	63	88	113	163	213
4	-182	-132	-82	-57	-32	-7	8	18	28	43	68	93	118	168	218
5	-177	-127	-77	-52	-27	-2	12	22	32	47	72	97	122	172	222
6	-173	-123	-73	-48	-23	2	17	27	37	52	77	102	127	177	227
7	-168	-118	-68	-43	-18	6	21	31	41	56	81	106	131	181	231
8	-164	-114	-64	-39	-14	11	26	36	46	61	86	111	136	186	236
10	-155	-105	-55	-30	-5	20	35	45	55	70	95	120	145	195	245
12	-146	-96	-46	-21	4	29	44	54	64	79	104	129	154	204	254
14	-137	-87	-37	-12	13	38	53	63	73	88	113	138	163	213	263
16	-128	-78	-28	-3	22	47	62	72	82	97	122	147	172	222	272
18	-119	-69	-19	6	31	56	71	81	91	106	131	156	181	231	281
20	-110	-60	-10	15	40	65	80	90	100	115	140	165	190	240	290
25	-87	-37	12	37	62	87	102	112	122	137	162	187	212	262	312
30	-65	-15	35	60	85	110	125	135	145	160	185	210	235	285	335
40	-20	30	80	105	130	155	170	180	190	205	230	255	280	330	380

Table 5-28

Separation, inch: 3.5

Leg Distance, inch: 40

SHIM REQUIRED FOR ALIGNMENT

Reverse Indicator Method

DriveN TIR MILS	DriveR TIR, MILS -40	-30	-20	-15	-10	-5	-2	0	2	5	10	15	20	30	40
-40	-437	-380	-322	-294	-265	-237	-220	-208	-197	-180	-151	-122	-94	-37	20
-30	-385	-327	-270	-242	-213	-185	-167	-156	-145	-127	-99	-70	-42	15	72
-25	-358	-301	-244	-216	-187	-158	-141	-130	-118	-101	-73	-44	-16	41	98
-20	-332	-275	-218	-190	-161	-132	-115	-104	-92	-75	-47	-18	10	67	124
-18	-322	-265	-208	-179	-151	-122	-105	-93	-82	-65	-36	-8	20	77	134
-16	-312	-254	-197	-169	-140	-112	-94	-83	-72	-54	-26	2	30	88	145
-14	-301	-244	-187	-158	-130	-101	-84	-73	-61	-44	-15	12	41	98	155
-12	-291	-234	-176	-148	-119	-91	-74	-62	-51	-34	-5	23	51	108	166
-10	-280	-223	-166	-137	-109	-80	-63	-52	-40	-23	5	33	62	119	176
-8	-270	-213	-156	-127	-98	-70	-53	-41	-30	-13	15	44	72	129	186
-7	-265	-207	-150	-122	-93	-65	-47	-36	-25	-7	20	49	77	134	192
-6	-259	-202	-145	-117	-88	-59	-42	-31	-19	-2	25	54	83	140	197
-5	-254	-197	-140	-111	-83	-54	-37	-26	-14	2	31	59	88	145	202
-4	-249	-192	-135	-106	-78	-49	-32	-20	-9	7	36	64	93	150	207
-3	-244	-187	-129	-101	-72	-44	-27	-15	-4	12	41	70	98	155	212
-2	-239	-181	-124	-96	-67	-39	-21	-10	1	18	46	75	103	161	218
-1	-233	-176	-119	-90	-62	-33	-16	-5	6	23	51	80	109	166	223
0	-228	-171	-114	-85	-57	-28	-11	0	11	28	57	85	114	171	228
1	-223	-166	-109	-80	-51	-23	-6	5	16	33	62	90	119	176	233
2	-218	-161	-103	-75	-46	-18	-1	10	21	39	67	96	124	181	239
3	-212	-155	-98	-70	-41	-12	4	15	27	44	72	101	129	187	244
4	-207	-150	-93	-64	-36	-7	9	20	32	49	78	106	135	192	249
5	-202	-145	-88	-59	-31	-2	14	26	37	54	83	111	140	197	254
6	-197	-140	-83	-54	-25	2	19	31	42	59	88	117	145	202	259
7	-192	-134	-77	-49	-20	7	25	36	47	65	93	122	150	207	265
8	-186	-129	-72	-44	-15	13	30	41	53	70	98	127	156	213	270
10	-176	-119	-62	-33	-5	23	40	52	63	80	109	137	166	223	280
12	-166	-108	-51	-23	5	34	51	62	74	91	119	148	176	234	291
14	-155	-98	-41	-12	15	44	61	73	84	101	130	158	187	244	301
16	-145	-88	-30	-2	26	54	72	83	94	112	140	169	197	254	312
18	-134	-77	-20	8	36	65	82	93	105	122	151	179	208	265	322
20	-124	-67	-10	18	47	75	92	104	115	132	161	190	218	275	332
25	-98	-41	16	44	73	101	118	130	141	158	187	216	244	301	358
30	-72	-15	42	70	99	127	145	156	167	185	213	242	270	327	385
40	-20	37	94	122	151	180	197	208	220	237	265	294	322	380	437

Table 5-29

Separation, inch: 3.5

Leg Distance, inch: 50

SHIM REQUIRED FOR ALIGNMENT

Reverse Indicator Method

riveN DriveR TIR, MILS

IR

ILS	-40	-30	-20	-15	-10	-5	-2	0	2	5	10	15	20	30	40
-40	-551	-480	-408	-372	-337	-301	-280	-265	-251	-230	-194	-158	-122	-51	20
-30	-485	-413	-342	-306	-270	-235	-213	-199	-185	-163	-127	-92	-56	15	86
-25	-451	-380	-308	-273	-237	-201	-180	-166	-151	-130	-94	-58	-23	48	119
-20	-418	-347	-275	-240	-204	-168	-147	-132	-118	-97	-61	-25	10	81	152
-18	-405	-333	-262	-226	-191	-155	-133	-119	-105	-83	-48	-12	23	94	166
-16	-392	-320	-249	-213	-177	-142	-120	-106	-92	-70	-34	0	36	108	179
-14	-378	-307	-235	-200	-164	-128	-107	-93	-78	-57	-21	14	49	121	192
-12	-365	-294	-222	-186	-151	-115	-94	-79	-65	-44	-8	27	63	134	206
-10	-352	-280	-209	-173	-137	-102	-80	-66	-52	-30	5	40	76	147	219
-8	-338	-267	-196	-160	-124	-88	-67	-53	-38	-17	18	54	89	161	232
-7	-332	-260	-189	-153	-117	-82	-60	-46	-32	-10	24	60	96	167	239
-6	-325	-254	-182	-147	-111	-75	-54	-39	-25	-4	31	67	103	174	245
-5	-318	-247	-176	-140	-104	-68	-47	-33	-18	2	38	73	109	181	252
-4	-312	-240	-169	-133	-98	-62	-40	-26	-12	9	44	80	116	187	259
-3	-305	-234	-162	-127	-91	-55	-34	-19	-5	15	51	87	122	194	265
-2	-299	-227	-156	-120	-84	-49	-27	-13	1	22	58	93	129	201	272
-1	-292	-220	-149	-113	-78	-42	-20	-6	7	29	64	100	136	207	279
0	-285	-214	-142	-107	-71	-35	-14	0	14	35	71	107	142	214	285
1	-279	-207	-136	-100	-64	-29	-7	6	20	42	78	113	149	220	292
2	-272	-201	-129	-93	-58	-22	-1	13	27	49	84	120	156	227	299
3	-265	-194	-122	-87	-51	-15	5	19	34	55	91	127	162	234	305
4	-259	-187	-116	-80	-44	-9	12	26	40	62	98	133	169	240	312
5	-252	-181	-109	-73	-38	-2	18	33	47	68	104	140	176	247	318
6	-245	-174	-103	-67	-31	4	25	39	54	75	111	147	182	254	325
7	-239	-167	-96	-60	-24	10	32	46	60	82	117	153	189	260	332
8	-232	-161	-89	-54	-18	17	38	53	67	88	124	160	196	267	338
10	-219	-147	-76	-40	-5	30	52	66	80	102	137	173	209	280	352
12	-206	-134	-63	-27	8	44	65	79	94	115	151	186	222	294	365
14	-192	-121	-49	-14	21	57	78	93	107	128	164	200	235	307	378
16	-179	-108	-36	0	34	70	92	106	120	142	177	213	249	320	392
18	-166	-94	-23	12	48	83	105	119	133	155	191	226	262	333	405
20	-152	-81	-10	25	61	97	118	132	147	168	204	240	275	347	418
25	-119	-48	23	58	94	130	151	166	180	201	237	273	308	380	451
30	-86	-15	56	92	127	163	185	199	213	235	270	306	342	413	485
40	-20	51	122	158	194	230	251	265	280	301	337	372	408	480	551

Table 5-30

Separation, inch: 3.5

Leg Distance, inch: 60

SHIM REQUIRED FOR ALIGNMENT

Reverse Indicator Method

DriveN TIR MILS	DriveR TIR, MILS -40	-30	-20	-15	-10	-5	-2	0	2	5	10	15	20	30	40
-40	-665	-580	-494	-451	-408	-365	-340	-322	-305	-280	-237	-194	-151	-65	20
-30	-585	-499	-413	-370	-327	-285	-259	-242	-225	-199	-156	-113	-70	15	100
-25	-544	-458	-373	-330	-287	-244	-218	-201	-184	-158	-116	-73	-30	55	141
-20	-504	-418	-332	-290	-247	-204	-178	-161	-144	-118	-75	-32	10	95	181
-18	-488	-402	-316	-273	-231	-188	-162	-145	-128	-102	-59	-16	26	111	197
-16	-472	-386	-300	-257	-214	-172	-146	-129	-112	-86	-43	0	42	128	213
-14	-455	-370	-284	-241	-198	-155	-130	-113	-95	-70	-27	15	58	144	229
-12	-439	-354	-268	-225	-182	-139	-114	-96	-79	-54	-11	31	74	160	246
-10	-423	-337	-252	-209	-166	-123	-97	-80	-63	-37	5	47	90	176	262
-8	-407	-321	-236	-193	-150	-107	-81	-64	-47	-21	21	64	106	192	278
-7	-399	-313	-227	-185	-142	-99	-73	-56	-39	-13	29	72	114	200	286
-6	-391	-305	-219	-177	-134	-91	-65	-48	-31	-5	37	80	123	208	294
-5	-383	-297	-211	-168	-126	-83	-57	-40	-23	2	45	88	131	216	302
-4	-375	-289	-203	-160	-118	-75	-49	-32	-15	10	53	96	139	224	310
-3	-367	-281	-195	-152	-109	-67	-41	-24	-7	18	61	104	147	232	318
-2	-359	-273	-187	-144	-101	-59	-33	-16	1	26	69	112	155	241	326
-1	-350	-265	-179	-136	-93	-50	-25	-8	9	34	77	120	163	249	334
0	-342	-257	-171	-128	-85	-42	-17	0	17	42	85	128	171	257	342
1	-334	-249	-163	-120	-77	-34	-9	8	25	50	93	136	179	265	350
2	-326	-241	-155	-112	-69	-26	-1	16	33	59	101	144	187	273	359
3	-318	-232	-147	-104	-61	-18	7	24	41	67	109	152	195	281	367
4	-310	-224	-139	-96	-53	-10	15	32	49	75	118	160	203	289	375
5	-302	-216	-131	-88	-45	-2	23	40	57	83	126	168	211	297	383
6	-294	-208	-123	-80	-37	5	31	48	65	91	134	177	219	305	391
7	-286	-200	-114	-72	-29	13	39	56	73	99	142	185	227	313	399
8	-278	-192	-106	-64	-21	21	47	64	81	107	150	193	236	321	407
10	-262	-176	-90	-47	-5	37	63	80	97	123	166	209	252	337	423
12	-246	-160	-74	-31	11	54	79	96	114	139	182	225	268	354	439
14	-229	-144	-58	-15	27	70	95	113	130	155	198	241	284	370	455
16	-213	-128	-42	0	43	86	112	129	146	172	214	257	300	386	472
18	-197	-111	-26	16	59	102	128	145	162	188	231	273	316	402	488
20	-181	-95	-10	32	75	118	144	161	178	204	247	290	332	418	504
25	-141	-55	30	73	116	158	184	201	218	244	287	330	373	458	544
30	-100	-15	70	113	156	199	225	242	259	285	327	370	413	499	585
40	-20	65	151	194	237	280	305	322	340	365	408	451	494	580	665

Table 5-31

Separation, inch: 4

Leg Distance, inch: 10

SHIM REQUIRED FOR ALIGNMENT

Reverse Indicator Method

DriveN TIR MILS	DriveR TIR, MILS -40	-30	-20	-15	-10	-5	-2	0	2	5	10	15	20	30	40
-40	-80	-67	-55	-48	-42	-36	-32	-30	-27	-23	-17	-11	-5	7	20
-30	-72	-60	-47	-41	-35	-28	-25	-22	-20	-16	-10	-3	2	15	27
-25	-68	-56	-43	-37	-31	-25	-21	-18	-16	-12	-6	0	6	18	31
-20	-65	-52	-40	-33	-27	-21	-17	-15	-12	-8	-2	3	10	22	35
-18	-63	-51	-38	-32	-26	-19	-16	-13	-11	-7	-1	5	11	24	36
-16	-62	-49	-37	-30	-24	-18	-14	-12	-9	-5	0	6	13	25	38
-14	-60	-48	-35	-29	-23	-16	-13	-10	-8	-4	2	8	14	27	39
-12	-59	-46	-34	-27	-21	-15	-11	-9	-6	-2	3	9	16	28	41
-10	-57	-45	-32	-26	-20	-13	-10	-7	-5	-1	5	11	17	30	42
-8	-56	-43	-31	-24	-18	-12	-8	-6	-3	0	6	12	19	31	44
-7	-55	-42	-30	-24	-17	-11	-7	-5	-2	1	7	13	19	32	44
-6	-54	-42	-29	-23	-17	-10	-7	-4	-2	1	8	14	20	33	45
-5	-53	-41	-28	-22	-16	-10	-6	-3	-1	2	8	15	21	33	46
-4	-53	-40	-28	-21	-15	-9	-5	-3	0	3	9	15	22	34	47
-3	-52	-39	-27	-21	-14	-8	-4	-2	0	4	10	16	22	35	47
-2	-51	-39	-26	-20	-14	-7	-4	-1	1	4	11	17	23	36	48
-1	-50	-38	-25	-19	-13	-7	-3	0	1	5	11	18	24	36	49
0	-50	-37	-25	-18	-12	-6	-2	0	2	6	12	18	25	37	50
1	-49	-36	-24	-18	-11	-5	-1	0	3	7	13	19	25	38	50
2	-48	-36	-23	-17	-11	-4	-1	1	4	7	14	20	26	39	51
3	-47	-35	-22	-16	-10	-4	0	2	4	8	14	21	27	39	52
4	-47	-34	-22	-15	-9	-3	0	3	5	9	15	21	28	40	53
5	-46	-33	-21	-15	-8	-2	1	3	6	10	16	22	28	41	53
6	-45	-33	-20	-14	-8	-1	2	4	7	10	17	23	29	42	54
7	-44	-32	-19	-13	-7	-1	2	5	7	11	17	24	30	42	55
8	-44	-31	-19	-12	-6	0	3	6	8	12	18	24	31	43	56
10	-42	-30	-17	-11	-5	1	5	7	10	13	20	26	32	45	57
12	-41	-28	-16	-9	-3	2	6	9	11	15	21	27	34	46	59
14	-39	-27	-14	-8	-2	4	8	10	13	16	23	29	35	48	60
16	-38	-25	-13	-6	0	5	9	12	14	18	24	30	37	49	62
18	-36	-24	-11	-5	1	7	11	13	16	19	26	32	38	51	63
20	-35	-22	-10	-3	2	8	12	15	17	21	27	33	40	52	65
25	-31	-18	-6	0	6	12	16	18	21	25	31	37	43	56	68
30	-27	-15	-2	3	10	16	20	22	25	28	35	41	47	60	72
40	-20	-7	5	11	17	23	27	30	32	36	42	48	55	67	80

Table 5-32

Separation, inch: 4

Leg Distance, inch: 15

SHIM REQUIRED FOR ALIGNMENT

Reverse Indicator Method

DriveN TIR MILS	DriveR TIR, MILS -40	-30	-20	-15	-10	-5	-2	0	2	5	10	15	20	30	40
-40	-130	-111	-92	-83	-73	-64	-58	-55	-51	-45	-36	-26	-17	1	20
-30	-116	-97	-78	-69	-60	-50	-45	-41	-37	-31	-22	-13	-3	15	33
-25	-109	-90	-71	-62	-53	-43	-38	-34	-30	-25	-15	-6	3	21	40
-20	-102	-83	-65	-55	-46	-36	-31	-27	-23	-18	-8	0	10	28	47
-18	-99	-81	-62	-52	-43	-34	-28	-24	-21	-15	-6	3	12	31	50
-16	-97	-78	-59	-50	-40	-31	-25	-22	-18	-12	-3	6	15	34	53
-14	-94	-75	-56	-47	-38	-28	-23	-19	-15	-9	0	8	18	37	55
-12	-91	-72	-54	-44	-35	-25	-20	-16	-12	-7	2	11	21	39	58
-10	-88	-70	-51	-41	-32	-23	-17	-13	-10	-4	5	14	23	42	61
-8	-86	-67	-48	-39	-29	-20	-14	-11	-7	-1	7	17	26	45	64
-7	-84	-65	-47	-37	-28	-19	-13	-9	-5	0	9	18	27	46	65
-6	-83	-64	-45	-36	-27	-17	-12	-8	-4	1	10	19	29	48	66
-5	-81	-63	-44	-35	-25	-16	-10	-6	-3	2	11	21	30	49	68
-4	-80	-61	-43	-33	-24	-14	-9	-5	-1	3	13	22	32	50	69
-3	-79	-60	-41	-32	-22	-13	-7	-4	0	5	14	24	33	52	70
-2	-77	-59	-40	-30	-21	-12	-6	-2	1	6	16	25	34	53	72
-1	-76	-57	-38	-29	-20	-10	-5	-1	2	8	17	26	36	54	73
0	-75	-56	-37	-28	-18	-9	-3	0	3	9	18	28	37	56	75
1	-73	-54	-36	-26	-17	-8	-2	1	5	10	20	29	38	57	76
2	-72	-53	-34	-25	-16	-6	-1	2	6	12	21	30	40	59	77
3	-70	-52	-33	-24	-14	-5	0	4	7	13	22	32	41	60	79
4	-69	-50	-32	-22	-13	-3	1	5	9	14	24	33	43	61	80
5	-68	-49	-30	-21	-11	-2	3	6	10	16	25	35	44	63	81
6	-66	-48	-29	-19	-10	-1	4	8	12	17	27	36	45	64	83
7	-65	-46	-27	-18	-9	0	5	9	13	19	28	37	47	65	84
8	-64	-45	-26	-17	-7	1	7	11	14	20	29	39	48	67	86
10	-61	-42	-23	-14	-5	4	10	13	17	23	32	41	51	70	88
12	-58	-39	-21	-11	-2	7	12	16	20	25	35	44	54	72	91
14	-55	-37	-18	-8	0	9	15	19	23	28	38	47	56	75	94
16	-53	-34	-15	-6	3	12	18	22	25	31	40	50	59	78	97
18	-50	-31	-12	-3	6	15	21	24	28	34	43	52	62	81	99
20	-47	-28	-10	0	8	18	23	27	31	36	46	55	65	83	102
25	-40	-21	-3	6	15	25	30	34	38	43	53	62	71	90	109
30	-33	-15	3	13	22	31	37	41	45	50	60	69	78	97	116
40	-20	-1	17	26	36	45	51	55	58	64	73	83	92	111	130

Table 5-33

Separation, inch: 4

Leg Distance, inch: 20

SHIM REQUIRED FOR ALIGNMENT

Reverse Indicator Method

DriveN TIR MILS	DriveR TIR, MILS -40	-30	-20	-15	-10	-5	-2	0	2	5	10	15	20	30	40
-40	-180	-155	-130	-117	-105	-92	-85	-80	-75	-67	-55	-42	-30	-5	20
-30	-160	-135	-110	-97	-85	-72	-65	-60	-55	-47	-35	-22	-10	15	40
-25	-150	-125	-100	-87	-75	-62	-55	-50	-45	-37	-25	-12	0	25	50
-20	-140	-115	-90	-77	-65	-52	-45	-40	-35	-27	-15	-2	10	35	60
-18	-136	-111	-86	-73	-61	-48	-41	-36	-31	-23	-11	1	14	39	64
-16	-132	-107	-82	-69	-57	-44	-37	-32	-27	-19	-7	5	18	43	68
-14	-128	-103	-78	-65	-53	-40	-33	-28	-23	-15	-3	9	22	47	72
-12	-124	-99	-74	-61	-49	-36	-29	-24	-19	-11	1	13	26	51	76
-10	-120	-95	-70	-57	-45	-32	-25	-20	-15	-7	5	17	30	55	80
-8	-116	-91	-66	-53	-41	-28	-21	-16	-11	-3	9	21	34	59	84
-7	-114	-89	-64	-51	-39	-26	-19	-14	-9	-1	11	23	36	61	86
-6	-112	-87	-62	-49	-37	-24	-17	-12	-7	0	13	25	38	63	88
-5	-110	-85	-60	-47	-35	-22	-15	-10	-5	2	15	27	40	65	90
-4	-108	-83	-58	-45	-33	-20	-13	-8	-3	4	17	29	42	67	92
-3	-106	-81	-56	-43	-31	-18	-11	-6	-1	6	19	31	44	69	94
-2	-104	-79	-54	-41	-29	-16	-9	-4	1	8	21	33	46	71	96
-1	-102	-77	-52	-39	-27	-14	-7	-2	3	10	23	35	48	73	98
0	-100	-75	-50	-37	-25	-12	-5	0	5	12	25	37	50	75	100
1	-98	-73	-48	-35	-23	-10	-3	2	7	14	27	39	52	77	102
2	-96	-71	-46	-33	-21	-8	-1	4	9	16	29	41	54	79	104
3	-94	-69	-44	-31	-19	-6	1	6	11	18	31	43	56	81	106
4	-92	-67	-42	-29	-17	-4	3	8	13	20	33	45	58	83	108
5	-90	-65	-40	-27	-15	-2	5	10	15	22	35	47	60	85	110
6	-88	-63	-38	-25	-13	0	7	12	17	24	37	49	62	87	112
7	-86	-61	-36	-23	-11	1	9	14	19	26	39	51	64	89	114
8	-84	-59	-34	-21	-9	3	11	16	21	28	41	53	66	91	116
10	-80	-55	-30	-17	-5	7	15	20	25	32	45	57	70	95	120
12	-76	-51	-26	-13	-1	11	19	24	29	36	49	61	74	99	124
14	-72	-47	-22	-9	3	15	23	28	33	40	53	65	78	103	128
16	-68	-43	-18	-5	7	19	27	32	37	44	57	69	82	107	132
18	-64	-39	-14	-1	11	23	31	36	41	48	61	73	86	111	136
20	-60	-35	-10	2	15	27	35	40	45	52	65	77	90	115	140
25	-50	-25	0	12	25	37	45	50	55	62	75	87	100	125	150
30	-40	-15	10	22	35	47	55	60	65	72	85	97	110	135	160
40	-20	5	30	42	55	67	75	80	85	92	105	117	130	155	180

Table 5-34

Separation, inch: 4

Leg Distance, inch: 25

SHIM REQUIRED FOR ALIGNMENT

Reverse Indicator Method

DriveN TIR MILS	DriveR TIR, MILS -40	-30	-20	-15	-10	-5	-2	0	2	5	10	15	20	30	40
-40	-230	-198	-167	-151	-136	-120	-111	-105	-98	-89	-73	-58	-42	-11	20
-30	-203	-172	-141	-125	-110	-94	-85	-78	-72	-63	-47	-31	-16	15	46
-25	-190	-159	-128	-112	-96	-81	-71	-65	-59	-50	-34	-18	-3	28	59
-20	-177	-146	-115	-99	-83	-68	-58	-52	-46	-36	-21	-5	10	41	72
-18	-172	-141	-109	-94	-78	-62	-53	-47	-41	-31	-16	0	15	46	77
-16	-167	-135	-104	-88	-73	-57	-48	-42	-35	-26	-10	4	20	51	83
-14	-161	-130	-99	-83	-68	-52	-43	-36	-30	-21	-5	10	25	57	88
-12	-156	-125	-94	-78	-62	-47	-37	-31	-25	-15	0	15	31	62	93
-10	-151	-120	-88	-73	-57	-41	-32	-26	-20	-10	5	20	36	67	98
-8	-146	-114	-83	-67	-52	-36	-27	-21	-14	-5	10	25	41	72	104
-7	-143	-112	-80	-65	-49	-34	-24	-18	-12	-2	12	28	44	75	106
-6	-140	-109	-78	-62	-47	-31	-22	-15	-9	0	15	31	46	78	109
-5	-138	-106	-75	-60	-44	-28	-19	-13	-6	2	18	33	49	80	111
-4	-135	-104	-73	-57	-41	-26	-16	-10	-4	5	20	36	52	83	114
-3	-132	-101	-70	-54	-39	-23	-14	-7	-1	7	23	39	54	85	117
-2	-130	-99	-67	-52	-36	-20	-11	-5	1	10	26	41	57	88	119
-1	-127	-96	-65	-49	-33	-18	-8	-2	3	13	28	44	59	91	122
0	-125	-93	-62	-46	-31	-15	-6	0	6	15	31	46	62	93	125
1	-122	-91	-59	-44	-28	-13	-3	2	8	18	33	49	65	96	127
2	-119	-88	-57	-41	-26	-10	-1	5	11	20	36	52	67	99	130
3	-117	-85	-54	-39	-23	-7	1	7	14	23	39	54	70	101	132
4	-114	-83	-52	-36	-20	-5	4	10	16	26	41	57	73	104	135
5	-111	-80	-49	-33	-18	-2	6	13	19	28	44	60	75	106	138
6	-109	-78	-46	-31	-15	0	9	15	22	31	47	62	78	109	140
7	-106	-75	-44	-28	-12	2	12	18	24	34	49	65	80	112	143
8	-104	-72	-41	-25	-10	5	14	21	27	36	52	67	83	114	146
10	-98	-67	-36	-20	-5	10	20	26	32	41	57	73	88	120	151
12	-93	-62	-31	-15	0	15	25	31	37	47	62	78	94	125	156
14	-88	-57	-25	-10	5	21	30	36	43	52	68	83	99	130	161
16	-83	-51	-20	-4	10	26	35	42	48	57	73	88	104	135	167
18	-77	-46	-15	0	16	31	41	47	53	62	78	94	109	141	172
20	-72	-41	-10	5	21	36	46	52	58	68	83	99	115	146	177
25	-59	-28	3	18	34	50	59	65	71	81	96	112	128	159	190
30	-46	-15	16	31	47	63	72	78	85	94	110	125	141	172	203
40	-20	11	42	58	73	89	98	105	111	120	136	151	167	198	230

Table 5-35

Separation, inch: 4

Leg Distance, inch: 30

SHIM REQUIRED FOR ALIGNMENT

Reverse Indicator Method

DriveN TIR MILS	DriveR TIR, MILS -40	-30	-20	-15	-10	-5	-2	0	2	5	10	15	20	30	40
-40	-280	-242	-205	-186	-167	-148	-137	-130	-122	-111	-92	-73	-55	-17	20
-30	-247	-210	-172	-153	-135	-116	-105	-97	-90	-78	-60	-41	-22	15	52
-25	-231	-193	-156	-137	-118	-100	-88	-81	-73	-62	-43	-25	-6	31	68
-20	-215	-177	-140	-121	-102	-83	-72	-65	-57	-46	-27	-8	10	47	85
-18	-208	-171	-133	-114	-96	-77	-66	-58	-51	-39	-21	-2	16	54	91
-16	-202	-164	-127	-108	-89	-70	-59	-52	-44	-33	-14	4	23	60	98
-14	-195	-158	-120	-101	-83	-64	-53	-45	-38	-26	-8	10	29	67	104
-12	-189	-151	-114	-95	-76	-57	-46	-39	-31	-20	-1	17	36	73	111
-10	-182	-145	-107	-88	-70	-51	-40	-32	-25	-13	5	23	42	80	117
-8	-176	-138	-101	-82	-63	-44	-33	-26	-18	-7	11	30	49	86	124
-7	-172	-135	-97	-79	-60	-41	-30	-22	-15	-4	14	33	52	89	127
-6	-169	-132	-94	-75	-57	-38	-27	-19	-12	0	18	36	55	93	130
-5	-166	-128	-91	-72	-53	-35	-23	-16	-8	2	21	40	58	96	133
-4	-163	-125	-88	-69	-50	-31	-20	-13	-5	5	24	43	62	99	137
-3	-159	-122	-84	-66	-47	-28	-17	-9	-2	9	27	46	65	102	140
-2	-156	-119	-81	-62	-44	-25	-14	-6	1	12	31	49	68	106	143
-1	-153	-115	-78	-59	-40	-22	-10	-3	4	15	34	53	71	109	146
0	-150	-112	-75	-56	-37	-18	-7	0	7	18	37	56	75	112	150
1	-146	-109	-71	-53	-34	-15	-4	3	10	22	40	59	78	115	153
2	-143	-106	-68	-49	-31	-12	-1	6	14	25	44	62	81	119	156
3	-140	-102	-65	-46	-27	-9	2	9	17	28	47	66	84	122	159
4	-137	-99	-62	-43	-24	-5	5	13	20	31	50	69	88	125	163
5	-133	-96	-58	-40	-21	-2	8	16	23	35	53	72	91	128	166
6	-130	-93	-55	-36	-18	0	12	19	27	38	57	75	94	132	169
7	-127	-89	-52	-33	-14	4	15	22	30	41	60	79	97	135	172
8	-124	-86	-49	-30	-11	7	18	26	33	44	63	82	101	138	176
10	-117	-80	-42	-23	-5	13	25	32	40	51	70	88	107	145	182
12	-111	-73	-36	-17	1	20	31	39	46	57	76	95	114	151	189
14	-104	-67	-29	-10	8	26	38	45	53	64	83	101	120	158	195
16	-98	-60	-23	-4	14	33	44	52	59	70	89	108	127	164	202
18	-91	-54	-16	2	21	39	51	58	66	77	96	114	133	171	208
20	-85	-47	-10	8	27	46	57	65	72	83	102	121	140	177	215
25	-68	-31	6	25	43	62	73	81	88	100	118	137	156	193	231
30	-52	-15	22	41	60	78	90	97	105	116	135	153	172	210	247
40	-20	17	55	73	92	111	122	130	137	148	167	186	205	242	280

Table 5-36

Separation, inch: 4

Leg Distance, inch: 35

SHIM REQUIRED FOR ALIGNMENT

Reverse Indicator Method

DriveN TIR MILS	DriveR TIR, MILS -40	-30	-20	-15	-10	-5	-2	0	2	5	10	15	20	30	40
-40	-330	-286	-242	-220	-198	-176	-163	-155	-146	-133	-111	-89	-67	-23	20
-30	-291	-247	-203	-181	-160	-138	-125	-116	-107	-94	-72	-50	-28	15	58
-25	-271	-228	-184	-162	-140	-118	-105	-96	-88	-75	-53	-31	-9	34	78
-20	-252	-208	-165	-143	-121	-99	-86	-77	-68	-55	-33	-11	10	53	97
-18	-244	-201	-157	-135	-113	-91	-78	-69	-61	-47	-26	-4	17	61	105
-16	-237	-193	-149	-127	-105	-83	-70	-62	-53	-40	-18	3	25	69	113
-14	-229	-185	-141	-119	-98	-76	-63	-54	-45	-32	-10	11	33	77	120
-12	-221	-177	-134	-112	-90	-68	-55	-46	-37	-24	-2	19	41	84	128
-10	-213	-170	-126	-104	-82	-60	-47	-38	-30	-16	5	26	48	92	136
-8	-206	-162	-118	-96	-74	-52	-39	-31	-22	-9	12	34	56	100	144
-7	-202	-158	-114	-92	-70	-49	-35	-27	-18	-5	16	38	60	104	147
-6	-198	-154	-110	-88	-67	-45	-32	-23	-14	-1	20	42	64	108	151
-5	-194	-150	-106	-85	-63	-41	-28	-19	-10	2	24	46	68	111	155
-4	-190	-146	-103	-81	-59	-37	-24	-15	-6	6	28	50	72	115	159
-3	-186	-142	-99	-77	-55	-33	-20	-11	-2	10	32	54	75	119	163
-2	-182	-139	-95	-73	-51	-29	-16	-7	1	14	36	57	79	123	167
-1	-178	-135	-91	-69	-47	-25	-12	-3	4	18	39	61	83	127	171
0	-175	-131	-87	-65	-43	-21	-8	0	8	21	43	65	87	131	175
1	-171	-127	-83	-61	-39	-18	-4	3	12	25	47	69	91	135	178
2	-167	-123	-79	-57	-36	-14	-1	7	16	29	51	73	95	139	182
3	-163	-119	-75	-54	-32	-10	2	11	20	33	55	77	99	142	186
4	-159	-115	-72	-50	-28	-6	6	15	24	37	59	81	103	146	190
5	-155	-111	-68	-46	-24	-2	10	19	28	41	63	85	106	150	194
6	-151	-108	-64	-42	-20	1	14	23	32	45	67	88	110	154	198
7	-147	-104	-60	-38	-16	5	18	27	35	49	70	92	114	158	202
8	-144	-100	-56	-34	-12	9	22	31	39	52	74	96	118	162	206
10	-136	-92	-48	-26	-5	16	30	38	47	60	82	104	126	170	213
12	-128	-84	-41	-19	2	24	37	46	55	68	90	112	134	177	221
14	-120	-77	-33	-11	10	32	45	54	63	76	98	119	141	185	229
16	-113	-69	-25	-3	18	40	53	62	70	83	105	127	149	193	237
18	-105	-61	-17	4	26	47	61	69	78	91	113	135	157	201	244
20	-97	-53	-10	11	33	55	68	77	86	99	121	143	165	208	252
25	-78	-34	9	31	53	75	88	96	105	118	140	162	184	228	271
30	-58	-15	28	50	72	94	107	116	125	138	160	181	203	247	291
40	-20	23	67	89	111	133	146	155	163	176	198	220	242	286	330

Table 5-37

Separation, inch: 4

Leg Distance, inch: 40

SHIM REQUIRED FOR ALIGNMENT

Reverse Indicator Method

riveN IR ILS	DriveR TIR, MILS -40	-30	-20	-15	-10	-5	-2	0	2	5	10	15	20	30	40
-40	-380	-330	-280	-255	-230	-205	-190	-180	-170	-155	-130	-105	-80	-30	20
-30	-335	-285	-235	-210	-185	-160	-145	-135	-125	-110	-85	-60	-35	15	65
-25	-312	-262	-212	-187	-162	-137	-122	-112	-102	-87	-62	-37	-12	37	87
-20	-290	-240	-190	-165	-140	-115	-100	-90	-80	-65	-40	-15	10	60	110
-18	-281	-231	-181	-156	-131	-106	-91	-81	-71	-56	-31	-6	19	69	119
-16	-272	-222	-172	-147	-122	-97	-82	-72	-62	-47	-22	3	28	78	128
-14	-263	-213	-163	-138	-113	-88	-73	-63	-53	-38	-13	12	37	87	137
-12	-254	-204	-154	-129	-104	-79	-64	-54	-44	-29	-4	21	46	96	146
-10	-245	-195	-145	-120	-95	-70	-55	-45	-35	-20	5	30	55	105	155
-8	-236	-186	-136	-111	-86	-61	-46	-36	-26	-11	14	39	64	114	164
-7	-231	-181	-131	-106	-81	-56	-41	-31	-21	-6	18	43	68	118	168
-6	-227	-177	-127	-102	-77	-52	-37	-27	-17	-2	23	48	73	123	173
-5	-222	-172	-122	-97	-72	-47	-32	-22	-12	2	27	52	77	127	177
-4	-218	-168	-118	-93	-68	-43	-28	-18	-8	7	32	57	82	132	182
-3	-213	-163	-113	-88	-63	-38	-23	-13	-3	11	36	61	86	136	186
-2	-209	-159	-109	-84	-59	-34	-19	-9	1	16	41	66	91	141	191
-1	-204	-154	-104	-79	-54	-29	-14	-4	5	20	45	70	95	145	195
0	-200	-150	-100	-75	-50	-25	-10	0	10	25	50	75	100	150	200
1	-195	-145	-95	-70	-45	-20	-5	4	14	29	54	79	104	154	204
2	-191	-141	-91	-66	-41	-16	-1	9	19	34	59	84	109	159	209
3	-186	-136	-86	-61	-36	-11	3	13	23	38	63	88	113	163	213
4	-182	-132	-82	-57	-32	-7	8	18	28	43	68	93	118	168	218
5	-177	-127	-77	-52	-27	-2	12	22	32	47	72	97	122	172	222
6	-173	-123	-73	-48	-23	2	17	27	37	52	77	102	127	177	227
7	-168	-118	-68	-43	-18	6	21	31	41	56	81	106	131	181	231
8	-164	-114	-64	-39	-14	11	26	36	46	61	86	111	136	186	236
10	-155	-105	-55	-30	-5	20	35	45	55	70	95	120	145	195	245
12	-146	-96	-46	-21	4	29	44	54	64	79	104	129	154	204	254
14	-137	-87	-37	-12	13	38	53	63	73	88	113	138	163	213	263
16	-128	-78	-28	-3	22	47	62	72	82	97	122	147	172	222	272
18	-119	-69	-19	6	31	56	71	81	91	106	131	156	181	231	281
20	-110	-60	-10	15	40	65	80	90	100	115	140	165	190	240	290
25	-87	-37	12	37	62	87	102	112	122	137	162	187	212	262	312
30	-65	-15	35	60	85	110	125	135	145	160	185	210	235	285	335
40	-20	30	80	105	130	155	170	180	190	205	230	255	280	330	380

Table 5-38

Separation, inch: 4

Leg Distance, inch: 50

SHIM REQUIRED FOR ALIGNMENT

Reverse Indicator Method

DriveN TIR MILS	DriveR TIR, MILS -40	-30	-20	-15	-10	-5	-2	0	2	5	10	15	20	30	40
-40	-480	-417	-355	-323	-292	-261	-242	-230	-217	-198	-167	-136	-105	-42	20
-30	-422	-360	-297	-266	-235	-203	-185	-172	-160	-141	-110	-78	-47	15	77
-25	-393	-331	-268	-237	-206	-175	-156	-143	-131	-112	-81	-50	-18	43	106
-20	-365	-302	-240	-208	-177	-146	-127	-115	-102	-83	-52	-21	10	72	135
-18	-353	-291	-228	-197	-166	-134	-116	-103	-91	-72	-41	-9	21	84	146
-16	-342	-279	-217	-185	-154	-123	-104	-92	-79	-60	-29	1	33	95	158
-14	-330	-268	-205	-174	-143	-111	-93	-80	-68	-49	-18	13	44	107	169
-12	-319	-256	-194	-162	-131	-100	-81	-69	-56	-37	-6	24	56	118	181
-10	-307	-245	-182	-151	-120	-88	-70	-57	-45	-26	5	36	67	130	192
-8	-296	-233	-171	-139	-108	-77	-58	-46	-33	-14	16	47	79	141	204
-7	-290	-227	-165	-134	-102	-71	-52	-40	-27	-9	22	53	84	147	209
-6	-284	-222	-159	-128	-97	-65	-47	-34	-22	-3	28	59	90	153	215
-5	-278	-216	-153	-122	-91	-60	-41	-28	-16	2	33	65	96	158	221
-4	-273	-210	-148	-116	-85	-54	-35	-23	-10	8	39	70	102	164	227
-3	-267	-204	-142	-111	-79	-48	-29	-17	-4	14	45	76	107	170	232
-2	-261	-199	-136	-105	-74	-42	-24	-11	1	19	51	82	113	176	238
-1	-255	-193	-130	-99	-68	-37	-18	-5	6	25	56	88	119	181	244
0	-250	-187	-125	-93	-62	-31	-12	0	12	31	62	93	125	187	250
1	-244	-181	-119	-88	-56	-25	-6	5	18	37	68	99	130	193	255
2	-238	-176	-113	-82	-51	-19	-1	11	24	42	74	105	136	199	261
3	-232	-170	-107	-76	-45	-14	4	17	29	48	79	111	142	204	267
4	-227	-164	-102	-70	-39	-8	10	23	35	54	85	116	148	210	273
5	-221	-158	-96	-65	-33	-2	16	28	41	60	91	122	153	216	278
6	-215	-153	-90	-59	-28	3	22	34	47	65	97	128	159	222	284
7	-209	-147	-84	-53	-22	9	27	40	52	71	102	134	165	227	290
8	-204	-141	-79	-47	-16	14	33	46	58	77	108	139	171	233	296
10	-192	-130	-67	-36	-5	26	45	57	70	88	120	151	182	245	307
12	-181	-118	-56	-24	6	37	56	69	81	100	131	162	194	256	319
14	-169	-107	-44	-13	18	49	68	80	93	111	143	174	205	268	330
16	-158	-95	-33	-1	29	60	79	92	104	123	154	185	217	279	342
18	-146	-84	-21	9	41	72	91	103	116	134	166	197	228	291	353
20	-135	-72	-10	21	52	83	102	115	127	146	177	208	240	302	365
25	-106	-43	18	50	81	112	131	143	156	175	206	237	268	331	393
30	-77	-15	47	78	110	141	160	172	185	203	235	266	297	360	422
40	-20	42	105	136	167	198	217	230	242	261	292	323	355	417	480

Table 5-39

Separation, inch: 4

Leg Distance, inch: 60

SHIM REQUIRED FOR ALIGNMENT

Reverse Indicator Method

riveN IR ILS	DriveR TIR, MILS -40	-30	-20	-15	-10	-5	-2	0	2	5	10	15	20	30	40
-40	-580	-505	-430	-392	-355	-317	-295	-280	-265	-242	-205	-167	-130	-55	20
-30	-510	-435	-360	-322	-285	-247	-225	-210	-195	-172	-135	-97	-60	15	90
-25	-475	-400	-325	-287	-250	-212	-190	-175	-160	-137	-100	-62	-25	50	125
-20	-440	-365	-290	-252	-215	-177	-155	-140	-125	-102	-65	-27	10	85	160
-18	-426	-351	-276	-238	-201	-163	-141	-126	-111	-88	-51	-13	24	99	174
-16	-412	-337	-262	-224	-187	-149	-127	-112	-97	-74	-37	0	38	113	188
-14	-398	-323	-248	-210	-173	-135	-113	-98	-83	-60	-23	14	52	127	202
-12	-384	-309	-234	-196	-159	-121	-99	-84	-69	-46	-9	28	66	141	216
-10	-370	-295	-220	-182	-145	-107	-85	-70	-55	-32	5	42	80	155	230
-8	-356	-281	-206	-168	-131	-93	-71	-56	-41	-18	19	56	94	169	244
-7	-349	-274	-199	-161	-124	-86	-64	-49	-34	-11	26	63	101	176	251
-6	-342	-267	-192	-154	-117	-79	-57	-42	-27	-4	33	70	108	183	258
-5	-335	-260	-185	-147	-110	-72	-50	-35	-20	2	40	77	115	190	265
-4	-328	-253	-178	-140	-103	-65	-43	-28	-13	9	47	84	122	197	272
-3	-321	-246	-171	-133	-96	-58	-36	-21	-6	16	54	91	129	204	279
-2	-314	-239	-164	-126	-89	-51	-29	-14	1	23	61	98	136	211	286
-1	-307	-232	-157	-119	-82	-44	-22	-7	8	30	68	105	143	218	293
0	-300	-225	-150	-112	-75	-37	-15	0	15	37	75	112	150	225	300
1	-293	-218	-143	-105	-68	-30	-8	7	22	44	82	119	157	232	307
2	-286	-211	-136	-98	-61	-23	-1	14	29	51	89	126	164	239	314
3	-279	-204	-129	-91	-54	-16	6	21	36	58	96	133	171	246	321
4	-272	-197	-122	-84	-47	-9	13	28	43	65	103	140	178	253	328
5	-265	-190	-115	-77	-40	-2	20	35	50	72	110	147	185	260	335
6	-258	-183	-108	-70	-33	4	27	42	57	79	117	154	192	267	342
7	-251	-176	-101	-63	-26	11	34	49	64	86	124	161	199	274	349
8	-244	-169	-94	-56	-19	18	41	56	71	93	131	168	206	281	356
10	-230	-155	-80	-42	-5	32	55	70	85	107	145	182	220	295	370
12	-216	-141	-66	-28	9	46	69	84	99	121	159	196	234	309	384
14	-202	-127	-52	-14	23	60	83	98	113	135	173	210	248	323	398
16	-188	-113	-38	0	37	74	97	112	127	149	187	224	262	337	412
18	-174	-99	-24	13	51	88	111	126	141	163	201	238	276	351	426
20	-160	-85	-10	27	65	102	125	140	155	177	215	252	290	365	440
25	-125	-50	25	62	100	137	160	175	190	212	250	287	325	400	475
30	-90	-15	60	97	135	172	195	210	225	247	285	322	360	435	510
40	-20	55	130	167	205	242	265	280	295	317	355	392	430	505	580

Table 5-40

Separation, inch: 5

Leg Distance, inch: 15

SHIM REQUIRED FOR ALIGNMENT

Reverse Indicator Method

DriveN TIR MILS	DriveR TIR, MILS -40	-30	-20	-15	-10	-5	-2	0	2	5	10	15	20	30	40
-40	-100	-85	-70	-62	-55	-47	-43	-40	-37	-32	-25	-17	-10	5	20
-30	-90	-75	-60	-52	-45	-37	-33	-30	-27	-22	-15	-7	0	15	30
-25	-85	-70	-55	-47	-40	-32	-28	-25	-22	-17	-10	-2	5	20	35
-20	-80	-65	-50	-42	-35	-27	-23	-20	-17	-12	-5	2	10	25	40
-18	-78	-63	-48	-40	-33	-25	-21	-18	-15	-10	-3	4	12	27	42
-16	-76	-61	-46	-38	-31	-23	-19	-16	-13	-8	-1	6	14	29	44
-14	-74	-59	-44	-36	-29	-21	-17	-14	-11	-6	1	8	16	31	46
-12	-72	-57	-42	-34	-27	-19	-15	-12	-9	-4	3	10	18	33	48
-10	-70	-55	-40	-32	-25	-17	-13	-10	-7	-2	5	12	20	35	50
-8	-68	-53	-38	-30	-23	-15	-11	-8	-5	0	7	14	22	37	52
-7	-67	-52	-37	-29	-22	-14	-10	-7	-4	0	8	15	23	38	53
-6	-66	-51	-36	-28	-21	-13	-9	-6	-3	1	9	16	24	39	54
-5	-65	-50	-35	-27	-20	-12	-8	-5	-2	2	10	17	25	40	55
-4	-64	-49	-34	-26	-19	-11	-7	-4	-1	3	11	18	26	41	56
-3	-63	-48	-33	-25	-18	-10	-6	-3	0	4	12	19	27	42	57
-2	-62	-47	-32	-24	-17	-9	-5	-2	1	5	13	20	28	43	58
-1	-61	-46	-31	-23	-16	-8	-4	-1	2	6	14	21	29	44	59
0	-60	-45	-30	-22	-15	-7	-3	0	3	7	15	22	30	45	60
1	-59	-44	-29	-21	-14	-6	-2	1	4	8	16	23	31	46	61
2	-58	-43	-28	-20	-13	-5	-1	2	5	9	17	24	32	47	62
3	-57	-42	-27	-19	-12	-4	0	3	6	10	18	25	33	48	63
4	-56	-41	-26	-18	-11	-3	1	4	7	11	19	26	34	49	64
5	-55	-40	-25	-17	-10	-2	2	5	8	12	20	27	35	50	65
6	-54	-39	-24	-16	-9	-1	3	6	9	13	21	28	36	51	66
7	-53	-38	-23	-15	-8	0	4	7	10	14	22	29	37	52	67
8	-52	-37	-22	-14	-7	0	5	8	11	15	23	30	38	53	68
10	-50	-35	-20	-12	-5	2	7	10	13	17	25	32	40	55	70
12	-48	-33	-18	-10	-3	4	9	12	15	19	27	34	42	57	72
14	-46	-31	-16	-8	-1	6	11	14	17	21	29	36	44	59	74
16	-44	-29	-14	-6	1	8	13	16	19	23	31	38	46	61	76
18	-42	-27	-12	-4	3	10	15	18	21	25	33	40	48	63	78
20	-40	-25	-10	-2	5	12	17	20	23	27	35	42	50	65	80
25	-35	-20	-5	2	10	17	22	25	28	32	40	47	55	70	85
30	-30	-15	0	7	15	22	27	30	33	37	45	52	60	75	90
40	-20	-5	10	17	25	32	37	40	43	47	55	62	70	85	100

Table 5-41

Separation, inch: 5

Leg Distance, inch: 20

SHIM REQUIRED FOR ALIGNMENT

Reverse Indicator Method

riveN IR ILS	DriveR TIR, MILS -40	-30	-20	-15	-10	-5	-2	0	2	5	10	15	20	30	40
-40	-140	-120	-100	-90	-80	-70	-64	-60	-56	-50	-40	-30	-20	0	20
-30	-125	-105	-85	-75	-65	-55	-49	-45	-41	-35	-25	-15	-5	15	35
-25	-117	-97	-77	-67	-57	-47	-41	-37	-33	-27	-17	-7	2	22	42
-20	-110	-90	-70	-60	-50	-40	-34	-30	-26	-20	-10	0	10	30	50
-18	-107	-87	-67	-57	-47	-37	-31	-27	-23	-17	-7	3	13	33	53
-16	-104	-84	-64	-54	-44	-34	-28	-24	-20	-14	-4	6	16	36	56
-14	-101	-81	-61	-51	-41	-31	-25	-21	-17	-11	-1	9	19	39	59
-12	-98	-78	-58	-48	-38	-28	-22	-18	-14	-8	2	12	22	42	62
-10	-95	-75	-55	-45	-35	-25	-19	-15	-11	-5	5	15	25	45	65
-8	-92	-72	-52	-42	-32	-22	-16	-12	-8	-2	8	18	28	48	68
-7	-90	-70	-50	-40	-30	-20	-14	-10	-6	0	9	19	29	49	69
-6	-89	-69	-49	-39	-29	-19	-13	-9	-5	1	11	21	31	51	71
-5	-87	-67	-47	-37	-27	-17	-11	-7	-3	2	12	22	32	52	72
-4	-86	-66	-46	-36	-26	-16	-10	-6	-2	4	14	24	34	54	74
-3	-84	-64	-44	-34	-24	-14	-8	-4	0	5	15	25	35	55	75
-2	-83	-63	-43	-33	-23	-13	-7	-3	1	7	17	27	37	57	77
-1	-81	-61	-41	-31	-21	-11	-5	-1	2	8	18	28	38	58	78
0	-80	-60	-40	-30	-20	-10	-4	0	4	10	20	30	40	60	80
1	-78	-58	-38	-28	-18	-8	-2	1	5	11	21	31	41	61	81
2	-77	-57	-37	-27	-17	-7	-1	3	7	13	23	33	43	63	83
3	-75	-55	-35	-25	-15	-5	0	4	8	14	24	34	44	64	84
4	-74	-54	-34	-24	-14	-4	2	6	10	16	26	36	46	66	86
5	-72	-52	-32	-22	-12	-2	3	7	11	17	27	37	47	67	87
6	-71	-51	-31	-21	-11	-1	5	9	13	19	29	39	49	69	89
7	-69	-49	-29	-19	-9	0	6	10	14	20	30	40	50	70	90
8	-68	-48	-28	-18	-8	2	8	12	16	22	32	42	52	72	92
10	-65	-45	-25	-15	-5	5	11	15	19	25	35	45	55	75	95
12	-62	-42	-22	-12	-2	8	14	18	22	28	38	48	58	78	98
14	-59	-39	-19	-9	1	11	17	21	25	31	41	51	61	81	101
16	-56	-36	-16	-6	4	14	20	24	28	34	44	54	64	84	104
18	-53	-33	-13	-3	7	17	23	27	31	37	47	57	67	87	107
20	-50	-30	-10	0	10	20	26	30	34	40	50	60	70	90	110
25	-42	-22	-2	7	17	27	33	37	41	47	57	67	77	97	117
30	-35	-15	5	15	25	35	41	45	49	55	65	75	85	105	125
40	-20	0	20	30	40	50	56	60	64	70	80	90	100	120	140

Table 5-42

Separation, inch: 5

Leg Distance, inch: 25

SHIM REQUIRED FOR ALIGNMENT

Reverse Indicator Method

DriveN TIR MILS	DriveR TIR, MILS -40	-30	-20	-15	-10	-5	-2	0	2	5	10	15	20	30	40
-40	-180	-155	-130	-117	-105	-92	-85	-80	-75	-67	-55	-42	-30	-5	20
-30	-160	-135	-110	-97	-85	-72	-65	-60	-55	-47	-35	-22	-10	15	40
-25	-150	-125	-100	-87	-75	-62	-55	-50	-45	-37	-25	-12	0	25	50
-20	-140	-115	-90	-77	-65	-52	-45	-40	-35	-27	-15	-2	10	35	60
-18	-136	-111	-86	-73	-61	-48	-41	-36	-31	-23	-11	1	14	39	64
-16	-132	-107	-82	-69	-57	-44	-37	-32	-27	-19	-7	5	18	43	68
-14	-128	-103	-78	-65	-53	-40	-33	-28	-23	-15	-3	9	22	47	72
-12	-124	-99	-74	-61	-49	-36	-29	-24	-19	-11	1	13	26	51	76
-10	-120	-95	-70	-57	-45	-32	-25	-20	-15	-7	5	17	30	55	80
-8	-116	-91	-66	-53	-41	-28	-21	-16	-11	-3	9	21	34	59	84
-7	-114	-89	-64	-51	-39	-26	-19	-14	-9	-1	11	23	36	61	86
-6	-112	-87	-62	-49	-37	-24	-17	-12	-7	0	13	25	38	63	88
-5	-110	-85	-60	-47	-35	-22	-15	-10	-5	2	15	27	40	65	90
-4	-108	-83	-58	-45	-33	-20	-13	-8	-3	4	17	29	42	67	92
-3	-106	-81	-56	-43	-31	-18	-11	-6	-1	6	19	31	44	69	94
-2	-104	-79	-54	-41	-29	-16	-9	-4	1	8	21	33	46	71	96
-1	-102	-77	-52	-39	-27	-14	-7	-2	3	10	23	35	48	73	98
0	-100	-75	-50	-37	-25	-12	-5	0	5	12	25	37	50	75	100
1	-98	-73	-48	-35	-23	-10	-3	2	7	14	27	39	52	77	102
2	-96	-71	-46	-33	-21	-8	-1	4	9	16	29	41	54	79	104
3	-94	-69	-44	-31	-19	-6	1	6	11	18	31	43	56	81	106
4	-92	-67	-42	-29	-17	-4	3	8	13	20	33	45	58	83	108
5	-90	-65	-40	-27	-15	-2	5	10	15	22	35	47	60	85	110
6	-88	-63	-38	-25	-13	0	7	12	17	24	37	49	62	87	112
7	-86	-61	-36	-23	-11	1	9	14	19	26	39	51	64	89	114
8	-84	-59	-34	-21	-9	3	11	16	21	28	41	53	66	91	116
10	-80	-55	-30	-17	-5	7	15	20	25	32	45	57	70	95	120
12	-76	-51	-26	-13	-1	11	19	24	29	36	49	61	74	99	124
14	-72	-47	-22	-9	3	15	23	28	33	40	53	65	78	103	128
16	-68	-43	-18	-5	7	19	27	32	37	44	57	69	82	107	132
18	-64	-39	-14	-1	11	23	31	36	41	48	61	73	86	111	136
20	-60	-35	-10	2	15	27	35	40	45	52	65	77	90	115	140
25	-50	-25	0	12	25	37	45	50	55	62	75	87	100	125	150
30	-40	-15	10	22	35	47	55	60	65	72	85	97	110	135	160
40	-20	5	30	42	55	67	75	80	85	92	105	117	130	155	180

Table 5-43

Separation, inch: 5

Leg Distance, inch: 30

SHIM REQUIRED FOR ALIGNMENT

Reverse Indicator Method

iveN IR ILS	DriveR TIR, MILS -40	-30	-20	-15	-10	-5	-2	0	2	5	10	15	20	30	40
-40	-220	-190	-160	-145	-130	-115	-106	-100	-94	-85	-70	-55	-40	-10	20
-30	-195	-165	-135	-120	-105	-90	-81	-75	-69	-60	-45	-30	-15	15	45
-25	-182	-152	-122	-107	-92	-77	-68	-62	-56	-47	-32	-17	-2	27	57
-20	-170	-140	-110	-95	-80	-65	-56	-50	-44	-35	-20	-5	10	40	70
-18	-165	-135	-105	-90	-75	-60	-51	-45	-39	-30	-15	0	15	45	75
-16	-160	-130	-100	-85	-70	-55	-46	-40	-34	-25	-10	5	20	50	80
-14	-155	-125	-95	-80	-65	-50	-41	-35	-29	-20	-5	10	25	55	85
-12	-150	-120	-90	-75	-60	-45	-36	-30	-24	-15	0	15	30	60	90
-10	-145	-115	-85	-70	-55	-40	-31	-25	-19	-10	5	20	35	65	95
-8	-140	-110	-80	-65	-50	-35	-26	-20	-14	-5	10	25	40	70	100
-7	-137	-107	-77	-62	-47	-32	-23	-17	-11	-2	12	27	42	72	102
-6	-135	-105	-75	-60	-45	-30	-21	-15	-9	0	15	30	45	75	105
-5	-132	-102	-72	-57	-42	-27	-18	-12	-6	2	17	32	47	77	107
-4	-130	-100	-70	-55	-40	-25	-16	-10	-4	5	20	35	50	80	110
-3	-127	-97	-67	-52	-37	-22	-13	-7	-1	7	22	37	52	82	112
-2	-125	-95	-65	-50	-35	-20	-11	-5	1	10	25	40	55	85	115
-1	-122	-92	-62	-47	-32	-17	-8	-2	3	12	27	42	57	87	117
0	-120	-90	-60	-45	-30	-15	-6	0	6	15	30	45	60	90	120
1	-117	-87	-57	-42	-27	-12	-3	2	8	17	32	47	62	92	122
2	-115	-85	-55	-40	-25	-10	-1	5	11	20	35	50	65	95	125
3	-112	-82	-52	-37	-22	-7	1	7	13	22	37	52	67	97	127
4	-110	-80	-50	-35	-20	-5	4	10	16	25	40	55	70	100	130
5	-107	-77	-47	-32	-17	-2	6	12	18	27	42	57	72	102	132
6	-105	-75	-45	-30	-15	0	9	15	21	30	45	60	75	105	135
7	-102	-72	-42	-27	-12	2	11	17	23	32	47	62	77	107	137
8	-100	-70	-40	-25	-10	5	14	20	26	35	50	65	80	110	140
10	-95	-65	-35	-20	-5	10	19	25	31	40	55	70	85	115	145
12	-90	-60	-30	-15	0	15	24	30	36	45	60	75	90	120	150
14	-85	-55	-25	-10	5	20	29	35	41	50	65	80	95	125	155
16	-80	-50	-20	-5	10	25	34	40	46	55	70	85	100	130	160
18	-75	-45	-15	0	15	30	39	45	51	60	75	90	105	135	165
20	-70	-40	-10	5	20	35	44	50	56	65	80	95	110	140	170
25	-57	-27	2	17	32	47	56	62	68	77	92	107	122	152	182
30	-45	-15	15	30	45	60	69	75	81	90	105	120	135	165	195
40	-20	10	40	55	70	85	94	100	106	115	130	145	160	190	220

Table 5-44

Separation, inch: 5

Leg Distance, inch: 35

SHIM REQUIRED FOR ALIGNMENT

Reverse Indicator Method

DriveN TIR MILS	DriveR TIR, MILS -40	-30	-20	-15	-10	-5	-2	0	2	5	10	15	20	30	40
-40	-260	-225	-190	-172	-155	-137	-127	-120	-113	-102	-85	-67	-50	-15	20
-30	-230	-195	-160	-142	-125	-107	-97	-90	-83	-72	-55	-37	-20	15	50
-25	-215	-180	-145	-127	-110	-92	-82	-75	-68	-57	-40	-22	-5	30	65
-20	-200	-165	-130	-112	-95	-77	-67	-60	-53	-42	-25	-7	10	45	80
-18	-194	-159	-124	-106	-89	-71	-61	-54	-47	-36	-19	-1	16	51	86
-16	-188	-153	-118	-100	-83	-65	-55	-48	-41	-30	-13	4	22	57	92
-14	-182	-147	-112	-94	-77	-59	-49	-42	-35	-24	-7	10	28	63	98
-12	-176	-141	-106	-88	-71	-53	-43	-36	-29	-18	-1	16	34	69	104
-10	-170	-135	-100	-82	-65	-47	-37	-30	-23	-12	5	22	40	75	110
-8	-164	-129	-94	-76	-59	-41	-31	-24	-17	-6	11	28	46	81	116
-7	-161	-126	-91	-73	-56	-38	-28	-21	-14	-3	14	31	49	84	119
-6	-158	-123	-88	-70	-53	-35	-25	-18	-11	0	17	34	52	87	122
-5	-155	-120	-85	-67	-50	-32	-22	-15	-8	2	20	37	55	90	125
-4	-152	-117	-82	-64	-47	-29	-19	-12	-5	5	23	40	58	93	128
-3	-149	-114	-79	-61	-44	-26	-16	-9	-2	8	26	43	61	96	131
-2	-146	-111	-76	-58	-41	-23	-13	-6	1	11	29	46	64	99	134
-1	-143	-108	-73	-55	-38	-20	-10	-3	4	14	32	49	67	102	137
0	-140	-105	-70	-52	-35	-17	-7	0	7	17	35	52	70	105	140
1	-137	-102	-67	-49	-32	-14	-4	3	10	20	38	55	73	108	143
2	-134	-99	-64	-46	-29	-11	-1	6	13	23	41	58	76	111	146
3	-131	-96	-61	-43	-26	-8	2	9	16	26	44	61	79	114	149
4	-128	-93	-58	-40	-23	-5	5	12	19	29	47	64	82	117	152
5	-125	-90	-55	-37	-20	-2	8	15	22	32	50	67	85	120	155
6	-122	-87	-52	-34	-17	0	11	18	25	35	53	70	88	123	158
7	-119	-84	-49	-31	-14	3	14	21	28	38	56	73	91	126	161
8	-116	-81	-46	-28	-11	6	17	24	31	41	59	76	94	129	164
10	-110	-75	-40	-22	-5	12	23	30	37	47	65	82	100	135	170
12	-104	-69	-34	-16	1	18	29	36	43	53	71	88	106	141	176
14	-98	-63	-28	-10	7	24	35	42	49	59	77	94	112	147	182
16	-92	-57	-22	-4	13	30	41	48	55	65	83	100	118	153	188
18	-86	-51	-16	1	19	36	47	54	61	71	89	106	124	159	194
20	-80	-45	-10	7	25	42	53	60	67	77	95	112	130	165	200
25	-65	-30	5	22	40	57	68	75	82	92	110	127	145	180	215
30	-50	-15	20	37	55	72	83	90	97	107	125	142	160	195	230
40	-20	15	50	67	85	102	113	120	127	137	155	172	190	225	260

Table 5-45

Separation, inch: 5

Leg Distance, inch: 40

SHIM REQUIRED FOR ALIGNMENT

Reverse Indicator Method

DriveN TIR MILS	DriveR TIR, MILS -40	-30	-20	-15	-10	-5	-2	0	2	5	10	15	20	30	40
-40	-300	-260	-220	-200	-180	-160	-148	-140	-132	-120	-100	-80	-60	-20	20
-30	-265	-225	-185	-165	-145	-125	-113	-105	-97	-85	-65	-45	-25	15	55
-25	-247	-207	-167	-147	-127	-107	-95	-87	-79	-67	-47	-27	-7	32	72
-20	-230	-190	-150	-130	-110	-90	-78	-70	-62	-50	-30	-10	10	50	90
-18	-223	-183	-143	-123	-103	-83	-71	-63	-55	-43	-23	-3	17	57	97
-16	-216	-176	-136	-116	-96	-76	-64	-56	-48	-36	-16	4	24	64	104
-14	-209	-169	-129	-109	-89	-69	-57	-49	-41	-29	-9	11	31	71	111
-12	-202	-162	-122	-102	-82	-62	-50	-42	-34	-22	-2	18	38	78	118
-10	-195	-155	-115	-95	-75	-55	-43	-35	-27	-15	5	25	45	85	125
-8	-188	-148	-108	-88	-68	-48	-36	-28	-20	-8	12	32	52	92	132
-7	-184	-144	-104	-84	-64	-44	-32	-24	-16	-4	15	35	55	95	135
-6	-181	-141	-101	-81	-61	-41	-29	-21	-13	-1	19	39	59	99	139
-5	-177	-137	-97	-77	-57	-37	-25	-17	-9	2	22	42	62	102	142
-4	-174	-134	-94	-74	-54	-34	-22	-14	-6	6	26	46	66	106	146
-3	-170	-130	-90	-70	-50	-30	-18	-10	-2	9	29	49	69	109	149
-2	-167	-127	-87	-67	-47	-27	-15	-7	1	13	33	53	73	113	153
-1	-163	-123	-83	-63	-43	-23	-11	-3	4	16	36	56	76	116	156
0	-160	-120	-80	-60	-40	-20	-8	0	8	20	40	60	80	120	160
1	-156	-116	-76	-56	-36	-16	-4	3	11	23	43	63	83	123	163
2	-153	-113	-73	-53	-33	-13	-1	7	15	27	47	67	87	127	167
3	-149	-109	-69	-49	-29	-9	2	10	18	30	50	70	90	130	170
4	-146	-106	-66	-46	-26	-6	6	14	22	34	54	74	94	134	174
5	-142	-102	-62	-42	-22	-2	9	17	25	37	57	77	97	137	177
6	-139	-99	-59	-39	-19	1	13	21	29	41	61	81	101	141	181
7	-135	-95	-55	-35	-15	4	16	24	32	44	64	84	104	144	184
8	-132	-92	-52	-32	-12	8	20	28	36	48	68	88	108	148	188
10	-125	-85	-45	-25	-5	15	27	35	43	55	75	95	115	155	195
12	-118	-78	-38	-18	2	22	34	42	50	62	82	102	122	162	202
14	-111	-71	-31	-11	9	29	41	49	57	69	89	109	129	169	209
16	-104	-64	-24	-4	16	36	48	56	64	76	96	116	136	176	216
18	-97	-57	-17	3	23	43	55	63	71	83	103	123	143	183	223
20	-90	-50	-10	10	30	50	62	70	78	90	110	130	150	190	230
25	-72	-32	7	27	47	67	79	87	95	107	127	147	167	207	247
30	-55	-15	25	45	65	85	97	105	113	125	145	165	185	225	265
40	-20	20	60	80	100	120	132	140	148	160	180	200	220	260	300

Table 5-46

Separation, inch: 5

Leg Distance, inch: 50

SHIM REQUIRED FOR ALIGNMENT

Reverse Indicator Method

DriveN TIR MILS	DriveR TIR, MILS -40	-30	-20	-15	-10	-5	-2	0	2	5	10	15	20	30	40
-40	-380	-330	-280	-255	-230	-205	-190	-180	-170	-155	-130	-105	-80	-30	20
-30	-335	-285	-235	-210	-185	-160	-145	-135	-125	-110	-85	-60	-35	15	65
-25	-312	-262	-212	-187	-162	-137	-122	-112	-102	-87	-62	-37	-12	37	87
-20	-290	-240	-190	-165	-140	-115	-100	-90	-80	-65	-40	-15	10	60	110
-18	-281	-231	-181	-156	-131	-106	-91	-81	-71	-56	-31	-6	19	69	119
-16	-272	-222	-172	-147	-122	-97	-82	-72	-62	-47	-22	3	28	78	128
-14	-263	-213	-163	-138	-113	-88	-73	-63	-53	-38	-13	12	37	87	137
-12	-254	-204	-154	-129	-104	-79	-64	-54	-44	-29	-4	21	46	96	146
-10	-245	-195	-145	-120	-95	-70	-55	-45	-35	-20	5	30	55	105	155
-8	-236	-186	-136	-111	-86	-61	-46	-36	-26	-11	14	39	64	114	164
-7	-231	-181	-131	-106	-81	-56	-41	-31	-21	-6	18	43	68	118	168
-6	-227	-177	-127	-102	-77	-52	-37	-27	-17	-2	23	48	73	123	173
-5	-222	-172	-122	-97	-72	-47	-32	-22	-12	2	27	52	77	127	177
-4	-218	-168	-118	-93	-68	-43	-28	-18	-8	7	32	57	82	132	182
-3	-213	-163	-113	-88	-63	-38	-23	-13	-3	11	36	61	86	136	186
-2	-209	-159	-109	-84	-59	-34	-19	-9	1	16	41	66	91	141	191
-1	-204	-154	-104	-79	-54	-29	-14	-4	5	20	45	70	95	145	195
0	-200	-150	-100	-75	-50	-25	-10	0	10	25	50	75	100	150	200
1	-195	-145	-95	-70	-45	-20	-5	4	14	29	54	79	104	154	204
2	-191	-141	-91	-66	-41	-16	-1	9	19	34	59	84	109	159	209
3	-186	-136	-86	-61	-36	-11	3	13	23	38	63	88	113	163	213
4	-182	-132	-82	-57	-32	-7	8	18	28	43	68	93	118	168	218
5	-177	-127	-77	-52	-27	-2	12	22	32	47	72	97	122	172	222
6	-173	-123	-73	-48	-23	2	17	27	37	52	77	102	127	177	227
7	-168	-118	-68	-43	-18	6	21	31	41	56	81	106	131	181	231
8	-164	-114	-64	-39	-14	11	26	36	46	61	86	111	136	186	236
10	-155	-105	-55	-30	-5	20	35	45	55	70	95	120	145	195	245
12	-146	-96	-46	-21	4	29	44	54	64	79	104	129	154	204	254
14	-137	-87	-37	-12	13	38	53	63	73	88	113	138	163	213	263
16	-128	-78	-28	-3	22	47	62	72	82	97	122	147	172	222	272
18	-119	-69	-19	6	31	56	71	81	91	106	131	156	181	231	281
20	-110	-60	-10	15	40	65	80	90	100	115	140	165	190	240	290
25	-87	-37	12	37	62	87	102	112	122	137	162	187	212	262	312
30	-65	-15	35	60	85	110	125	135	145	160	185	210	235	285	335
40	-20	30	80	105	130	155	170	180	190	205	230	255	280	330	380

Table 5-47

Separation, inch: 5

Leg Distance, inch: 60

SHIM REQUIRED FOR ALIGNMENT

Reverse Indicator Method

DriveN TIR MILS	DriveR TIR, MILS -40	-30	-20	-15	-10	-5	-2	0	2	5	10	15	20	30	40
-40	-460	-400	-340	-310	-280	-250	-232	-220	-208	-190	-160	-130	-100	-40	20
-30	-405	-345	-285	-255	-225	-195	-177	-165	-153	-135	-105	-75	-45	15	75
-25	-377	-317	-257	-227	-197	-167	-149	-137	-125	-107	-77	-47	-17	42	102
-20	-350	-290	-230	-200	-170	-140	-122	-110	-98	-80	-50	-20	10	70	130
-18	-339	-279	-219	-189	-159	-129	-111	-99	-87	-69	-39	-9	21	81	141
-16	-328	-268	-208	-178	-148	-118	-100	-88	-76	-58	-28	2	32	92	152
-14	-317	-257	-197	-167	-137	-107	-89	-77	-65	-47	-17	13	43	103	163
-12	-306	-246	-186	-156	-126	-96	-78	-66	-54	-36	-6	24	54	114	174
-10	-295	-235	-175	-145	-115	-85	-67	-55	-43	-25	5	35	65	125	185
-8	-284	-224	-164	-134	-104	-74	-56	-44	-32	-14	16	46	76	136	196
-7	-278	-218	-158	-128	-98	-68	-50	-38	-26	-8	21	51	81	141	201
-6	-273	-213	-153	-123	-93	-63	-45	-33	-21	-3	27	57	87	147	207
-5	-267	-207	-147	-117	-87	-57	-39	-27	-15	2	32	62	92	152	212
-4	-262	-202	-142	-112	-82	-52	-34	-22	-10	8	38	68	98	158	218
-3	-256	-196	-136	-106	-76	-46	-28	-16	-4	13	43	73	103	163	223
-2	-251	-191	-131	-101	-71	-41	-23	-11	1	19	49	79	109	169	229
-1	-245	-185	-125	-95	-65	-35	-17	-5	6	24	54	84	114	174	234
0	-240	-180	-120	-90	-60	-30	-12	0	12	30	60	90	120	180	240
1	-234	-174	-114	-84	-54	-24	-6	5	17	35	65	95	125	185	245
2	-229	-169	-109	-79	-49	-19	-1	11	23	41	71	101	131	191	251
3	-223	-163	-103	-73	-43	-13	4	16	28	46	76	106	136	196	256
4	-218	-158	-98	-68	-38	-8	10	22	34	52	82	112	142	202	262
5	-212	-152	-92	-62	-32	-2	15	27	39	57	87	117	147	207	267
6	-207	-147	-87	-57	-27	3	21	33	45	63	93	123	153	213	273
7	-201	-141	-81	-51	-21	8	26	38	50	68	98	128	158	218	278
8	-196	-136	-76	-46	-16	14	32	44	56	74	104	134	164	224	284
10	-185	-125	-65	-35	-5	25	43	55	67	85	115	145	175	235	295
12	-174	-114	-54	-24	6	36	54	66	78	96	126	156	186	246	306
14	-163	-103	-43	-13	17	47	65	77	89	107	137	167	197	257	317
16	-152	-92	-32	-2	28	58	76	88	100	118	148	178	208	268	328
18	-141	-81	-21	9	39	69	87	99	111	129	159	189	219	279	339
20	-130	-70	-10	20	50	80	98	110	122	140	170	200	230	290	350
25	-102	-42	17	47	77	107	125	137	149	167	197	227	257	317	377
30	-75	-15	45	75	105	135	153	165	177	195	225	255	285	345	405
40	-20	40	100	130	160	190	208	220	232	250	280	310	340	400	460

Table 5-48

Separation, inch: 5

Leg Distance, inch: 70

SHIM REQUIRED FOR ALIGNMENT

Reverse Indicator Method

DriveN TIR MILS	DriveR TIR, MILS -40	-30	-20	-15	-10	-5	-2	0	2	5	10	15	20	30	40
-40	-540	-470	-400	-365	-330	-295	-274	-260	-246	-225	-190	-155	-120	-50	20
-30	-475	-405	-335	-300	-265	-230	-209	-195	-181	-160	-125	-90	-55	15	85
-25	-442	-372	-302	-267	-232	-197	-176	-162	-148	-127	-92	-57	-22	47	117
-20	-410	-340	-270	-235	-200	-165	-144	-130	-116	-95	-60	-25	10	80	150
-18	-397	-327	-257	-222	-187	-152	-131	-117	-103	-82	-47	-12	23	93	163
-16	-384	-314	-244	-209	-174	-139	-118	-104	-90	-69	-34	1	36	106	176
-14	-371	-301	-231	-196	-161	-126	-105	-91	-77	-56	-21	14	49	119	189
-12	-358	-288	-218	-183	-148	-113	-92	-78	-64	-43	-8	27	62	132	202
-10	-345	-275	-205	-170	-135	-100	-79	-65	-51	-30	5	40	75	145	215
-8	-332	-262	-192	-157	-122	-87	-66	-52	-38	-17	18	53	88	158	228
-7	-325	-255	-185	-150	-115	-80	-59	-45	-31	-10	24	59	94	164	234
-6	-319	-249	-179	-144	-109	-74	-53	-39	-25	-4	31	66	101	171	241
-5	-312	-242	-172	-137	-102	-67	-46	-32	-18	2	37	72	107	177	247
-4	-306	-236	-166	-131	-96	-61	-40	-26	-12	9	44	79	114	184	254
-3	-299	-229	-159	-124	-89	-54	-33	-19	-5	15	50	85	120	190	260
-2	-293	-223	-153	-118	-83	-48	-27	-13	1	22	57	92	127	197	267
-1	-286	-216	-146	-111	-76	-41	-20	-6	7	28	63	98	133	203	273
0	-280	-210	-140	-105	-70	-35	-14	0	14	35	70	105	140	210	280
1	-273	-203	-133	-98	-63	-28	-7	6	20	41	76	111	146	216	286
2	-267	-197	-127	-92	-57	-22	-1	13	27	48	83	118	153	223	293
3	-260	-190	-120	-85	-50	-15	5	19	33	54	89	124	159	229	299
4	-254	-184	-114	-79	-44	-9	12	26	40	61	96	131	166	236	306
5	-247	-177	-107	-72	-37	-2	18	32	46	67	102	137	172	242	312
6	-241	-171	-101	-66	-31	4	25	39	53	74	109	144	179	249	319
7	-234	-164	-94	-59	-24	10	31	45	59	80	115	150	185	255	325
8	-228	-158	-88	-53	-18	17	38	52	66	87	122	157	192	262	332
10	-215	-145	-75	-40	-5	30	51	65	79	100	135	170	205	275	345
12	-202	-132	-62	-27	8	43	64	78	92	113	148	183	218	288	358
14	-189	-119	-49	-14	21	56	77	91	105	126	161	196	231	301	371
16	-176	-106	-36	-1	34	69	90	104	118	139	174	209	244	314	384
18	-163	-93	-23	12	47	82	103	117	131	152	187	222	257	327	397
20	-150	-80	-10	25	60	95	116	130	144	165	200	235	270	340	410
25	-117	-47	22	57	92	127	148	162	176	197	232	267	302	372	442
30	-85	-15	55	90	125	160	181	195	209	230	265	300	335	405	475
40	-20	50	120	155	190	225	246	260	274	295	330	365	400	470	540

Table 5-49

Separation, inch: 6

Leg Distance, inch: 15

SHIM REQUIRED FOR ALIGNMENT

Reverse Indicator Method

DriveN TIR MILS	DriveR TIR, MILS -40	-30	-20	-15	-10	-5	-2	0	2	5	10	15	20	30	40
-40	-80	-67	-55	-48	-42	-36	-32	-30	-27	-23	-17	-11	-5	7	20
-30	-72	-60	-47	-41	-35	-28	-25	-22	-20	-16	-10	-3	2	15	27
-25	-68	-56	-43	-37	-31	-25	-21	-18	-16	-12	-6	0	6	18	31
-20	-65	-52	-40	-33	-27	-21	-17	-15	-12	-8	-2	3	10	22	35
-18	-63	-51	-38	-32	-26	-19	-16	-13	-11	-7	-1	5	11	24	36
-16	-62	-49	-37	-30	-24	-18	-14	-12	-9	-5	0	6	13	25	38
-14	-60	-48	-35	-29	-23	-16	-13	-10	-8	-4	2	8	14	27	39
-12	-59	-46	-34	-27	-21	-15	-11	-9	-6	-2	3	9	16	28	41
-10	-57	-45	-32	-26	-20	-13	-10	-7	-5	-1	5	11	17	30	42
-8	-56	-43	-31	-24	-18	-12	-8	-6	-3	0	6	12	19	31	44
-7	-55	-42	-30	-24	-17	-11	-7	-5	-2	1	7	13	19	32	44
-6	-54	-42	-29	-23	-17	-10	-7	-4	-2	1	8	14	20	33	45
-5	-53	-41	-28	-22	-16	-10	-6	-3	-1	2	8	15	21	33	46
-4	-53	-40	-28	-21	-15	-9	-5	-3	0	3	9	15	22	34	47
-3	-52	-39	-27	-21	-14	-8	-4	-2	0	4	10	16	22	35	47
-2	-51	-39	-26	-20	-14	-7	-4	-1	1	4	11	17	23	36	48
-1	-50	-38	-25	-19	-13	-7	-3	0	1	5	11	18	24	36	49
0	-50	-37	-25	-18	-12	-6	-2	0	2	6	12	18	25	37	50
1	-49	-36	-24	-18	-11	-5	-1	0	3	7	13	19	25	38	50
2	-48	-36	-23	-17	-11	-4	-1	1	4	7	14	20	26	39	51
3	-47	-35	-22	-16	-10	-4	0	2	4	8	14	21	27	39	52
4	-47	-34	-22	-15	-9	-3	0	3	5	9	15	21	28	40	53
5	-46	-33	-21	-15	-8	-2	1	3	6	10	16	22	28	41	53
6	-45	-33	-20	-14	-8	-1	2	4	7	10	17	23	29	42	54
7	-44	-32	-19	-13	-7	-1	2	5	7	11	17	24	30	42	55
8	-44	-31	-19	-12	-6	0	3	6	8	12	18	24	31	43	56
10	-42	-30	-17	-11	-5	1	5	7	10	13	20	26	32	45	57
12	-41	-28	-16	-9	-3	2	6	9	11	15	21	27	34	46	59
14	-39	-27	-14	-8	-2	4	8	10	13	16	23	29	35	48	60
16	-38	-25	-13	-6	0	5	9	12	14	18	24	30	37	49	62
18	-36	-24	-11	-5	1	7	11	13	16	19	26	32	38	51	63
20	-35	-22	-10	-3	2	8	12	15	17	21	27	33	40	52	65
25	-31	-18	-6	0	6	12	16	18	21	25	31	37	43	56	68
30	-27	-15	-2	3	10	16	20	22	25	28	35	41	47	60	72
40	-20	-7	5	11	17	23	27	30	32	36	42	48	55	67	80

Table 5-50

Separation, inch: 6

Leg Distance, inch: 20

SHIM REQUIRED FOR ALIGNMENT

Reverse Indicator Method

DriveN TIR MILS	DriveR TIR, MILS -40	-30	-20	-15	-10	-5	-2	0	2	5	10	15	20	30	40
-40	-113	-96	-80	-71	-63	-55	-50	-46	-43	-38	-30	-21	-13	3	20
-30	-101	-85	-68	-60	-51	-43	-38	-35	-31	-26	-18	-10	-1	15	31
-25	-95	-79	-62	-54	-45	-37	-32	-29	-25	-20	-12	-4	4	20	37
-20	-90	-73	-56	-48	-40	-31	-26	-23	-20	-15	-6	1	10	26	43
-18	-87	-71	-54	-46	-37	-29	-24	-21	-17	-12	-4	4	12	29	45
-16	-85	-68	-52	-43	-35	-27	-22	-18	-15	-10	-2	6	14	31	48
-14	-83	-66	-49	-41	-33	-24	-19	-16	-13	-8	0	8	17	33	50
-12	-80	-64	-47	-39	-30	-22	-17	-14	-10	-5	2	11	19	36	52
-10	-78	-61	-45	-36	-28	-20	-15	-11	-8	-3	5	13	21	38	55
-8	-76	-59	-42	-34	-26	-17	-12	-9	-6	-1	7	15	24	40	57
-7	-74	-58	-41	-33	-24	-16	-11	-8	-4	0	8	16	25	41	58
-6	-73	-57	-40	-32	-23	-15	-10	-7	-3	1	9	18	26	43	59
-5	-72	-55	-39	-30	-22	-14	-9	-5	-2	2	10	19	27	44	60
-4	-71	-54	-38	-29	-21	-13	-8	-4	-1	3	12	20	28	45	62
-3	-70	-53	-36	-28	-20	-11	-6	-3	0	4	13	21	29	46	63
-2	-69	-52	-35	-27	-19	-10	-5	-2	1	6	14	22	31	47	64
-1	-67	-51	-34	-26	-17	-9	-4	-1	2	7	15	23	32	48	65
0	-66	-50	-33	-25	-16	-8	-3	0	3	8	16	25	33	50	66
1	-65	-48	-32	-23	-15	-7	-2	1	4	9	17	26	34	51	67
2	-64	-47	-31	-22	-14	-6	-1	2	5	10	19	27	35	52	69
3	-63	-46	-29	-21	-13	-4	0	3	6	11	20	28	36	53	70
4	-62	-45	-28	-20	-12	-3	1	4	8	13	21	29	38	54	71
5	-60	-44	-27	-19	-10	-2	2	5	9	14	22	30	39	55	72
6	-59	-43	-26	-18	-9	-1	3	7	10	15	23	32	40	57	73
7	-58	-41	-25	-16	-8	0	4	8	11	16	24	33	41	58	74
8	-57	-40	-24	-15	-7	1	6	9	12	17	26	34	42	59	76
10	-55	-38	-21	-13	-5	3	8	11	15	20	28	36	45	61	78
12	-52	-36	-19	-11	-2	5	10	14	17	22	30	39	47	64	80
14	-50	-33	-17	-8	0	8	13	16	19	24	33	41	49	66	83
16	-48	-31	-14	-6	2	10	15	18	22	27	35	43	52	68	85
18	-45	-29	-12	-4	4	12	17	21	24	29	37	46	54	71	87
20	-43	-26	-10	-1	6	15	20	23	26	31	40	48	56	73	90
25	-37	-20	-4	4	12	20	25	29	32	37	45	54	62	79	95
30	-31	-15	1	10	18	26	31	35	38	43	51	60	68	85	101
40	-20	-3	13	21	30	38	43	46	50	55	63	71	80	96	113

Table 5-51

Separation, inch: 6

Leg Distance, inch: 25

SHIM REQUIRED FOR ALIGNMENT

Reverse Indicator Method

DriveN TIR MILS	DriveR TIR, MILS -40	-30	-20	-15	-10	-5	-2	0	2	5	10	15	20	30	40
-40	-146	-125	-105	-94	-84	-73	-67	-63	-59	-52	-42	-32	-21	0	20
-30	-130	-110	-89	-78	-68	-57	-51	-47	-43	-37	-26	-16	-5	15	35
-25	-122	-102	-81	-70	-60	-50	-43	-39	-35	-29	-18	-8	2	22	43
-20	-115	-94	-73	-62	-52	-42	-35	-31	-27	-21	-10	0	10	30	51
-18	-111	-91	-70	-59	-49	-38	-32	-28	-24	-18	-7	2	13	34	54
-16	-108	-87	-67	-56	-46	-35	-29	-25	-21	-14	-4	5	16	37	58
-14	-105	-84	-63	-53	-43	-32	-26	-22	-18	-11	-1	9	19	40	61
-12	-102	-81	-60	-50	-39	-29	-23	-19	-14	-8	1	12	22	43	64
-10	-99	-78	-57	-47	-36	-26	-20	-15	-11	-5	5	15	25	46	67
-8	-96	-75	-54	-43	-33	-23	-16	-12	-8	-2	8	18	29	49	70
-7	-94	-73	-52	-42	-31	-21	-15	-11	-6	0	9	20	30	51	72
-6	-92	-72	-51	-40	-30	-19	-13	-9	-5	0	11	21	32	53	73
-5	-91	-70	-49	-39	-28	-18	-12	-7	-3	2	12	23	33	54	75
-4	-89	-68	-48	-37	-27	-16	-10	-6	-2	4	14	24	35	56	77
-3	-88	-67	-46	-36	-25	-15	-8	-4	0	5	16	26	36	57	78
-2	-86	-65	-44	-34	-24	-13	-7	-3	1	7	17	28	38	59	80
-1	-84	-64	-43	-32	-22	-12	-5	-1	2	8	19	29	40	60	81
0	-83	-62	-41	-31	-20	-10	-4	0	4	10	20	31	41	62	83
1	-81	-60	-40	-29	-19	-8	-2	1	5	12	22	32	43	64	84
2	-80	-59	-38	-28	-17	-7	-1	3	7	13	24	34	44	65	86
3	-78	-57	-36	-26	-16	-5	0	4	8	15	25	36	46	67	88
4	-77	-56	-35	-24	-14	-4	2	6	10	16	27	37	48	68	89
5	-75	-54	-33	-23	-12	-2	3	7	12	18	28	39	49	70	91
6	-73	-53	-32	-21	-11	0	5	9	13	19	30	40	51	72	92
7	-72	-51	-30	-20	-9	0	6	11	15	21	31	42	52	73	94
8	-70	-49	-29	-18	-8	2	8	12	16	23	33	43	54	75	96
10	-67	-46	-25	-15	-5	5	11	15	20	26	36	47	57	78	99
12	-64	-43	-22	-12	-1	8	14	19	23	29	39	50	60	81	102
14	-61	-40	-19	-9	1	11	18	22	26	32	43	53	63	84	105
16	-58	-37	-16	-5	4	14	21	25	29	35	46	56	67	87	108
18	-54	-34	-13	-2	7	18	24	28	32	38	49	59	70	91	111
20	-51	-30	-10	0	10	21	27	31	35	42	52	62	73	94	115
25	-43	-22	-2	8	18	29	35	39	43	50	60	70	81	102	122
30	-35	-15	5	16	26	37	43	47	51	57	68	78	89	110	130
40	-20	0	21	32	42	52	59	63	67	73	84	94	105	125	146

Table 5-52

Separation, inch: 6

Leg Distance, inch: 30

SHIM REQUIRED FOR ALIGNMENT

Reverse Indicator Method

DriveN TIR MILS	DriveR TIR, MILS -40	-30	-20	-15	-10	-5	-2	0	2	5	10	15	20	30	40
-40	-180	-155	-130	-117	-105	-92	-85	-80	-75	-67	-55	-42	-30	-5	20
-30	-160	-135	-110	-97	-85	-72	-65	-60	-55	-47	-35	-22	-10	15	40
-25	-150	-125	-100	-87	-75	-62	-55	-50	-45	-37	-25	-12	0	25	50
-20	-140	-115	-90	-77	-65	-52	-45	-40	-35	-27	-15	-2	10	35	60
-18	-136	-111	-86	-73	-61	-48	-41	-36	-31	-23	-11	1	14	39	64
-16	-132	-107	-82	-69	-57	-44	-37	-32	-27	-19	-7	5	18	43	68
-14	-128	-103	-78	-65	-53	-40	-33	-28	-23	-15	-3	9	22	47	72
-12	-124	-99	-74	-61	-49	-36	-29	-24	-19	-11	1	13	26	51	76
-10	-120	-95	-70	-57	-45	-32	-25	-20	-15	-7	5	17	30	55	80
-8	-116	-91	-66	-53	-41	-28	-21	-16	-11	-3	9	21	34	59	84
-7	-114	-89	-64	-51	-39	-26	-19	-14	-9	-1	11	23	36	61	86
-6	-112	-87	-62	-49	-37	-24	-17	-12	-7	0	13	25	38	63	88
-5	-110	-85	-60	-47	-35	-22	-15	-10	-5	2	15	27	40	65	90
-4	-108	-83	-58	-45	-33	-20	-13	-8	-3	4	17	29	42	67	92
-3	-106	-81	-56	-43	-31	-18	-11	-6	-1	6	19	31	44	69	94
-2	-104	-79	-54	-41	-29	-16	-9	-4	1	8	21	33	46	71	96
-1	-102	-77	-52	-39	-27	-14	-7	-2	3	10	23	35	48	73	98
0	-100	-75	-50	-37	-25	-12	-5	0	5	12	25	37	50	75	100
1	-98	-73	-48	-35	-23	-10	-3	2	7	14	27	39	52	77	102
2	-96	-71	-46	-33	-21	-8	-1	4	9	16	29	41	54	79	104
3	-94	-69	-44	-31	-19	-6	1	6	11	18	31	43	56	81	106
4	-92	-67	-42	-29	-17	-4	3	8	13	20	33	45	58	83	108
5	-90	-65	-40	-27	-15	-2	5	10	15	22	35	47	60	85	110
6	-88	-63	-38	-25	-13	0	7	12	17	24	37	49	62	87	112
7	-86	-61	-36	-23	-11	1	9	14	19	26	39	51	64	89	114
8	-84	-59	-34	-21	-9	3	11	16	21	28	41	53	66	91	116
10	-80	-55	-30	-17	-5	7	15	20	25	32	45	57	70	95	120
12	-76	-51	-26	-13	-1	11	19	24	29	36	49	61	74	99	124
14	-72	-47	-22	-9	3	15	23	28	33	40	53	65	78	103	128
16	-68	-43	-18	-5	7	19	27	32	37	44	57	69	82	107	132
18	-64	-39	-14	-1	11	23	31	36	41	48	61	73	86	111	136
20	-60	-35	-10	2	15	27	35	40	45	52	65	77	90	115	140
25	-50	-25	0	12	25	37	45	50	55	62	75	87	100	125	150
30	-40	-15	10	22	35	47	55	60	65	72	85	97	110	135	160
40	-20	5	30	42	55	67	75	80	85	92	105	117	130	155	180

Table 5-53

Separation, inch: 6

Leg Distance, inch: 35

SHIM REQUIRED FOR ALIGNMENT

Reverse Indicator Method

riveN DriveR TIR, MILS

IR

ILS	-40	-30	-20	-15	-10	-5	-2	0	2	5	10	15	20	30	40
-40	-213	-184	-155	-140	-125	-111	-102	-96	-90	-82	-67	-52	-38	-9	20
-30	-189	-160	-130	-116	-101	-87	-78	-72	-66	-57	-43	-28	-14	15	44
-25	-177	-147	-118	-104	-89	-75	-66	-60	-54	-45	-31	-16	-2	27	56
-20	-165	-135	-106	-92	-77	-62	-54	-48	-42	-33	-19	-4	10	39	68
-18	-160	-131	-101	-87	-72	-58	-49	-43	-37	-28	-14	0	14	44	73
-16	-155	-126	-97	-82	-67	-53	-44	-38	-32	-24	-9	5	19	48	78
-14	-150	-121	-92	-77	-63	-48	-39	-33	-28	-19	-4	9	24	53	82
-12	-145	-116	-87	-72	-58	-43	-34	-29	-23	-14	0	14	29	58	87
-10	-140	-111	-82	-67	-53	-38	-30	-24	-18	-9	5	19	34	63	92
-8	-136	-106	-77	-63	-48	-33	-25	-19	-13	-4	9	24	39	68	97
-7	-133	-104	-75	-60	-46	-31	-22	-16	-11	-2	12	26	41	70	99
-6	-131	-102	-72	-58	-43	-29	-20	-14	-8	0	14	29	43	73	102
-5	-128	-99	-70	-55	-41	-26	-17	-12	-6	2	17	31	46	75	104
-4	-126	-97	-68	-53	-38	-24	-15	-9	-3	4	19	34	48	77	107
-3	-123	-94	-65	-51	-36	-21	-13	-7	-1	7	21	36	51	80	109
-2	-121	-92	-63	-48	-34	-19	-10	-4	1	9	24	38	53	82	111
-1	-119	-89	-60	-46	-31	-17	-8	-2	3	12	26	41	55	85	114
0	-116	-87	-58	-43	-29	-14	-5	0	5	14	29	43	58	87	116
1	-114	-85	-55	-41	-26	-12	-3	2	8	17	31	46	60	89	119
2	-111	-82	-53	-38	-24	-9	-1	4	10	19	34	48	63	92	121
3	-109	-80	-51	-36	-21	-7	1	7	13	21	36	51	65	94	123
4	-107	-77	-48	-34	-19	-4	3	9	15	24	38	53	68	97	126
5	-104	-75	-46	-31	-17	-2	6	12	17	26	41	55	70	99	128
6	-102	-73	-43	-29	-14	0	8	14	20	29	43	58	72	102	131
7	-99	-70	-41	-26	-12	2	11	16	22	31	46	60	75	104	133
8	-97	-68	-39	-24	-9	4	13	19	25	33	48	63	77	106	136
10	-92	-63	-34	-19	-5	9	18	24	30	38	53	67	82	111	140
12	-87	-58	-29	-14	0	14	23	29	34	43	58	72	87	116	145
14	-82	-53	-24	-9	4	19	28	33	39	48	63	77	92	121	150
16	-78	-48	-19	-5	9	24	32	38	44	53	67	82	97	126	155
18	-73	-44	-14	0	14	28	37	43	49	58	72	87	101	131	160
20	-68	-39	-10	4	19	33	42	48	54	62	77	92	106	135	165
25	-56	-27	2	16	31	45	54	60	66	75	89	104	118	147	177
30	-44	-15	14	28	43	57	66	72	78	87	101	116	130	160	189
40	-20	9	38	52	67	82	90	96	102	111	125	140	155	184	213

Table 5-54

Separation, inch: 6

Leg Distance, inch: 40

SHIM REQUIRED FOR ALIGNMENT

Reverse Indicator Method

DriveN TIR MILS	DriveR TIR, MILS														
	-40	-30	-20	-15	-10	-5	-2	0	2	5	10	15	20	30	40
-40	-246	-213	-180	-163	-146	-130	-120	-113	-106	-96	-80	-63	-46	-13	20
-30	-218	-185	-151	-135	-118	-101	-91	-85	-78	-68	-51	-35	-18	15	48
-25	-204	-170	-137	-120	-104	-87	-77	-70	-64	-54	-37	-20	-4	29	62
-20	-190	-156	-123	-106	-90	-73	-63	-56	-50	-40	-23	-6	10	43	76
-18	-184	-151	-117	-101	-84	-67	-57	-51	-44	-34	-17	-1	15	49	82
-16	-178	-145	-112	-95	-78	-62	-52	-45	-38	-28	-12	4	21	54	88
-14	-173	-139	-106	-89	-73	-56	-46	-39	-33	-23	-6	10	27	60	93
-12	-167	-134	-100	-84	-67	-50	-40	-34	-27	-17	0	16	32	66	99
-10	-161	-128	-95	-78	-61	-45	-35	-28	-21	-11	5	21	38	71	105
-8	-156	-122	-89	-72	-56	-39	-29	-22	-16	-6	10	27	44	77	110
-7	-153	-119	-86	-69	-53	-36	-26	-19	-13	-3	13	30	46	80	113
-6	-150	-117	-83	-67	-50	-33	-23	-17	-10	0	16	33	49	83	116
-5	-147	-114	-80	-64	-47	-30	-20	-14	-7	2	19	35	52	85	119
-4	-144	-111	-78	-61	-44	-28	-18	-11	-4	5	22	38	55	88	122
-3	-141	-108	-75	-58	-41	-25	-15	-8	-1	8	24	41	58	91	124
-2	-139	-105	-72	-55	-39	-22	-12	-5	1	11	27	44	61	94	127
-1	-136	-102	-69	-52	-36	-19	-9	-2	3	13	30	47	63	97	130
0	-133	-100	-66	-50	-33	-16	-6	0	6	16	33	50	66	100	133
1	-130	-97	-63	-47	-30	-13	-3	2	9	19	36	52	69	102	136
2	-127	-94	-61	-44	-27	-11	-1	5	12	22	39	55	72	105	139
3	-124	-91	-58	-41	-24	-8	1	8	15	25	41	58	75	108	141
4	-122	-88	-55	-38	-22	-5	4	11	18	28	44	61	78	111	144
5	-119	-85	-52	-35	-19	-2	7	14	20	30	47	64	80	114	147
6	-116	-83	-49	-33	-16	0	10	17	23	33	50	67	83	117	150
7	-113	-80	-46	-30	-13	3	13	19	26	36	53	69	86	119	153
8	-110	-77	-44	-27	-10	6	16	22	29	39	56	72	89	122	156
10	-105	-71	-38	-21	-5	11	21	28	35	45	61	78	95	128	161
12	-99	-66	-32	-16	0	17	27	34	40	50	67	84	100	134	167
14	-93	-60	-27	-10	6	23	33	39	46	56	73	89	106	139	173
16	-88	-54	-21	-4	12	28	38	45	52	62	78	95	112	145	178
18	-82	-49	-15	1	17	34	44	51	57	67	84	101	117	151	184
20	-76	-43	-10	6	23	40	50	56	63	73	90	106	123	156	190
25	-62	-29	4	20	37	54	64	70	77	87	104	120	137	170	204
30	-48	-15	18	35	51	68	78	85	91	101	118	135	151	185	218
40	-20	13	46	63	80	96	106	113	120	130	146	163	180	213	246

Table 5-55

Separation, inch: 6

Leg Distance, inch: 50

SHIM REQUIRED FOR ALIGNMENT

Reverse Indicator Method

iveN IR ILS	DriveR TIR, MILS														
	-40	-30	-20	-15	-10	-5	-2	0	2	5	10	15	20	30	40
-40	-313	-271	-230	-209	-188	-167	-155	-146	-138	-125	-105	-84	-63	-21	20
-30	-276	-235	-193	-172	-151	-130	-118	-110	-101	-89	-68	-47	-26	15	56
-25	-258	-216	-175	-154	-133	-112	-100	-91	-83	-70	-50	-29	-8	33	75
-20	-240	-198	-156	-135	-115	-94	-81	-73	-65	-52	-31	-10	10	51	93
-18	-232	-191	-149	-128	-107	-86	-74	-66	-57	-45	-24	-3	17	59	100
-16	-225	-183	-142	-121	-100	-79	-67	-58	-50	-37	-17	3	24	66	108
-14	-218	-176	-134	-113	-93	-72	-59	-51	-43	-30	-9	11	32	73	115
-12	-210	-169	-127	-106	-85	-64	-52	-44	-35	-23	-2	18	39	81	122
-10	-203	-161	-120	-99	-78	-57	-45	-36	-28	-15	5	25	46	88	130
-8	-196	-154	-112	-91	-71	-50	-37	-29	-21	-8	12	33	54	95	137
-7	-192	-150	-109	-88	-67	-46	-34	-25	-17	-4	16	36	57	99	141
-6	-188	-147	-105	-84	-63	-42	-30	-22	-13	-1	19	40	61	103	144
-5	-185	-143	-101	-80	-60	-39	-26	-18	-10	2	23	44	65	106	148
-4	-181	-139	-98	-77	-56	-35	-23	-14	-6	6	27	47	68	110	152
-3	-177	-136	-94	-73	-52	-31	-19	-11	-2	9	30	51	72	114	155
-2	-174	-132	-90	-69	-49	-28	-15	-7	1	13	34	55	76	117	159
-1	-170	-128	-87	-66	-45	-24	-12	-3	4	17	38	58	79	121	163
0	-166	-125	-83	-62	-41	-20	-8	0	8	20	41	62	83	125	166
1	-163	-121	-79	-58	-38	-17	-4	3	12	24	45	66	87	128	170
2	-159	-117	-76	-55	-34	-13	-1	7	15	28	49	69	90	132	174
3	-155	-114	-72	-51	-30	-9	2	11	19	31	52	73	94	136	177
4	-152	-110	-68	-47	-27	-6	6	14	23	35	56	77	98	139	181
5	-148	-106	-65	-44	-23	-2	10	18	26	39	60	80	101	143	185
6	-144	-103	-61	-40	-19	1	13	22	30	42	63	84	105	147	188
7	-141	-99	-57	-36	-16	4	17	25	34	46	67	88	109	150	192
8	-137	-95	-54	-33	-12	8	21	29	37	50	71	91	112	154	196
10	-130	-88	-46	-25	-5	15	28	36	45	57	78	99	120	161	203
12	-122	-81	-39	-18	2	23	35	44	52	64	85	106	127	169	210
14	-115	-73	-32	-11	9	30	43	51	59	72	93	113	134	176	218
16	-108	-66	-24	-3	17	37	50	58	67	79	100	121	142	183	225
18	-100	-59	-17	3	24	45	57	66	74	86	107	128	149	191	232
20	-93	-51	-10	10	31	52	65	73	81	94	115	135	156	198	240
25	-75	-33	8	29	50	70	83	91	100	112	133	154	175	216	258
30	-56	-15	26	47	68	89	101	110	118	130	151	172	193	235	276
40	-20	21	63	84	105	125	138	146	155	167	188	209	230	271	313

Table 5-56

Separation, inch: 6

Leg Distance, inch: 60

SHIM REQUIRED FOR ALIGNMENT

Reverse Indicator Method

DriveN TIR MILS	DriveR TIR, MILS -40	-30	-20	-15	-10	-5	-2	0	2	5	10	15	20	30	40
-40	-380	-330	-280	-255	-230	-205	-190	-180	-170	-155	-130	-105	-80	-30	20
-30	-335	-285	-235	-210	-185	-160	-145	-135	-125	-110	-85	-60	-35	15	65
-25	-312	-262	-212	-187	-162	-137	-122	-112	-102	-87	-62	-37	-12	37	87
-20	-290	-240	-190	-165	-140	-115	-100	-90	-80	-65	-40	-15	10	60	110
-18	-281	-231	-181	-156	-131	-106	-91	-81	-71	-56	-31	-6	19	69	119
-16	-272	-222	-172	-147	-122	-97	-82	-72	-62	-47	-22	3	28	78	128
-14	-263	-213	-163	-138	-113	-88	-73	-63	-53	-38	-13	12	37	87	137
-12	-254	-204	-154	-129	-104	-79	-64	-54	-44	-29	-4	21	46	96	146
-10	-245	-195	-145	-120	-95	-70	-55	-45	-35	-20	5	30	55	105	155
-8	-236	-186	-136	-111	-86	-61	-46	-36	-26	-11	14	39	64	114	164
-7	-231	-181	-131	-106	-81	-56	-41	-31	-21	-6	18	43	68	118	168
-6	-227	-177	-127	-102	-77	-52	-37	-27	-17	-2	23	48	73	123	173
-5	-222	-172	-122	-97	-72	-47	-32	-22	-12	2	27	52	77	127	177
-4	-218	-168	-118	-93	-68	-43	-28	-18	-8	7	32	57	82	132	182
-3	-213	-163	-113	-88	-63	-38	-23	-13	-3	11	36	61	86	136	186
-2	-209	-159	-109	-84	-59	-34	-19	-9	1	16	41	66	91	141	191
-1	-204	-154	-104	-79	-54	-29	-14	-4	5	20	45	70	95	145	195
0	-200	-150	-100	-75	-50	-25	-10	0	10	25	50	75	100	150	200
1	-195	-145	-95	-70	-45	-20	-5	4	14	29	54	79	104	154	204
2	-191	-141	-91	-66	-41	-16	-1	9	19	34	59	84	109	159	209
3	-186	-136	-86	-61	-36	-11	3	13	23	38	63	88	113	163	213
4	-182	-132	-82	-57	-32	-7	8	18	28	43	68	93	118	168	218
5	-177	-127	-77	-52	-27	-2	12	22	32	47	72	97	122	172	222
6	-173	-123	-73	-48	-23	2	17	27	37	52	77	102	127	177	227
7	-168	-118	-68	-43	-18	6	21	31	41	56	81	106	131	181	231
8	-164	-114	-64	-39	-14	11	26	36	46	61	86	111	136	186	236
10	-155	-105	-55	-30	-5	20	35	45	55	70	95	120	145	195	245
12	-146	-96	-46	-21	4	29	44	54	64	79	104	129	154	204	254
14	-137	-87	-37	-12	13	38	53	63	73	88	113	138	163	213	263
16	-128	-78	-28	-3	22	47	62	72	82	97	122	147	172	222	272
18	-119	-69	-19	6	31	56	71	81	91	106	131	156	181	231	281
20	-110	-60	-10	15	40	65	80	90	100	115	140	165	190	240	290
25	-87	-37	12	37	62	87	102	112	122	137	162	187	212	262	312
30	-65	-15	35	60	85	110	125	135	145	160	185	210	235	285	335
40	-20	30	80	105	130	155	170	180	190	205	230	255	280	330	380

Table 5-57

Separation, inch: 6

Leg Distance, inch: 70

SHIM REQUIRED FOR ALIGNMENT

Reverse Indicator Method

-iveN IR ILS	DriveR TIR, MILS -40	-30	-20	-15	-10	-5	-2	0	2	5	10	15	20	30	40
-40	-446	-388	-330	-300	-271	-242	-225	-213	-201	-184	-155	-125	-96	-38	20
-30	-393	-335	-276	-247	-218	-189	-171	-160	-148	-130	-101	-72	-43	15	73
-25	-366	-308	-250	-220	-191	-162	-145	-133	-121	-104	-75	-45	-16	41	100
-20	-340	-281	-223	-194	-165	-135	-118	-106	-95	-77	-48	-19	10	68	126
-18	-329	-271	-212	-183	-154	-125	-107	-96	-84	-66	-37	-8	20	79	137
-16	-318	-260	-202	-172	-143	-114	-97	-85	-73	-56	-27	2	31	89	148
-14	-308	-249	-191	-162	-133	-103	-86	-74	-63	-45	-16	12	42	100	158
-12	-297	-239	-180	-151	-122	-93	-75	-64	-52	-34	-5	23	52	111	169
-10	-286	-228	-170	-140	-111	-82	-65	-53	-41	-24	5	34	63	121	180
-8	-276	-217	-159	-130	-101	-71	-54	-42	-31	-13	15	44	74	132	190
-7	-270	-212	-154	-124	-95	-66	-49	-37	-25	-8	21	50	79	137	196
-6	-265	-207	-148	-119	-90	-61	-43	-32	-20	-2	26	55	84	143	201
-5	-260	-201	-143	-114	-85	-55	-38	-26	-15	2	31	60	90	148	206
-4	-254	-196	-138	-108	-79	-50	-33	-21	-9	7	37	66	95	153	212
-3	-249	-191	-132	-103	-74	-45	-27	-16	-4	13	42	71	100	159	217
-2	-244	-185	-127	-98	-69	-39	-22	-10	1	18	47	76	106	164	222
-1	-238	-180	-122	-92	-63	-34	-17	-5	6	23	53	82	111	169	228
0	-233	-175	-116	-87	-58	-29	-11	0	11	29	58	87	116	175	233
1	-228	-169	-111	-82	-53	-23	-6	5	17	34	63	92	122	180	238
2	-222	-164	-106	-76	-47	-18	-1	10	22	39	69	98	127	185	244
3	-217	-159	-100	-71	-42	-13	4	16	27	45	74	103	132	191	249
4	-212	-153	-95	-66	-37	-7	9	21	33	50	79	108	138	196	254
5	-206	-148	-90	-60	-31	-2	15	26	38	55	85	114	143	201	260
6	-201	-143	-84	-55	-26	2	20	32	43	61	90	119	148	207	265
7	-196	-137	-79	-50	-21	8	25	37	49	66	95	124	154	212	270
8	-190	-132	-74	-44	-15	13	31	42	54	71	101	130	159	217	276
10	-180	-121	-63	-34	-5	24	41	53	65	82	111	140	170	228	286
12	-169	-111	-52	-23	5	34	52	64	75	93	122	151	180	239	297
14	-158	-100	-42	-12	16	45	63	74	86	103	133	162	191	249	308
16	-148	-89	-31	-2	27	56	73	85	97	114	143	172	202	260	318
18	-137	-79	-20	8	37	66	84	96	107	125	154	183	212	271	329
20	-126	-68	-10	19	48	77	95	106	118	135	165	194	223	281	340
25	-100	-41	16	45	75	104	121	133	145	162	191	220	250	308	366
30	-73	-15	43	72	101	130	148	160	171	189	218	247	276	335	393
40	-20	38	96	125	155	184	201	213	225	242	271	300	330	388	446

Table 5-58

Separation, inch: 7

Leg Distance, inch: 15

SHIM REQUIRED FOR ALIGNMENT

Reverse Indicator Method

DriveN TIR MILS	DriveR TIR, MILS														
	-40	-30	-20	-15	-10	-5	-2	0	2	5	10	15	20	30	40
-40	-65	-55	-44	-38	-33	-28	-25	-22	-20	-17	-12	-6	-1	9	20
-30	-60	-49	-38	-33	-27	-22	-19	-17	-15	-11	-6	-1	4	15	25
-25	-57	-46	-35	-30	-25	-19	-16	-14	-12	-8	-3	1	7	17	28
-20	-54	-43	-32	-27	-22	-16	-13	-11	-9	-6	0	4	10	20	31
-18	-53	-42	-31	-26	-21	-15	-12	-10	-8	-4	0	5	11	21	32
-16	-52	-41	-30	-25	-19	-14	-11	-9	-7	-3	1	6	12	23	33
-14	-50	-40	-29	-24	-18	-13	-10	-8	-5	-2	2	8	13	24	34
-12	-49	-39	-28	-22	-17	-12	-9	-6	-4	-1	3	9	14	25	36
-10	-48	-37	-27	-21	-16	-11	-7	-5	-3	0	5	10	15	26	37
-8	-47	-36	-26	-20	-15	-9	-6	-4	-2	0	6	11	16	27	38
-7	-46	-36	-25	-20	-14	-9	-6	-4	-1	1	6	12	17	28	38
-6	-46	-35	-24	-19	-14	-8	-5	-3	-1	1	7	12	18	28	39
-5	-45	-35	-24	-18	-13	-8	-5	-2	0	2	7	13	18	29	40
-4	-45	-34	-23	-18	-13	-7	-4	-2	0	3	8	13	19	29	40
-3	-44	-33	-23	-17	-12	-7	-3	-1	0	3	9	14	19	30	41
-2	-44	-33	-22	-17	-11	-6	-3	-1	1	4	9	14	20	31	41
-1	-43	-32	-22	-16	-11	-5	-2	0	1	4	10	15	20	31	42
0	-42	-32	-21	-16	-10	-5	-2	0	2	5	10	16	21	32	42
1	-42	-31	-20	-15	-10	-4	-1	0	2	5	11	16	22	32	43
2	-41	-31	-20	-14	-9	-4	-1	1	3	6	11	17	22	33	44
3	-41	-30	-19	-14	-9	-3	0	1	3	7	12	17	23	33	44
4	-40	-29	-19	-13	-8	-3	0	2	4	7	13	18	23	34	45
5	-40	-29	-18	-13	-7	-2	0	2	5	8	13	18	24	35	45
6	-39	-28	-18	-12	-7	-1	1	3	5	8	14	19	24	35	46
7	-38	-28	-17	-12	-6	-1	1	4	6	9	14	20	25	36	46
8	-38	-27	-16	-11	-6	0	2	4	6	9	15	20	26	36	47
10	-37	-26	-15	-10	-5	0	3	5	7	11	16	21	27	37	48
12	-36	-25	-14	-9	-3	1	4	6	9	12	17	22	28	39	49
14	-34	-24	-13	-8	-2	2	5	8	10	13	18	24	29	40	50
16	-33	-23	-12	-6	-1	3	7	9	11	14	19	25	30	41	52
18	-32	-21	-11	-5	0	4	8	10	12	15	21	26	31	42	53
20	-31	-20	-10	-4	0	6	9	11	13	16	22	27	32	43	54
25	-28	-17	-7	-1	3	8	12	14	16	19	25	30	35	46	57
30	-25	-15	-4	1	6	11	15	17	19	22	27	33	38	49	60
40	-20	-9	1	6	12	17	20	22	25	28	33	38	44	55	65

Table 5-59

Separation, inch: 7

Leg Distance, inch: 20

SHIM REQUIRED FOR ALIGNMENT

Reverse Indicator Method

iveN R LS	DriveR TIR, MILS														
	-40	-30	-20	-15	-10	-5	-2	0	2	5	10	15	20	30	40
40	-94	-80	-65	-58	-51	-44	-40	-37	-34	-30	-22	-15	-8	5	20
30	-85	-70	-56	-49	-42	-35	-30	-27	-25	-20	-13	-6	0	15	29
25	-80	-66	-51	-44	-37	-30	-26	-23	-20	-16	-8	-1	5	19	33
20	-75	-61	-47	-40	-32	-25	-21	-18	-15	-11	-4	2	10	24	38
18	-73	-59	-45	-38	-31	-23	-19	-16	-13	-9	-2	4	11	26	40
16	-72	-57	-43	-36	-29	-22	-17	-14	-12	-7	0	6	13	28	42
14	-70	-55	-41	-34	-27	-20	-15	-13	-10	-5	1	8	15	29	44
12	-68	-54	-39	-32	-25	-18	-14	-11	-8	-4	3	10	17	31	46
10	-66	-52	-37	-30	-23	-16	-12	-9	-6	-2	5	12	19	33	47
-8	-64	-50	-36	-28	-21	-14	-10	-7	-4	0	6	14	21	35	49
-7	-63	-49	-35	-27	-20	-13	-9	-6	-3	0	7	14	22	36	50
-6	-62	-48	-34	-27	-19	-12	-8	-5	-2	1	8	15	23	37	51
-5	-61	-47	-33	-26	-18	-11	-7	-4	-1	2	9	16	23	38	52
-4	-60	-46	-32	-25	-18	-10	-6	-3	0	3	10	17	24	39	53
-3	-59	-45	-31	-24	-17	-9	-5	-2	0	4	11	18	25	40	54
-2	-59	-44	-30	-23	-16	-9	-4	-1	1	5	12	19	26	41	55
-1	-58	-43	-29	-22	-15	-8	-3	0	1	6	13	20	27	41	56
0	-57	-42	-28	-21	-14	-7	-2	0	2	7	14	21	28	42	57
1	-56	-41	-27	-20	-13	-6	-1	0	3	8	15	22	29	43	58
2	-55	-41	-26	-19	-12	-5	-1	1	4	9	16	23	30	44	59
3	-54	-40	-25	-18	-11	-4	0	2	5	9	17	24	31	45	59
4	-53	-39	-24	-17	-10	-3	0	3	6	10	18	25	32	46	60
5	-52	-38	-23	-16	-9	-2	1	4	7	11	18	26	33	47	61
6	-51	-37	-23	-15	-8	-1	2	5	8	12	19	27	34	48	62
7	-50	-36	-22	-14	-7	0	3	6	9	13	20	27	35	49	63
8	-49	-35	-21	-14	-6	0	4	7	10	14	21	28	36	50	64
10	-47	-33	-19	-12	-5	2	6	9	12	16	23	30	37	52	66
12	-46	-31	-17	-10	-3	4	8	11	14	18	25	32	39	54	68
14	-44	-29	-15	-8	-1	5	10	13	15	20	27	34	41	55	70
16	-42	-28	-13	-6	0	7	12	14	17	22	29	36	43	57	72
18	-40	-26	-11	-4	2	9	13	16	19	23	31	38	45	59	73
20	-38	-24	-10	-2	4	11	15	18	21	25	32	40	47	61	75
25	-33	-19	-5	1	8	16	20	23	26	30	37	44	51	66	80
30	-29	-15	0	6	13	20	25	27	30	35	42	49	56	70	85
40	-20	-5	8	15	22	30	34	37	40	44	51	58	65	80	94

Table 5-60

Separation, inch: 7

Leg Distance, inch: 25

SHIM REQUIRED FOR ALIGNMENT

Reverse Indicator Method

DriveN TIR MILS	DriveR TIR, MILS -40	-30	-20	-15	-10	-5	-2	0	2	5	10	15	20	30	40
-40	-122	-105	-87	-78	-69	-60	-55	-51	-47	-42	-33	-24	-15	2	20
-30	-110	-92	-74	-65	-56	-47	-42	-38	-35	-29	-20	-11	-2	15	32
-25	-103	-85	-67	-58	-50	-41	-35	-32	-28	-23	-14	-5	3	21	39
-20	-97	-79	-61	-52	-43	-34	-29	-25	-22	-16	-7	1	10	27	45
-18	-94	-76	-58	-49	-41	-32	-26	-23	-19	-14	-5	3	12	30	48
-16	-92	-74	-56	-47	-38	-29	-24	-20	-17	-11	-2	6	15	33	50
-14	-89	-71	-53	-44	-35	-26	-21	-18	-14	-9	0	8	17	35	53
-12	-86	-69	-51	-42	-33	-24	-19	-15	-11	-6	2	11	20	38	56
-10	-84	-66	-48	-39	-30	-21	-16	-12	-9	-3	5	13	22	40	58
-8	-81	-63	-46	-37	-28	-19	-13	-10	-6	-1	7	16	25	43	61
-7	-80	-62	-44	-35	-26	-17	-12	-9	-5	0	8	17	26	44	62
-6	-79	-61	-43	-34	-25	-16	-11	-7	-4	1	10	19	28	45	63
-5	-77	-60	-42	-33	-24	-15	-10	-6	-2	2	11	20	29	47	65
-4	-76	-58	-40	-31	-23	-14	-8	-5	-1	3	12	21	30	48	66
-3	-75	-57	-39	-30	-21	-12	-7	-3	0	5	14	22	31	49	67
-2	-74	-56	-38	-29	-20	-11	-6	-2	1	6	15	24	33	51	68
-1	-72	-54	-37	-28	-19	-10	-4	-1	2	7	16	25	34	52	70
0	-71	-53	-35	-26	-17	-8	-3	0	3	8	17	26	35	53	71
1	-70	-52	-34	-25	-16	-7	-2	1	4	10	19	28	37	54	72
2	-68	-51	-33	-24	-15	-6	-1	2	6	11	20	29	38	56	74
3	-67	-49	-31	-22	-14	-5	0	3	7	12	21	30	39	57	75
4	-66	-48	-30	-21	-12	-3	1	5	8	14	23	31	40	58	76
5	-65	-47	-29	-20	-11	-2	2	6	10	15	24	33	42	60	77
6	-63	-45	-28	-19	-10	-1	4	7	11	16	25	34	43	61	79
7	-62	-44	-26	-17	-8	0	5	9	12	17	26	35	44	62	80
8	-61	-43	-25	-16	-7	1	6	10	13	19	28	37	46	63	81
10	-58	-40	-22	-13	-5	3	9	12	16	21	30	39	48	66	84
12	-56	-38	-20	-11	-2	6	11	15	19	24	33	42	51	69	86
14	-53	-35	-17	-8	0	9	14	18	21	26	35	44	53	71	89
16	-50	-33	-15	-6	2	11	17	20	24	29	38	47	56	74	92
18	-48	-30	-12	-3	5	14	19	23	26	32	41	49	58	76	94
20	-45	-27	-10	-1	7	16	22	25	29	34	43	52	61	79	97
25	-39	-21	-3	5	14	23	28	32	35	41	50	58	67	85	103
30	-32	-15	2	11	20	29	35	38	42	47	56	65	74	92	110
40	-20	-2	15	24	33	42	47	51	55	60	69	78	87	105	122

Table 5-61

Separation, inch: 7

Leg Distance, inch: 30

SHIM REQUIRED FOR ALIGNMENT

Reverse Indicator Method

riveN IR ILS	DriveR TIR, MILS														
	-40	-30	-20	-15	-10	-5	-2	0	2	5	10	15	20	30	40
-40	-151	-130	-108	-97	-87	-76	-70	-65	-61	-55	-44	-33	-22	-1	20
-30	-135	-113	-92	-81	-70	-60	-53	-49	-45	-38	-27	-17	-6	15	36
-25	-126	-105	-83	-73	-62	-51	-45	-41	-36	-30	-19	-8	1	23	44
-20	-118	-97	-75	-65	-54	-43	-37	-32	-28	-22	-11	0	10	31	52
-18	-115	-93	-72	-61	-51	-40	-33	-29	-25	-18	-8	2	13	34	56
-16	-112	-90	-69	-58	-47	-37	-30	-26	-22	-15	-4	5	16	38	59
-14	-108	-87	-65	-55	-44	-33	-27	-23	-18	-12	-1	9	19	41	62
-12	-105	-84	-62	-51	-41	-30	-24	-19	-15	-9	1	12	23	44	66
-10	-102	-80	-59	-48	-37	-27	-20	-16	-12	-5	5	15	26	47	69
-8	-98	-77	-56	-45	-34	-23	-17	-13	-8	-2	8	19	29	51	72
-7	-97	-75	-54	-43	-32	-22	-15	-11	-7	0	9	20	31	52	74
-6	-95	-74	-52	-42	-31	-20	-14	-9	-5	0	11	22	33	54	75
-5	-93	-72	-51	-40	-29	-18	-12	-8	-3	2	13	23	34	56	77
-4	-92	-70	-49	-38	-28	-17	-10	-6	-2	4	14	25	36	57	79
-3	-90	-69	-47	-37	-26	-15	-9	-4	0	5	16	27	37	59	80
-2	-89	-67	-46	-35	-24	-14	-7	-3	1	7	18	28	39	61	82
-1	-87	-65	-44	-33	-23	-12	-5	-1	2	9	19	30	41	62	84
0	-85	-64	-42	-32	-21	-10	-4	0	4	10	21	32	42	64	85
1	-84	-62	-41	-30	-19	-9	-2	1	5	12	23	33	44	65	87
2	-82	-61	-39	-28	-18	-7	-1	3	7	14	24	35	46	67	89
3	-80	-59	-37	-27	-16	-5	0	4	9	15	26	37	47	69	90
4	-79	-57	-36	-25	-14	-4	2	6	10	17	28	38	49	70	92
5	-77	-56	-34	-23	-13	-2	3	8	12	18	29	40	51	72	93
6	-75	-54	-33	-22	-11	0	5	9	14	20	31	42	52	74	95
7	-74	-52	-31	-20	-9	0	7	11	15	22	32	43	54	75	97
8	-72	-51	-29	-19	-8	2	8	13	17	23	34	45	56	77	98
10	-69	-47	-26	-15	-5	5	12	16	20	27	37	48	59	80	102
12	-66	-44	-23	-12	-1	9	15	19	24	30	41	51	62	84	105
14	-62	-41	-19	-9	1	12	18	23	27	33	44	55	65	87	108
16	-59	-38	-16	-5	4	15	22	26	30	37	47	58	69	90	112
18	-56	-34	-13	-2	8	18	25	29	33	40	51	61	72	93	115
20	-52	-31	-10	0	11	22	28	32	37	43	54	65	75	97	118
25	-44	-23	-1	8	19	30	36	41	45	51	62	73	83	105	126
30	-36	-15	6	17	27	38	45	49	53	60	70	81	92	113	135
40	-20	1	22	33	44	55	61	65	70	76	87	97	108	130	151

Table 5-62

Separation, inch: 7

Leg Distance, inch: 35

SHIM REQUIRED FOR ALIGNMENT

Reverse Indicator Method

DriveN TIR MILS	DriveR TIR, MILS -40	-30	-20	-15	-10	-5	-2	0	2	5	10	15	20	30	40
-40	-180	-155	-130	-117	-105	-92	-85	-80	-75	-67	-55	-42	-30	-5	20
-30	-160	-135	-110	-97	-85	-72	-65	-60	-55	-47	-35	-22	-10	15	40
-25	-150	-125	-100	-87	-75	-62	-55	-50	-45	-37	-25	-12	0	25	50
-20	-140	-115	-90	-77	-65	-52	-45	-40	-35	-27	-15	-2	10	35	60
-18	-136	-111	-86	-73	-61	-48	-41	-36	-31	-23	-11	1	14	39	64
-16	-132	-107	-82	-69	-57	-44	-37	-32	-27	-19	-7	5	18	43	68
-14	-128	-103	-78	-65	-53	-40	-33	-28	-23	-15	-3	9	22	47	72
-12	-124	-99	-74	-61	-49	-36	-29	-24	-19	-11	1	13	26	51	76
-10	-120	-95	-70	-57	-45	-32	-25	-20	-15	-7	5	17	30	55	80
-8	-116	-91	-66	-53	-41	-28	-21	-16	-11	-3	9	21	34	59	84
-7	-114	-89	-64	-51	-39	-26	-19	-14	-9	-1	11	23	36	61	86
-6	-112	-87	-62	-49	-37	-24	-17	-12	-7	0	13	25	38	63	88
-5	-110	-85	-60	-47	-35	-22	-15	-10	-5	2	15	27	40	65	90
-4	-108	-83	-58	-45	-33	-20	-13	-8	-3	4	17	29	42	67	92
-3	-106	-81	-56	-43	-31	-18	-11	-6	-1	6	19	31	44	69	94
-2	-104	-79	-54	-41	-29	-16	-9	-4	1	8	21	33	46	71	96
-1	-102	-77	-52	-39	-27	-14	-7	-2	3	10	23	35	48	73	98
0	-100	-75	-50	-37	-25	-12	-5	0	5	12	25	37	50	75	100
1	-98	-73	-48	-35	-23	-10	-3	2	7	14	27	39	52	77	102
2	-96	-71	-46	-33	-21	-8	-1	4	9	16	29	41	54	79	104
3	-94	-69	-44	-31	-19	-6	1	6	11	18	31	43	56	81	106
4	-92	-67	-42	-29	-17	-4	3	8	13	20	33	45	58	83	108
5	-90	-65	-40	-27	-15	-2	5	10	15	22	35	47	60	85	110
6	-88	-63	-38	-25	-13	0	7	12	17	24	37	49	62	87	112
7	-86	-61	-36	-23	-11	1	9	14	19	26	39	51	64	89	114
8	-84	-59	-34	-21	-9	3	11	16	21	28	41	53	66	91	116
10	-80	-55	-30	-17	-5	7	15	20	25	32	45	57	70	95	120
12	-76	-51	-26	-13	-1	11	19	24	29	36	49	61	74	99	124
14	-72	-47	-22	-9	3	15	23	28	33	40	53	65	78	103	128
16	-68	-43	-18	-5	7	19	27	32	37	44	57	69	82	107	132
18	-64	-39	-14	-1	11	23	31	36	41	48	61	73	86	111	136
20	-60	-35	-10	2	15	27	35	40	45	52	65	77	90	115	140
25	-50	-25	0	12	25	37	45	50	55	62	75	87	100	125	150
30	-40	-15	10	22	35	47	55	60	65	72	85	97	110	135	160
40	-20	5	30	42	55	67	75	80	85	92	105	117	130	155	180

Table 5-63

Separation, inch: 7

Leg Distance, inch: 40

SHIM REQUIRED FOR ALIGNMENT

Reverse Indicator Method

DriveN TIR, MILS	DriveR TIR, MILS														
	-40	-30	-20	-15	-10	-5	-2	0	2	5	10	15	20	30	40
-40	-208	-180	-151	-137	-122	-108	-100	-94	-88	-80	-65	-51	-37	-8	20
-30	-185	-156	-127	-113	-99	-85	-76	-70	-65	-56	-42	-27	-13	15	43
-25	-173	-144	-116	-101	-87	-73	-64	-58	-53	-44	-30	-16	-1	26	55
-20	-161	-132	-104	-90	-75	-61	-52	-47	-41	-32	-18	-4	10	38	67
-18	-156	-128	-99	-85	-71	-56	-48	-42	-36	-28	-13	0	14	43	71
-16	-152	-123	-94	-80	-66	-52	-43	-37	-32	-23	-9	5	19	48	76
-14	-147	-118	-90	-75	-61	-47	-38	-33	-27	-18	-4	9	24	52	81
-12	-142	-114	-85	-71	-56	-42	-34	-28	-22	-14	0	14	28	57	86
-10	-137	-109	-80	-66	-52	-37	-29	-23	-17	-9	5	19	33	62	90
-8	-133	-104	-76	-61	-47	-33	-24	-18	-13	-4	9	24	38	66	95
-7	-130	-102	-73	-59	-45	-30	-22	-16	-10	-2	12	26	40	69	97
-6	-128	-99	-71	-57	-42	-28	-19	-14	-8	0	14	28	43	71	100
-5	-126	-97	-68	-54	-40	-26	-17	-11	-6	2	16	31	45	73	102
-4	-123	-95	-66	-52	-38	-23	-15	-9	-3	4	19	33	47	76	104
-3	-121	-92	-64	-49	-35	-21	-12	-7	-1	7	21	35	50	78	107
-2	-119	-90	-61	-47	-33	-19	-10	-4	1	9	23	38	52	81	109
-1	-116	-88	-59	-45	-30	-16	-8	-2	3	11	26	40	54	83	111
0	-114	-85	-57	-42	-28	-14	-5	0	5	14	28	42	57	85	114
1	-111	-83	-54	-40	-26	-11	-3	2	8	16	30	45	59	88	116
2	-109	-81	-52	-38	-23	-9	-1	4	10	19	33	47	61	90	119
3	-107	-78	-50	-35	-21	-7	1	7	12	21	35	49	64	92	121
4	-104	-76	-47	-33	-19	-4	3	9	15	23	38	52	66	95	123
5	-102	-73	-45	-31	-16	-2	6	11	17	26	40	54	68	97	126
6	-100	-71	-43	-28	-14	0	8	14	19	28	42	57	71	99	128
7	-97	-69	-40	-26	-12	2	10	16	22	30	45	59	73	102	130
8	-95	-66	-38	-24	-9	4	13	18	24	33	47	61	76	104	133
10	-90	-62	-33	-19	-5	9	17	23	29	37	52	66	80	109	137
12	-86	-57	-28	-14	0	14	22	28	34	42	56	71	85	114	142
14	-81	-52	-24	-9	4	18	27	33	38	47	61	75	90	118	147
16	-76	-48	-19	-5	9	23	32	37	43	52	66	80	94	123	152
18	-71	-43	-14	0	13	28	36	42	48	56	71	85	99	128	156
20	-67	-38	-10	4	18	32	41	47	52	61	75	90	104	132	161
25	-55	-26	1	16	30	44	53	58	64	73	87	101	116	144	173
30	-43	-15	13	27	42	56	65	70	76	85	99	113	127	156	185
40	-20	8	37	51	65	80	88	94	100	108	122	137	151	180	208

Table 5-64

Separation, inch: 7

Leg Distance, inch: 50

SHIM REQUIRED FOR ALIGNMENT

Reverse Indicator Method

DriveN TIR MILS	DriveR TIR, MILS -40	-30	-20	-15	-10	-5	-2	0	2	5	10	15	20	30	40
-40	-265	-230	-194	-176	-158	-140	-130	-122	-115	-105	-87	-69	-51	-15	20
-30	-235	-199	-163	-145	-127	-110	-99	-92	-85	-74	-56	-38	-20	15	50
-25	-219	-183	-148	-130	-112	-94	-83	-76	-69	-58	-41	-23	-5	30	66
-20	-204	-168	-132	-115	-97	-79	-68	-61	-54	-43	-25	-7	10	45	81
-18	-198	-162	-126	-108	-91	-73	-62	-55	-48	-37	-19	-1	16	51	87
-16	-192	-156	-120	-102	-84	-67	-56	-49	-42	-31	-13	4	22	58	93
-14	-185	-150	-114	-96	-78	-60	-50	-43	-35	-25	-7	10	28	64	99
-12	-179	-144	-108	-90	-72	-54	-44	-36	-29	-19	-1	16	34	70	106
-10	-173	-137	-102	-84	-66	-48	-37	-30	-23	-12	5	22	40	76	112
-8	-167	-131	-96	-78	-60	-42	-31	-24	-17	-6	11	29	46	82	118
-7	-164	-128	-92	-75	-57	-39	-28	-21	-14	-3	14	32	49	85	121
-6	-161	-125	-89	-72	-54	-36	-25	-18	-11	0	17	35	53	88	124
-5	-158	-122	-86	-68	-51	-33	-22	-15	-8	2	20	38	56	91	127
-4	-155	-119	-83	-65	-48	-30	-19	-12	-5	5	23	41	59	94	130
-3	-152	-116	-80	-62	-44	-27	-16	-9	-2	8	26	44	62	97	133
-2	-149	-113	-77	-59	-41	-24	-13	-6	1	11	29	47	65	101	136
-1	-145	-110	-74	-56	-38	-20	-10	-3	4	14	32	50	68	104	139
0	-142	-107	-71	-53	-35	-17	-7	0	7	17	35	53	71	107	142
1	-139	-104	-68	-50	-32	-14	-4	3	10	20	38	56	74	110	145
2	-136	-101	-65	-47	-29	-11	-1	6	13	24	41	59	77	113	149
3	-133	-97	-62	-44	-26	-8	2	9	16	27	44	62	80	116	152
4	-130	-94	-59	-41	-23	-5	5	12	19	30	48	65	83	119	155
5	-127	-91	-56	-38	-20	-2	8	15	22	33	51	68	86	122	158
6	-124	-88	-53	-35	-17	0	11	18	25	36	54	72	89	125	161
7	-121	-85	-49	-32	-14	3	14	21	28	39	57	75	92	128	164
8	-118	-82	-46	-29	-11	6	17	24	31	42	60	78	96	131	167
10	-112	-76	-40	-22	-5	12	23	30	37	48	66	84	102	137	173
12	-106	-70	-34	-16	1	19	29	36	44	54	72	90	108	144	179
14	-99	-64	-28	-10	7	25	35	43	50	60	78	96	114	150	185
16	-93	-58	-22	-4	13	31	42	49	56	67	84	102	120	156	192
18	-87	-51	-16	1	19	37	48	55	62	73	91	108	126	162	198
20	-81	-45	-10	7	25	43	54	61	68	79	97	115	132	168	204
25	-66	-30	5	23	41	58	69	76	83	94	112	130	148	183	219
30	-50	-15	20	38	56	74	85	92	99	110	127	145	163	199	235
40	-20	15	51	69	87	105	115	122	130	140	158	176	194	230	265

Table 5-65

Separation, inch: 7

Leg Distance, inch: 60

SHIM REQUIRED FOR ALIGNMENT

Reverse Indicator Method

riveN DriveR TIR, MILS

IR

ILS	-40	-30	-20	-15	-10	-5	-2	0	2	5	10	15	20	30	40
-40	-322	-280	-237	-215	-194	-172	-160	-151	-142	-130	-108	-87	-65	-22	20
-30	-285	-242	-199	-177	-156	-135	-122	-113	-105	-92	-70	-49	-27	15	57
-25	-266	-223	-180	-158	-137	-116	-103	-94	-86	-73	-51	-30	-8	33	76
-20	-247	-204	-161	-140	-118	-97	-84	-75	-67	-54	-32	-11	10	52	95
-18	-239	-196	-153	-132	-111	-89	-76	-68	-59	-46	-25	-3	17	60	103
-16	-232	-189	-146	-124	-103	-82	-69	-60	-52	-39	-17	3	25	68	110
-14	-224	-181	-138	-117	-95	-74	-61	-53	-44	-31	-10	11	32	75	118
-12	-216	-174	-131	-109	-88	-66	-54	-45	-36	-24	-2	18	40	83	126
-10	-209	-166	-123	-102	-80	-59	-46	-37	-29	-16	5	26	47	90	133
-8	-201	-158	-116	-94	-73	-51	-38	-30	-21	-8	12	34	55	98	141
-7	-197	-155	-112	-90	-69	-47	-35	-26	-17	-5	16	37	59	102	144
-6	-194	-151	-108	-87	-65	-44	-31	-22	-14	-1	20	41	63	105	148
-5	-190	-147	-104	-83	-61	-40	-27	-18	-10	2	23	45	66	109	152
-4	-186	-143	-100	-79	-58	-36	-23	-15	-6	6	27	49	70	113	156
-3	-182	-139	-97	-75	-54	-32	-19	-11	-2	10	31	52	74	117	160
-2	-179	-136	-93	-71	-50	-29	-16	-7	1	13	35	56	78	121	163
-1	-175	-132	-89	-68	-46	-25	-12	-3	4	17	39	60	81	124	167
0	-171	-128	-85	-64	-42	-21	-8	0	8	21	42	64	85	128	171
1	-167	-124	-81	-60	-39	-17	-4	3	12	25	46	68	89	132	175
2	-163	-121	-78	-56	-35	-13	-1	7	16	29	50	71	93	136	179
3	-160	-117	-74	-52	-31	-10	2	11	19	32	54	75	97	139	182
4	-156	-113	-70	-49	-27	-6	6	15	23	36	58	79	100	143	186
5	-152	-109	-66	-45	-23	-2	10	18	27	40	61	83	104	147	190
6	-148	-105	-63	-41	-20	1	14	22	31	44	65	87	108	151	194
7	-144	-102	-59	-37	-16	5	17	26	35	47	69	90	112	155	197
8	-141	-98	-55	-34	-12	8	21	30	38	51	73	94	116	158	201
10	-133	-90	-47	-26	-5	16	29	37	46	59	80	102	123	166	209
12	-126	-83	-40	-18	2	24	36	45	54	66	88	109	131	174	216
14	-118	-75	-32	-11	10	31	44	53	61	74	95	117	138	181	224
16	-110	-68	-25	-3	17	39	52	60	69	82	103	124	146	189	232
18	-103	-60	-17	3	25	46	59	68	76	89	111	132	153	196	239
20	-95	-52	-10	11	32	54	67	75	84	97	118	140	161	204	247
25	-76	-33	8	30	51	73	86	94	103	116	137	158	180	223	266
30	-57	-15	27	49	70	92	105	113	122	135	156	177	199	242	285
40	-20	22	65	87	108	130	142	151	160	172	194	215	237	280	322

Table 5-66

Separation, inch: 7

Leg Distance, inch: 70

SHIM REQUIRED FOR ALIGNMENT

Reverse Indicator Method

DriveN TIR MILS	DriveR TIR, MILS -40	-30	-20	-15	-10	-5	-2	0	2	5	10	15	20	30	40
-40	-380	-330	-280	-255	-230	-205	-190	-180	-170	-155	-130	-105	-80	-30	20
-30	-335	-285	-235	-210	-185	-160	-145	-135	-125	-110	-85	-60	-35	15	65
-25	-312	-262	-212	-187	-162	-137	-122	-112	-102	-87	-62	-37	-12	37	87
-20	-290	-240	-190	-165	-140	-115	-100	-90	-80	-65	-40	-15	10	60	110
-18	-281	-231	-181	-156	-131	-106	-91	-81	-71	-56	-31	-6	19	69	119
-16	-272	-222	-172	-147	-122	-97	-82	-72	-62	-47	-22	3	28	78	128
-14	-263	-213	-163	-138	-113	-88	-73	-63	-53	-38	-13	12	37	87	137
-12	-254	-204	-154	-129	-104	-79	-64	-54	-44	-29	-4	21	46	96	146
-10	-245	-195	-145	-120	-95	-70	-55	-45	-35	-20	5	30	55	105	155
-8	-236	-186	-136	-111	-86	-61	-46	-36	-26	-11	14	39	64	114	164
-7	-231	-181	-131	-106	-81	-56	-41	-31	-21	-6	18	43	68	118	168
-6	-227	-177	-127	-102	-77	-52	-37	-27	-17	-2	23	48	73	123	173
-5	-222	-172	-122	-97	-72	-47	-32	-22	-12	2	27	52	77	127	177
-4	-218	-168	-118	-93	-68	-43	-28	-18	-8	7	32	57	82	132	182
-3	-213	-163	-113	-88	-63	-38	-23	-13	-3	11	36	61	86	136	186
-2	-209	-159	-109	-84	-59	-34	-19	-9	1	16	41	66	91	141	191
-1	-204	-154	-104	-79	-54	-29	-14	-4	5	20	45	70	95	145	195
0	-200	-150	-100	-75	-50	-25	-10	0	10	25	50	75	100	150	200
1	-195	-145	-95	-70	-45	-20	-5	4	14	29	54	79	104	154	204
2	-191	-141	-91	-66	-41	-16	-1	9	19	34	59	84	109	159	209
3	-186	-136	-86	-61	-36	-11	3	13	23	38	63	88	113	163	213
4	-182	-132	-82	-57	-32	-7	8	18	28	43	68	93	118	168	218
5	-177	-127	-77	-52	-27	-2	12	22	32	47	72	97	122	172	222
6	-173	-123	-73	-48	-23	2	17	27	37	52	77	102	127	177	227
7	-168	-118	-68	-43	-18	6	21	31	41	56	81	106	131	181	231
8	-164	-114	-64	-39	-14	11	26	36	46	61	86	111	136	186	236
10	-155	-105	-55	-30	-5	20	35	45	55	70	95	120	145	195	245
12	-146	-96	-46	-21	4	29	44	54	64	79	104	129	154	204	254
14	-137	-87	-37	-12	13	38	53	63	73	88	113	138	163	213	263
16	-128	-78	-28	-3	22	47	62	72	82	97	122	147	172	222	272
18	-119	-69	-19	6	31	56	71	81	91	106	131	156	181	231	281
20	-110	-60	-10	15	40	65	80	90	100	115	140	165	190	240	290
25	-87	-37	12	37	62	87	102	112	122	137	162	187	212	262	312
30	-65	-15	35	60	85	110	125	135	145	160	185	210	235	285	335
40	-20	30	80	105	130	155	170	180	190	205	230	255	280	330	380

Table 5-67

Separation, inch: 8

Leg Distance, inch: 25

SHIM REQUIRED FOR ALIGNMENT

Reverse Indicator Method

DriveN TIR MILS	DriveR TIR, MILS -40	-30	-20	-15	-10	-5	-2	0	2	5	10	15	20	30	40
-40	-105	-89	-73	-65	-58	-50	-45	-42	-39	-34	-26	-19	-11	4	20
-30	-94	-78	-63	-55	-47	-39	-35	-31	-28	-24	-16	-8	0	15	30
-25	-89	-73	-57	-50	-42	-34	-29	-26	-23	-18	-10	-3	4	20	35
-20	-83	-68	-52	-44	-36	-29	-24	-21	-18	-13	-5	2	10	25	41
-18	-81	-66	-50	-42	-34	-26	-22	-19	-16	-11	-3	4	12	27	43
-16	-79	-63	-48	-40	-32	-24	-20	-17	-13	-9	-1	6	14	29	45
-14	-77	-61	-46	-38	-30	-22	-18	-14	-11	-7	0	8	16	32	47
-12	-75	-59	-44	-36	-28	-20	-15	-12	-9	-4	2	10	18	34	49
-10	-73	-57	-41	-34	-26	-18	-13	-10	-7	-2	5	12	20	36	51
-8	-71	-55	-39	-31	-24	-16	-11	-8	-5	0	7	14	22	38	54
-7	-69	-54	-38	-30	-23	-15	-10	-7	-4	0	8	16	23	39	55
-6	-68	-53	-37	-29	-22	-14	-9	-6	-3	1	9	17	24	40	56
-5	-67	-52	-36	-28	-20	-13	-8	-5	-2	2	10	18	25	41	57
-4	-66	-51	-35	-27	-19	-12	-7	-4	-1	3	11	19	27	42	58
-3	-65	-50	-34	-26	-18	-11	-6	-3	0	4	12	20	28	43	59
-2	-64	-49	-33	-25	-17	-9	-5	-2	1	5	13	21	29	44	60
-1	-63	-47	-32	-24	-16	-8	-4	-1	2	6	14	22	30	45	61
0	-62	-46	-31	-23	-15	-7	-3	0	3	7	15	23	31	46	62
1	-61	-45	-30	-22	-14	-6	-2	1	4	8	16	24	32	47	63
2	-60	-44	-29	-21	-13	-5	-1	2	5	9	17	25	33	49	64
3	-59	-43	-28	-20	-12	-4	0	3	6	11	18	26	34	50	65
4	-58	-42	-27	-19	-11	-3	1	4	7	12	19	27	35	51	66
5	-57	-41	-25	-18	-10	-2	2	5	8	13	20	28	36	52	67
6	-56	-40	-24	-17	-9	-1	3	6	9	14	22	29	37	53	68
7	-55	-39	-23	-16	-8	0	4	7	10	15	23	30	38	54	69
8	-54	-38	-22	-14	-7	0	5	8	11	16	24	31	39	55	71
10	-51	-36	-20	-12	-5	2	7	10	13	18	26	34	41	57	73
12	-49	-34	-18	-10	-2	4	9	12	15	20	28	36	44	59	75
14	-47	-32	-16	-8	0	7	11	14	18	22	30	38	46	61	77
16	-45	-29	-14	-6	1	9	13	17	20	24	32	40	48	63	79
18	-43	-27	-12	-4	3	11	16	19	22	26	34	42	50	66	81
20	-41	-25	-10	-2	5	13	18	21	24	29	36	44	52	68	83
25	-35	-20	-4	3	10	18	23	26	29	34	42	50	57	73	89
30	-30	-15	0	8	16	24	28	31	35	39	47	55	63	78	94
40	-20	-4	11	19	26	34	39	42	45	50	58	65	73	89	105

Table 5-68

Separation, inch: 8

Leg Distance, inch: 30

SHIM REQUIRED FOR ALIGNMENT

Reverse Indicator Method

DriveN TIR MILS	DriveR TIR, MILS -40	-30	-20	-15	-10	-5	-2	0	2	5	10	15	20	30	40
-40	-130	-111	-92	-83	-73	-64	-58	-55	-51	-45	-36	-26	-17	1	20
-30	-116	-97	-78	-69	-60	-50	-45	-41	-37	-31	-22	-13	-3	15	33
-25	-109	-90	-71	-62	-53	-43	-38	-34	-30	-25	-15	-6	3	21	40
-20	-102	-83	-65	-55	-46	-36	-31	-27	-23	-18	-8	0	10	28	47
-18	-99	-81	-62	-52	-43	-34	-28	-24	-21	-15	-6	3	12	31	50
-16	-97	-78	-59	-50	-40	-31	-25	-22	-18	-12	-3	6	15	34	53
-14	-94	-75	-56	-47	-38	-28	-23	-19	-15	-9	0	8	18	37	55
-12	-91	-72	-54	-44	-35	-25	-20	-16	-12	-7	2	11	21	39	58
-10	-88	-70	-51	-41	-32	-23	-17	-13	-10	-4	5	14	23	42	61
-8	-86	-67	-48	-39	-29	-20	-14	-11	-7	-1	7	17	26	45	64
-7	-84	-65	-47	-37	-28	-19	-13	-9	-5	0	9	18	27	46	65
-6	-83	-64	-45	-36	-27	-17	-12	-8	-4	1	10	19	29	48	66
-5	-81	-63	-44	-35	-25	-16	-10	-6	-3	2	11	21	30	49	68
-4	-80	-61	-43	-33	-24	-14	-9	-5	-1	3	13	22	32	50	69
-3	-79	-60	-41	-32	-22	-13	-7	-4	0	5	14	24	33	52	70
-2	-77	-59	-40	-30	-21	-12	-6	-2	1	6	16	25	34	53	72
-1	-76	-57	-38	-29	-20	-10	-5	-1	2	8	17	26	36	54	73
0	-75	-56	-37	-28	-18	-9	-3	0	3	9	18	28	37	56	75
1	-73	-54	-36	-26	-17	-8	-2	1	5	10	20	29	38	57	76
2	-72	-53	-34	-25	-16	-6	-1	2	6	12	21	30	40	59	77
3	-70	-52	-33	-24	-14	-5	0	4	7	13	22	32	41	60	79
4	-69	-50	-32	-22	-13	-3	1	5	9	14	24	33	43	61	80
5	-68	-49	-30	-21	-11	-2	3	6	10	16	25	35	44	63	81
6	-66	-48	-29	-19	-10	-1	4	8	12	17	27	36	45	64	83
7	-65	-46	-27	-18	-9	0	5	9	13	19	28	37	47	65	84
8	-64	-45	-26	-17	-7	1	7	11	14	20	29	39	48	67	86
10	-61	-42	-23	-14	-5	4	10	13	17	23	32	41	51	70	88
12	-58	-39	-21	-11	-2	7	12	16	20	25	35	44	54	72	91
14	-55	-37	-18	-8	0	9	15	19	23	28	38	47	56	75	94
16	-53	-34	-15	-6	3	12	18	22	25	31	40	50	59	78	97
18	-50	-31	-12	-3	6	15	21	24	28	34	43	52	62	81	99
20	-47	-28	-10	0	8	18	23	27	31	36	46	55	65	83	102
25	-40	-21	-3	6	15	25	30	34	38	43	53	62	71	90	109
30	-33	-15	3	13	22	31	37	41	45	50	60	69	78	97	116
40	-20	-1	17	26	36	45	51	55	58	64	73	83	92	111	130

Table 5-69

Separation, inch: 8

Leg Distance, inch: 35

SHIM REQUIRED FOR ALIGNMENT

Reverse Indicator Method

DriveR TIR, MILS

riveN IR ILS	-40	-30	-20	-15	-10	-5	-2	0	2	5	10	15	20	30	40
-40	-155	-133	-111	-100	-89	-78	-71	-67	-63	-56	-45	-34	-23	-1	20
-30	-138	-116	-94	-83	-72	-61	-55	-50	-46	-39	-28	-17	-6	15	36
-25	-129	-107	-85	-75	-64	-53	-46	-42	-37	-31	-20	-9	1	23	45
-20	-121	-99	-77	-66	-55	-44	-38	-33	-29	-22	-11	0	10	31	53
-18	-117	-96	-74	-63	-52	-41	-34	-30	-26	-19	-8	2	13	35	57
-16	-114	-92	-70	-59	-48	-37	-31	-27	-22	-16	-5	5	16	38	60
-14	-111	-89	-67	-56	-45	-34	-28	-23	-19	-12	-1	9	20	42	63
-12	-107	-85	-64	-53	-42	-31	-24	-20	-15	-9	1	12	23	45	67
-10	-104	-82	-60	-49	-38	-27	-21	-16	-12	-5	5	15	26	48	70
-8	-101	-79	-57	-46	-35	-24	-17	-13	-9	-2	8	19	30	52	74
-7	-99	-77	-55	-44	-33	-22	-16	-11	-7	0	10	21	31	53	75
-6	-97	-75	-53	-42	-32	-21	-14	-10	-5	0	11	22	33	55	77
-5	-95	-74	-52	-41	-30	-19	-12	-8	-4	2	13	24	35	57	79
-4	-94	-72	-50	-39	-28	-17	-11	-6	-2	4	15	26	37	58	80
-3	-92	-70	-48	-37	-26	-16	-9	-5	0	5	16	27	38	60	82
-2	-90	-69	-47	-36	-25	-14	-7	-3	1	7	18	29	40	62	84
-1	-89	-67	-45	-34	-23	-12	-6	-1	2	9	20	31	42	63	85
0	-87	-65	-43	-32	-21	-10	-4	0	4	10	21	32	43	65	87
1	-85	-63	-42	-31	-20	-9	-2	1	6	12	23	34	45	67	89
2	-84	-62	-40	-29	-18	-7	-1	3	7	14	25	36	47	69	90
3	-82	-60	-38	-27	-16	-5	0	5	9	16	26	37	48	70	92
4	-80	-58	-37	-26	-15	-4	2	6	11	17	28	39	50	72	94
5	-79	-57	-35	-24	-13	-2	4	8	12	19	30	41	52	74	95
6	-77	-55	-33	-22	-11	0	5	10	14	21	32	42	53	75	97
7	-75	-53	-31	-21	-10	0	7	11	16	22	33	44	55	77	99
8	-74	-52	-30	-19	-8	2	9	13	17	24	35	46	57	79	101
10	-70	-48	-26	-15	-5	5	12	16	21	27	38	49	60	82	104
12	-67	-45	-23	-12	-1	9	15	20	24	31	42	53	64	85	107
14	-63	-42	-20	-9	1	12	19	23	28	34	45	56	67	89	111
16	-60	-38	-16	-5	5	16	22	27	31	37	48	59	70	92	114
18	-57	-35	-13	-2	8	19	26	30	34	41	52	63	74	96	117
20	-53	-31	-10	0	11	22	29	33	38	44	55	66	77	99	121
25	-45	-23	-1	9	20	31	37	42	46	53	64	75	85	107	129
30	-36	-15	6	17	28	39	46	50	55	61	72	83	94	116	138
40	-20	1	23	34	45	56	63	67	71	78	89	100	111	133	155

Table 5-70

Separation, inch: 8

Leg Distance, inch: 40

SHIM REQUIRED FOR ALIGNMENT

Reverse Indicator Method

DriveN TIR MILS	DriveR TIR, MILS -40	-30	-20	-15	-10	-5	-2	0	2	5	10	15	20	30	40
-40	-180	-155	-130	-117	-105	-92	-85	-80	-75	-67	-55	-42	-30	-5	20
-30	-160	-135	-110	-97	-85	-72	-65	-60	-55	-47	-35	-22	-10	15	40
-25	-150	-125	-100	-87	-75	-62	-55	-50	-45	-37	-25	-12	0	25	50
-20	-140	-115	-90	-77	-65	-52	-45	-40	-35	-27	-15	-2	10	35	60
-18	-136	-111	-86	-73	-61	-48	-41	-36	-31	-23	-11	1	14	39	64
-16	-132	-107	-82	-69	-57	-44	-37	-32	-27	-19	-7	5	18	43	68
-14	-128	-103	-78	-65	-53	-40	-33	-28	-23	-15	-3	9	22	47	72
-12	-124	-99	-74	-61	-49	-36	-29	-24	-19	-11	1	13	26	51	76
-10	-120	-95	-70	-57	-45	-32	-25	-20	-15	-7	5	17	30	55	80
-8	-116	-91	-66	-53	-41	-28	-21	-16	-11	-3	9	21	34	59	84
-7	-114	-89	-64	-51	-39	-26	-19	-14	-9	-1	11	23	36	61	86
-6	-112	-87	-62	-49	-37	-24	-17	-12	-7	0	13	25	38	63	88
-5	-110	-85	-60	-47	-35	-22	-15	-10	-5	2	15	27	40	65	90
-4	-108	-83	-58	-45	-33	-20	-13	-8	-3	4	17	29	42	67	92
-3	-106	-81	-56	-43	-31	-18	-11	-6	-1	6	19	31	44	69	94
-2	-104	-79	-54	-41	-29	-16	-9	-4	1	8	21	33	46	71	96
-1	-102	-77	-52	-39	-27	-14	-7	-2	3	10	23	35	48	73	98
0	-100	-75	-50	-37	-25	-12	-5	0	5	12	25	37	50	75	100
1	-98	-73	-48	-35	-23	-10	-3	2	7	14	27	39	52	77	102
2	-96	-71	-46	-33	-21	-8	-1	4	9	16	29	41	54	79	104
3	-94	-69	-44	-31	-19	-6	1	6	11	18	31	43	56	81	106
4	-92	-67	-42	-29	-17	-4	3	8	13	20	33	45	58	83	108
5	-90	-65	-40	-27	-15	-2	5	10	15	22	35	47	60	85	110
6	-88	-63	-38	-25	-13	0	7	12	17	24	37	49	62	87	112
7	-86	-61	-36	-23	-11	1	9	14	19	26	39	51	64	89	114
8	-84	-59	-34	-21	-9	3	11	16	21	28	41	53	66	91	116
10	-80	-55	-30	-17	-5	7	15	20	25	32	45	57	70	95	120
12	-76	-51	-26	-13	-1	11	19	24	29	36	49	61	74	99	124
14	-72	-47	-22	-9	3	15	23	28	33	40	53	65	78	103	128
16	-68	-43	-18	-5	7	19	27	32	37	44	57	69	82	107	132
18	-64	-39	-14	-1	11	23	31	36	41	48	61	73	86	111	136
20	-60	-35	-10	2	15	27	35	40	45	52	65	77	90	115	140
25	-50	-25	0	12	25	37	45	50	55	62	75	87	100	125	150
30	-40	-15	10	22	35	47	55	60	65	72	85	97	110	135	160
40	-20	5	30	42	55	67	75	80	85	92	105	117	130	155	180

Table 5-71

Separation, inch: 8

Leg Distance, inch: 50

SHIM REQUIRED FOR ALIGNMENT

Reverse Indicator Method

DriveN TIR MILS	DriveR TIR, MILS -40	-30	-20	-15	-10	-5	-2	0	2	5	10	15	20	30	40
-40	-230	-198	-167	-151	-136	-120	-111	-105	-98	-89	-73	-58	-42	-11	20
-30	-203	-172	-141	-125	-110	-94	-85	-78	-72	-63	-47	-31	-16	15	46
-25	-190	-159	-128	-112	-96	-81	-71	-65	-59	-50	-34	-18	-3	28	59
-20	-177	-146	-115	-99	-83	-68	-58	-52	-46	-36	-21	-5	10	41	72
-18	-172	-141	-109	-94	-78	-62	-53	-47	-41	-31	-16	0	15	46	77
-16	-167	-135	-104	-88	-73	-57	-48	-42	-35	-26	-10	4	20	51	83
-14	-161	-130	-99	-83	-68	-52	-43	-36	-30	-21	-5	10	25	57	88
-12	-156	-125	-94	-78	-62	-47	-37	-31	-25	-15	0	15	31	62	93
-10	-151	-120	-88	-73	-57	-41	-32	-26	-20	-10	5	20	36	67	98
-8	-146	-114	-83	-67	-52	-36	-27	-21	-14	-5	10	25	41	72	104
-7	-143	-112	-80	-65	-49	-34	-24	-18	-12	-2	12	28	44	75	106
-6	-140	-109	-78	-62	-47	-31	-22	-15	-9	0	15	31	46	78	109
-5	-138	-106	-75	-60	-44	-28	-19	-13	-6	2	18	33	49	80	111
-4	-135	-104	-73	-57	-41	-26	-16	-10	-4	5	20	36	52	83	114
-3	-132	-101	-70	-54	-39	-23	-14	-7	-1	7	23	39	54	85	117
-2	-130	-99	-67	-52	-36	-20	-11	-5	1	10	26	41	57	88	119
-1	-127	-96	-65	-49	-33	-18	-8	-2	3	13	28	44	59	91	122
0	-125	-93	-62	-46	-31	-15	-6	0	6	15	31	46	62	93	125
1	-122	-91	-59	-44	-28	-13	-3	2	8	18	33	49	65	96	127
2	-119	-88	-57	-41	-26	-10	-1	5	11	20	36	52	67	99	130
3	-117	-85	-54	-39	-23	-7	1	7	14	23	39	54	70	101	132
4	-114	-83	-52	-36	-20	-5	4	10	16	26	41	57	73	104	135
5	-111	-80	-49	-33	-18	-2	6	13	19	28	44	60	75	106	138
6	-109	-78	-46	-31	-15	0	9	15	22	31	47	62	78	109	140
7	-106	-75	-44	-28	-12	2	12	18	24	34	49	65	80	112	143
8	-104	-72	-41	-25	-10	5	14	21	27	36	52	67	83	114	146
10	-98	-67	-36	-20	-5	10	20	26	32	41	57	73	88	120	151
12	-93	-62	-31	-15	0	15	25	31	37	47	62	78	94	125	156
14	-88	-57	-25	-10	5	21	30	36	43	52	68	83	99	130	161
16	-83	-51	-20	-4	10	26	35	42	48	57	73	88	104	135	167
18	-77	-46	-15	0	16	31	41	47	53	62	78	94	109	141	172
20	-72	-41	-10	5	21	36	46	52	58	68	83	99	115	146	177
25	-59	-28	3	18	34	50	59	65	71	81	96	112	128	159	190
30	-46	-15	16	31	47	63	72	78	85	94	110	125	141	172	203
40	-20	11	42	58	73	89	98	105	111	120	136	151	167	198	230

Table 5-72

Separation, inch: 8

Leg Distance, inch: 60

SHIM REQUIRED FOR ALIGNMENT

Reverse Indicator Method

DriveN TIR MILS	DriveR TIR, MILS -40	-30	-20	-15	-10	-5	-2	0	2	5	10	15	20	30	40
-40	-280	-242	-205	-186	-167	-148	-137	-130	-122	-111	-92	-73	-55	-17	20
-30	-247	-210	-172	-153	-135	-116	-105	-97	-90	-78	-60	-41	-22	15	52
-25	-231	-193	-156	-137	-118	-100	-88	-81	-73	-62	-43	-25	-6	31	68
-20	-215	-177	-140	-121	-102	-83	-72	-65	-57	-46	-27	-8	10	47	85
-18	-208	-171	-133	-114	-96	-77	-66	-58	-51	-39	-21	-2	16	54	91
-16	-202	-164	-127	-108	-89	-70	-59	-52	-44	-33	-14	4	23	60	98
-14	-195	-158	-120	-101	-83	-64	-53	-45	-38	-26	-8	10	29	67	104
-12	-189	-151	-114	-95	-76	-57	-46	-39	-31	-20	-1	17	36	73	111
-10	-182	-145	-107	-88	-70	-51	-40	-32	-25	-13	5	23	42	80	117
-8	-176	-138	-101	-82	-63	-44	-33	-26	-18	-7	11	30	49	86	124
-7	-172	-135	-97	-79	-60	-41	-30	-22	-15	-4	14	33	52	89	127
-6	-169	-132	-94	-75	-57	-38	-27	-19	-12	0	18	36	55	93	130
-5	-166	-128	-91	-72	-53	-35	-23	-16	-8	2	21	40	58	96	133
-4	-163	-125	-88	-69	-50	-31	-20	-13	-5	5	24	43	62	99	137
-3	-159	-122	-84	-66	-47	-28	-17	-9	-2	9	27	46	65	102	140
-2	-156	-119	-81	-62	-44	-25	-14	-6	1	12	31	49	68	106	143
-1	-153	-115	-78	-59	-40	-22	-10	-3	4	15	34	53	71	109	146
0	-150	-112	-75	-56	-37	-18	-7	0	7	18	37	56	75	112	150
1	-146	-109	-71	-53	-34	-15	-4	3	10	22	40	59	78	115	153
2	-143	-106	-68	-49	-31	-12	-1	6	14	25	44	62	81	119	156
3	-140	-102	-65	-46	-27	-9	2	9	17	28	47	66	84	122	159
4	-137	-99	-62	-43	-24	-5	5	13	20	31	50	69	88	125	163
5	-133	-96	-58	-40	-21	-2	8	16	23	35	53	72	91	128	166
6	-130	-93	-55	-36	-18	0	12	19	27	38	57	75	94	132	169
7	-127	-89	-52	-33	-14	4	15	22	30	41	60	79	97	135	172
8	-124	-86	-49	-30	-11	7	18	26	33	44	63	82	101	138	176
10	-117	-80	-42	-23	-5	13	25	32	40	51	70	88	107	145	182
12	-111	-73	-36	-17	1	20	31	39	46	57	76	95	114	151	189
14	-104	-67	-29	-10	8	26	38	45	53	64	83	101	120	158	195
16	-98	-60	-23	-4	14	33	44	52	59	70	89	108	127	164	202
18	-91	-54	-16	2	21	39	51	58	66	77	96	114	133	171	208
20	-85	-47	-10	8	27	46	57	65	72	83	102	121	140	177	215
25	-68	-31	6	25	43	62	73	81	88	100	118	137	156	193	231
30	-52	-15	22	41	60	78	90	97	105	116	135	153	172	210	247
40	-20	17	55	73	92	111	122	130	137	148	167	186	205	242	280

Table 5-73

Separation, inch: 8

Leg Distance, inch: 70

SHIM REQUIRED FOR ALIGNMENT

Reverse Indicator Method

riveN IR ILS	DriveR TIR, MILS -40	-30	-20	-15	-10	-5	-2	0	2	5	10	15	20	30	40
-40	-330	-286	-242	-220	-198	-176	-163	-155	-146	-133	-111	-89	-67	-23	20
-30	-291	-247	-203	-181	-160	-138	-125	-116	-107	-94	-72	-50	-28	15	58
-25	-271	-228	-184	-162	-140	-118	-105	-96	-88	-75	-53	-31	-9	34	78
-20	-252	-208	-165	-143	-121	-99	-86	-77	-68	-55	-33	-11	10	53	97
-18	-244	-201	-157	-135	-113	-91	-78	-69	-61	-47	-26	-4	17	61	105
-16	-237	-193	-149	-127	-105	-83	-70	-62	-53	-40	-18	3	25	69	113
-14	-229	-185	-141	-119	-98	-76	-63	-54	-45	-32	-10	11	33	77	120
-12	-221	-177	-134	-112	-90	-68	-55	-46	-37	-24	-2	19	41	84	128
-10	-213	-170	-126	-104	-82	-60	-47	-38	-30	-16	5	26	48	92	136
-8	-206	-162	-118	-96	-74	-52	-39	-31	-22	-9	12	34	56	100	144
-7	-202	-158	-114	-92	-70	-49	-35	-27	-18	-5	16	38	60	104	147
-6	-198	-154	-110	-88	-67	-45	-32	-23	-14	-1	20	42	64	108	151
-5	-194	-150	-106	-85	-63	-41	-28	-19	-10	2	24	46	68	111	155
-4	-190	-146	-103	-81	-59	-37	-24	-15	-6	6	28	50	72	115	159
-3	-186	-142	-99	-77	-55	-33	-20	-11	-2	10	32	54	75	119	163
-2	-182	-139	-95	-73	-51	-29	-16	-7	1	14	36	57	79	123	167
-1	-178	-135	-91	-69	-47	-25	-12	-3	4	18	39	61	83	127	171
0	-175	-131	-87	-65	-43	-21	-8	0	8	21	43	65	87	131	175
1	-171	-127	-83	-61	-39	-18	-4	3	12	25	47	69	91	135	178
2	-167	-123	-79	-57	-36	-14	-1	7	16	29	51	73	95	139	182
3	-163	-119	-75	-54	-32	-10	2	11	20	33	55	77	99	142	186
4	-159	-115	-72	-50	-28	-6	6	15	24	37	59	81	103	146	190
5	-155	-111	-68	-46	-24	-2	10	19	28	41	63	85	106	150	194
6	-151	-108	-64	-42	-20	1	14	23	32	45	67	88	110	154	198
7	-147	-104	-60	-38	-16	5	18	27	35	49	70	92	114	158	202
8	-144	-100	-56	-34	-12	9	22	31	39	52	74	96	118	162	206
10	-136	-92	-48	-26	-5	16	30	38	47	60	82	104	126	170	213
12	-128	-84	-41	-19	2	24	37	46	55	68	90	112	134	177	221
14	-120	-77	-33	-11	10	32	45	54	63	76	98	119	141	185	229
16	-113	-69	-25	-3	18	40	53	62	70	83	105	127	149	193	237
18	-105	-61	-17	4	26	47	61	69	78	91	113	135	157	201	244
20	-97	-53	-10	11	33	55	68	77	86	99	121	143	165	208	252
25	-78	-34	9	31	53	75	88	96	105	118	140	162	184	228	271
30	-58	-15	28	50	72	94	107	116	125	138	160	181	203	247	291
40	-20	23	67	89	111	133	146	155	163	176	198	220	242	286	330

Table 5-74

Separation, inch: 8

Leg Distance, inch: 80

SHIM REQUIRED FOR ALIGNMENT

Reverse Indicator Method

DriveN TIR MILS	DriveR TIR, MILS -40	-30	-20	-15	-10	-5	-2	0	2	5	10	15	20	30	40
-40	-380	-330	-280	-255	-230	-205	-190	-180	-170	-155	-130	-105	-80	-30	20
-30	-335	-285	-235	-210	-185	-160	-145	-135	-125	-110	-85	-60	-35	15	65
-25	-312	-262	-212	-187	-162	-137	-122	-112	-102	-87	-62	-37	-12	37	87
-20	-290	-240	-190	-165	-140	-115	-100	-90	-80	-65	-40	-15	10	60	110
-18	-281	-231	-181	-156	-131	-106	-91	-81	-71	-56	-31	-6	19	69	119
-16	-272	-222	-172	-147	-122	-97	-82	-72	-62	-47	-22	3	28	78	128
-14	-263	-213	-163	-138	-113	-88	-73	-63	-53	-38	-13	12	37	87	137
-12	-254	-204	-154	-129	-104	-79	-64	-54	-44	-29	-4	21	46	96	146
-10	-245	-195	-145	-120	-95	-70	-55	-45	-35	-20	5	30	55	105	155
-8	-236	-186	-136	-111	-86	-61	-46	-36	-26	-11	14	39	64	114	164
-7	-231	-181	-131	-106	-81	-56	-41	-31	-21	-6	18	43	68	118	168
-6	-227	-177	-127	-102	-77	-52	-37	-27	-17	-2	23	48	73	123	173
-5	-222	-172	-122	-97	-72	-47	-32	-22	-12	2	27	52	77	127	177
-4	-218	-168	-118	-93	-68	-43	-28	-18	-8	7	32	57	82	132	182
-3	-213	-163	-113	-88	-63	-38	-23	-13	-3	11	36	61	86	136	186
-2	-209	-159	-109	-84	-59	-34	-19	-9	1	16	41	66	91	141	191
-1	-204	-154	-104	-79	-54	-29	-14	-4	5	20	45	70	95	145	195
0	-200	-150	-100	-75	-50	-25	-10	0	10	25	50	75	100	150	200
1	-195	-145	-95	-70	-45	-20	-5	4	14	29	54	79	104	154	204
2	-191	-141	-91	-66	-41	-16	-1	9	19	34	59	84	109	159	209
3	-186	-136	-86	-61	-36	-11	3	13	23	38	63	88	113	163	213
4	-182	-132	-82	-57	-32	-7	8	18	28	43	68	93	118	168	218
5	-177	-127	-77	-52	-27	-2	12	22	32	47	72	97	122	172	222
6	-173	-123	-73	-48	-23	2	17	27	37	52	77	102	127	177	227
7	-168	-118	-68	-43	-18	6	21	31	41	56	81	106	131	181	231
8	-164	-114	-64	-39	-14	11	26	36	46	61	86	111	136	186	236
10	-155	-105	-55	-30	-5	20	35	45	55	70	95	120	145	195	245
12	-146	-96	-46	-21	4	29	44	54	64	79	104	129	154	204	254
14	-137	-87	-37	-12	13	38	53	63	73	88	113	138	163	213	263
16	-128	-78	-28	-3	22	47	62	72	82	97	122	147	172	222	272
18	-119	-69	-19	6	31	56	71	81	91	106	131	156	181	231	281
20	-110	-60	-10	15	40	65	80	90	100	115	140	165	190	240	290
25	-87	-37	12	37	62	87	102	112	122	137	162	187	212	262	312
30	-65	-15	35	60	85	110	125	135	145	160	185	210	235	285	335
40	-20	30	80	105	130	155	170	180	190	205	230	255	280	330	380

Table 5-75

Separation, inch: 8

Leg Distance, inch: 90

SHIM REQUIRED FOR ALIGNMENT

Reverse Indicator Method

iveN R LS	DriveR TIR, MILS -40	-30	-20	-15	-10	-5	-2	0	2	5	10	15	20	30	40
40	-430	-373	-317	-289	-261	-233	-216	-205	-193	-176	-148	-120	-92	-36	20
30	-378	-322	-266	-238	-210	-181	-165	-153	-142	-125	-97	-69	-41	15	71
25	-353	-296	-240	-212	-184	-156	-139	-128	-116	-100	-71	-43	-15	40	96
20	-327	-271	-215	-186	-158	-130	-113	-102	-91	-74	-46	-18	10	66	122
18	-317	-261	-204	-176	-148	-120	-103	-92	-81	-64	-36	-7	20	76	132
16	-307	-250	-194	-166	-138	-110	-93	-82	-70	-53	-25	2	30	86	143
14	-296	-240	-184	-156	-128	-99	-83	-71	-60	-43	-15	12	40	97	153
12	-286	-230	-174	-145	-117	-89	-72	-61	-50	-33	-5	22	51	107	163
10	-276	-220	-163	-135	-107	-79	-62	-51	-40	-23	5	33	61	117	173
-8	-266	-209	-153	-125	-97	-69	-52	-41	-29	-12	15	43	71	127	184
-7	-260	-204	-148	-120	-92	-64	-47	-35	-24	-7	20	48	76	132	189
-6	-255	-199	-143	-115	-87	-58	-42	-30	-19	-2	25	53	81	138	194
-5	-250	-194	-138	-110	-81	-53	-36	-25	-14	2	30	58	86	143	199
-4	-245	-189	-133	-104	-76	-48	-31	-20	-9	7	35	63	92	148	204
-3	-240	-184	-127	-99	-71	-43	-26	-15	-4	12	40	69	97	153	209
-2	-235	-179	-122	-94	-66	-38	-21	-10	1	17	46	74	102	158	214
-1	-230	-173	-117	-89	-61	-33	-16	-5	6	23	51	79	107	163	219
0	-225	-168	-112	-84	-56	-28	-11	0	11	28	56	84	112	168	225
1	-219	-163	-107	-79	-51	-23	-6	5	16	33	61	89	117	173	230
2	-214	-158	-102	-74	-46	-17	-1	10	21	38	66	94	122	179	235
3	-209	-153	-97	-69	-40	-12	4	15	26	43	71	99	127	184	240
4	-204	-148	-92	-63	-35	-7	9	20	31	48	76	104	133	189	245
5	-199	-143	-86	-58	-30	-2	14	25	36	53	81	110	138	194	250
6	-194	-138	-81	-53	-25	2	19	30	42	58	87	115	143	199	255
7	-189	-132	-76	-48	-20	7	24	35	47	64	92	120	148	204	260
8	-184	-127	-71	-43	-15	12	29	41	52	69	97	125	153	209	266
10	-173	-117	-61	-33	-5	23	40	51	62	79	107	135	163	220	276
12	-163	-107	-51	-22	5	33	50	61	72	89	117	145	174	230	286
14	-153	-97	-40	-12	15	43	60	71	83	99	128	156	184	240	296
16	-143	-86	-30	-2	25	53	70	82	93	110	138	166	194	250	307
18	-132	-76	-20	7	36	64	81	92	103	120	148	176	204	261	317
20	-122	-66	-10	18	46	74	91	102	113	130	158	186	215	271	327
25	-96	-40	15	43	71	100	116	128	139	156	184	212	240	296	353
30	-71	-15	41	69	97	125	142	153	165	181	210	238	266	322	378
40	-20	36	92	120	148	176	193	205	216	233	261	289	317	373	430

Table 5-76

Separation, inch: 8

Leg Distance, inch: 100

SHIM REQUIRED FOR ALIGNMENT

Reverse Indicator Method

DriveN TIR MILS	DriveR TIR, MILS -40	-30	-20	-15	-10	-5	-2	0	2	5	10	15	20	30	40
-40	-480	-417	-355	-323	-292	-261	-242	-230	-217	-198	-167	-136	-105	-42	20
-30	-422	-360	-297	-266	-235	-203	-185	-172	-160	-141	-110	-78	-47	15	77
-25	-393	-331	-268	-237	-206	-175	-156	-143	-131	-112	-81	-50	-18	43	106
-20	-365	-302	-240	-208	-177	-146	-127	-115	-102	-83	-52	-21	10	72	135
-18	-353	-291	-228	-197	-166	-134	-116	-103	-91	-72	-41	-9	21	84	146
-16	-342	-279	-217	-185	-154	-123	-104	-92	-79	-60	-29	1	33	95	158
-14	-330	-268	-205	-174	-143	-111	-93	-80	-68	-49	-18	13	44	107	169
-12	-319	-256	-194	-162	-131	-100	-81	-69	-56	-37	-6	24	56	118	181
-10	-307	-245	-182	-151	-120	-88	-70	-57	-45	-26	5	36	67	130	192
-8	-296	-233	-171	-139	-108	-77	-58	-46	-33	-14	16	47	79	141	204
-7	-290	-227	-165	-134	-102	-71	-52	-40	-27	-9	22	53	84	147	209
-6	-284	-222	-159	-128	-97	-65	-47	-34	-22	-3	28	59	90	153	215
-5	-278	-216	-153	-122	-91	-60	-41	-28	-16	2	33	65	96	158	221
-4	-273	-210	-148	-116	-85	-54	-35	-23	-10	8	39	70	102	164	227
-3	-267	-204	-142	-111	-79	-48	-29	-17	-4	14	45	76	107	170	232
-2	-261	-199	-136	-105	-74	-42	-24	-11	1	19	51	82	113	176	238
-1	-255	-193	-130	-99	-68	-37	-18	-5	6	25	56	88	119	181	244
0	-250	-187	-125	-93	-62	-31	-12	0	12	31	62	93	125	187	250
1	-244	-181	-119	-88	-56	-25	-6	5	18	37	68	99	130	193	255
2	-238	-176	-113	-82	-51	-19	-1	11	24	42	74	105	136	199	261
3	-232	-170	-107	-76	-45	-14	4	17	29	48	79	111	142	204	267
4	-227	-164	-102	-70	-39	-8	10	23	35	54	85	116	148	210	273
5	-221	-158	-96	-65	-33	-2	16	28	41	60	91	122	153	216	278
6	-215	-153	-90	-59	-28	3	22	34	47	65	97	128	159	222	284
7	-209	-147	-84	-53	-22	9	27	40	52	71	102	134	165	227	290
8	-204	-141	-79	-47	-16	14	33	46	58	77	108	139	171	233	296
10	-192	-130	-67	-36	-5	26	45	57	70	88	120	151	182	245	307
12	-181	-118	-56	-24	6	37	56	69	81	100	131	162	194	256	319
14	-169	-107	-44	-13	18	49	68	80	93	111	143	174	205	268	330
16	-158	-95	-33	-1	29	60	79	92	104	123	154	185	217	279	342
18	-146	-84	-21	9	41	72	91	103	116	134	166	197	228	291	353
20	-135	-72	-10	21	52	83	102	115	127	146	177	208	240	302	365
25	-106	-43	18	50	81	112	131	143	156	175	206	237	268	331	393
30	-77	-15	47	78	110	141	160	172	185	203	235	266	297	360	422
40	-20	42	105	136	167	198	217	230	242	261	292	323	355	417	480

Table 5-77

Separation, inch: 9

Leg Distance, inch: 25

SHIM REQUIRED FOR ALIGNMENT

Reverse Indicator Method

DriveN TIR MILS	DriveR TIR, MILS -40	-30	-20	-15	-10	-5	-2	0	2	5	10	15	20	30	40
-40	-91	-77	-63	-56	-49	-42	-38	-35	-32	-28	-21	-14	-7	6	20
-30	-82	-68	-54	-47	-40	-33	-29	-26	-23	-19	-12	-5	1	15	28
-25	-77	-63	-50	-43	-36	-29	-25	-22	-19	-15	-8	-1	5	19	33
-20	-73	-59	-45	-38	-31	-24	-20	-17	-15	-10	-3	3	10	23	37
-18	-71	-57	-43	-36	-29	-22	-18	-16	-13	-9	-2	4	11	25	39
-16	-69	-55	-42	-35	-28	-21	-17	-14	-11	-7	0	6	13	27	41
-14	-68	-54	-40	-33	-26	-19	-15	-12	-9	-5	1	8	15	29	43
-12	-66	-52	-38	-31	-24	-17	-13	-10	-7	-3	3	10	17	31	44
-10	-64	-50	-36	-29	-22	-15	-11	-8	-6	-1	5	11	18	32	46
-8	-62	-48	-34	-27	-21	-14	-9	-7	-4	0	6	13	20	34	48
-7	-61	-47	-34	-27	-20	-13	-9	-6	-3	0	7	14	21	35	49
-6	-60	-47	-33	-26	-19	-12	-8	-5	-2	1	8	15	22	36	50
-5	-60	-46	-32	-25	-18	-11	-7	-4	-1	2	9	16	23	37	51
-4	-59	-45	-31	-24	-17	-10	-6	-3	0	3	10	17	24	38	52
-3	-58	-44	-30	-23	-16	-9	-5	-2	0	4	11	18	25	39	52
-2	-57	-43	-29	-22	-15	-8	-4	-1	1	5	12	19	26	39	53
-1	-56	-42	-28	-21	-14	-7	-3	0	1	6	13	19	26	40	54
0	-55	-41	-27	-20	-13	-6	-2	0	2	6	13	20	27	41	55
1	-54	-40	-26	-19	-13	-6	-1	0	3	7	14	21	28	42	56
2	-53	-39	-26	-19	-12	-5	-1	1	4	8	15	22	29	43	57
3	-52	-39	-25	-18	-11	-4	0	2	5	9	16	23	30	44	58
4	-52	-38	-24	-17	-10	-3	0	3	6	10	17	24	31	45	59
5	-51	-37	-23	-16	-9	-2	1	4	7	11	18	25	32	46	60
6	-50	-36	-22	-15	-8	-1	2	5	8	12	19	26	33	47	60
7	-49	-35	-21	-14	-7	0	3	6	9	13	20	27	34	47	61
8	-48	-34	-20	-13	-6	0	4	7	9	14	21	27	34	48	62
10	-46	-32	-18	-11	-5	1	6	8	11	15	22	29	36	50	64
12	-44	-31	-17	-10	-3	3	7	10	13	17	24	31	38	52	66
14	-43	-29	-15	-8	-1	5	9	12	15	19	26	33	40	54	68
16	-41	-27	-13	-6	0	7	11	14	17	21	28	35	42	55	69
18	-39	-25	-11	-4	2	9	13	16	18	22	29	36	43	57	71
20	-37	-23	-10	-3	3	10	15	17	20	24	31	38	45	59	73
25	-33	-19	-5	1	8	15	19	22	25	29	36	43	50	63	77
30	-28	-15	-1	5	12	19	23	26	29	33	40	47	54	68	82
40	-20	-6	7	14	21	28	32	35	38	42	49	56	63	77	91

Table 5-78

Separation, inch: 9

Leg Distance, inch: 30

SHIM REQUIRED FOR ALIGNMENT

Reverse Indicator Method

DriveN TIR MILS	DriveR TIR, MILS -40	-30	-20	-15	-10	-5	-2	0	2	5	10	15	20	30	40
-40	-113	-96	-80	-71	-63	-55	-50	-46	-43	-38	-30	-21	-13	3	20
-30	-101	-85	-68	-60	-51	-43	-38	-35	-31	-26	-18	-10	-1	15	31
-25	-95	-79	-62	-54	-45	-37	-32	-29	-25	-20	-12	-4	4	20	37
-20	-90	-73	-56	-48	-40	-31	-26	-23	-20	-15	-6	1	10	26	43
-18	-87	-71	-54	-46	-37	-29	-24	-21	-17	-12	-4	4	12	29	45
-16	-85	-68	-52	-43	-35	-27	-22	-18	-15	-10	-2	6	14	31	48
-14	-83	-66	-49	-41	-33	-24	-19	-16	-13	-8	0	8	17	33	50
-12	-80	-64	-47	-39	-30	-22	-17	-14	-10	-5	2	11	19	36	52
-10	-78	-61	-45	-36	-28	-20	-15	-11	-8	-3	5	13	21	38	55
-8	-76	-59	-42	-34	-26	-17	-12	-9	-6	-1	7	15	24	40	57
-7	-74	-58	-41	-33	-24	-16	-11	-8	-4	0	8	16	25	41	58
-6	-73	-57	-40	-32	-23	-15	-10	-7	-3	1	9	18	26	43	59
-5	-72	-55	-39	-30	-22	-14	-9	-5	-2	2	10	19	27	44	60
-4	-71	-54	-38	-29	-21	-13	-8	-4	-1	3	12	20	28	45	62
-3	-70	-53	-36	-28	-20	-11	-6	-3	0	4	13	21	29	46	63
-2	-69	-52	-35	-27	-19	-10	-5	-2	1	6	14	22	31	47	64
-1	-67	-51	-34	-26	-17	-9	-4	-1	2	7	15	23	32	48	65
0	-66	-50	-33	-25	-16	-8	-3	0	3	8	16	25	33	50	66
1	-65	-48	-32	-23	-15	-7	-2	1	4	9	17	26	34	51	67
2	-64	-47	-31	-22	-14	-6	-1	2	5	10	19	27	35	52	69
3	-63	-46	-29	-21	-13	-4	0	3	6	11	20	28	36	53	70
4	-62	-45	-28	-20	-12	-3	1	4	8	13	21	29	38	54	71
5	-60	-44	-27	-19	-10	-2	2	5	9	14	22	30	39	55	72
6	-59	-43	-26	-18	-9	-1	3	7	10	15	23	32	40	57	73
7	-58	-41	-25	-16	-8	0	4	8	11	16	24	33	41	58	74
8	-57	-40	-24	-15	-7	1	6	9	12	17	26	34	42	59	76
10	-55	-38	-21	-13	-5	3	8	11	15	20	28	36	45	61	78
12	-52	-36	-19	-11	-2	5	10	14	17	22	30	39	47	64	80
14	-50	-33	-17	-8	0	8	13	16	19	24	33	41	49	66	83
16	-48	-31	-14	-6	2	10	15	18	22	27	35	43	52	68	85
18	-45	-29	-12	-4	4	12	17	21	24	29	37	46	54	71	87
20	-43	-26	-10	-1	6	15	20	23	26	31	40	48	56	73	90
25	-37	-20	-4	4	12	20	25	29	32	37	45	54	62	79	95
30	-31	-15	1	10	18	26	31	35	38	43	51	60	68	85	101
40	-20	-3	13	21	30	38	43	46	50	55	63	71	80	96	113

Table 5-79

Separation, inch: 9

Leg Distance, inch: 35

SHIM REQUIRED FOR ALIGNMENT

Reverse Indicator Method

riveN DriveR TIR, MILS
IR

ILS	-40	-30	-20	-15	-10	-5	-2	0	2	5	10	15	20	30	40
-40	-135	-116	-96	-86	-77	-67	-61	-57	-53	-48	-38	-28	-18	0	20
-30	-121	-101	-82	-72	-62	-53	-47	-43	-39	-33	-23	-14	-4	15	34
-25	-113	-94	-75	-65	-55	-45	-40	-36	-32	-26	-16	-6	2	22	41
-20	-106	-87	-67	-58	-48	-38	-32	-28	-25	-19	-9	0	10	29	48
-18	-103	-84	-64	-55	-45	-35	-29	-26	-22	-16	-6	3	12	32	51
-16	-100	-81	-62	-52	-42	-32	-27	-23	-19	-13	-3	6	15	35	54
-14	-98	-78	-59	-49	-39	-29	-24	-20	-16	-10	0	8	18	38	57
-12	-95	-75	-56	-46	-36	-27	-21	-17	-13	-7	2	11	21	41	60
-10	-92	-72	-53	-43	-33	-24	-18	-14	-10	-4	5	14	24	43	63
-8	-89	-69	-50	-40	-31	-21	-15	-11	-7	-1	7	17	27	46	66
-7	-87	-68	-49	-39	-29	-19	-14	-10	-6	0	9	19	28	48	67
-6	-86	-67	-47	-37	-28	-18	-12	-8	-4	1	10	20	30	49	69
-5	-85	-65	-46	-36	-26	-16	-11	-7	-3	2	12	21	31	51	70
-4	-83	-64	-44	-34	-25	-15	-9	-5	-1	3	13	23	33	52	72
-3	-82	-62	-43	-33	-23	-14	-8	-4	0	5	15	24	34	54	73
-2	-80	-61	-41	-32	-22	-12	-6	-2	1	6	16	26	36	55	74
-1	-79	-59	-40	-30	-20	-11	-5	-1	2	8	18	27	37	56	76
0	-77	-58	-38	-29	-19	-9	-3	0	3	9	19	29	38	58	77
1	-76	-56	-37	-27	-18	-8	-2	1	5	11	20	30	40	59	79
2	-74	-55	-36	-26	-16	-6	-1	2	6	12	22	32	41	61	80
3	-73	-54	-34	-24	-15	-5	0	4	8	14	23	33	43	62	82
4	-72	-52	-33	-23	-13	-3	1	5	9	15	25	34	44	64	83
5	-70	-51	-31	-21	-12	-2	3	7	11	16	26	36	46	65	85
6	-69	-49	-30	-20	-10	-1	4	8	12	18	28	37	47	67	86
7	-67	-48	-28	-19	-9	0	6	10	14	19	29	39	49	68	87
8	-66	-46	-27	-17	-7	1	7	11	15	21	31	40	50	69	89
10	-63	-43	-24	-14	-5	4	10	14	18	24	33	43	53	72	92
12	-60	-41	-21	-11	-2	7	13	17	21	27	36	46	56	75	95
14	-57	-38	-18	-8	0	10	16	20	24	29	39	49	59	78	98
16	-54	-35	-15	-6	3	13	19	23	27	32	42	52	62	81	100
18	-51	-32	-12	-3	6	16	22	26	29	35	45	55	64	84	103
20	-48	-29	-10	0	9	19	25	28	32	38	48	58	67	87	106
25	-41	-22	-2	6	16	26	32	36	40	45	55	65	75	94	113
30	-34	-15	4	14	23	33	39	43	47	53	62	72	82	101	121
40	-20	0	18	28	38	48	53	57	61	67	77	86	96	116	135

Table 5-80

Separation, inch: 9

Leg Distance, inch: 40

SHIM REQUIRED FOR ALIGNMENT

Reverse Indicator Method

DriveN TIR MILS	DriveR TIR, MILS -40	-30	-20	-15	-10	-5	-2	0	2	5	10	15	20	30	40
-40	-157	-135	-113	-102	-91	-80	-73	-68	-64	-57	-46	-35	-24	-2	20
-30	-140	-118	-96	-85	-73	-62	-56	-51	-47	-40	-29	-18	-7	15	37
-25	-131	-109	-87	-76	-65	-54	-47	-43	-38	-31	-20	-9	1	23	45
-20	-123	-101	-78	-67	-56	-45	-38	-34	-30	-23	-12	-1	10	32	54
-18	-119	-97	-75	-64	-53	-42	-35	-31	-26	-19	-8	2	13	35	57
-16	-116	-94	-72	-60	-49	-38	-32	-27	-23	-16	-5	5	16	39	61
-14	-113	-90	-68	-57	-46	-35	-28	-24	-19	-13	-1	9	20	42	64
-12	-109	-87	-65	-54	-42	-31	-25	-20	-16	-9	1	12	23	46	68
-10	-106	-83	-61	-50	-39	-28	-21	-17	-12	-6	5	16	27	49	71
-8	-102	-80	-58	-47	-36	-24	-18	-13	-9	-2	8	19	30	52	75
-7	-100	-78	-56	-45	-34	-23	-16	-12	-7	0	10	21	32	54	76
-6	-99	-77	-54	-43	-32	-21	-14	-10	-5	0	11	23	34	56	78
-5	-97	-75	-53	-41	-30	-19	-13	-8	-4	2	13	24	35	58	80
-4	-95	-73	-51	-40	-29	-18	-11	-6	-2	4	15	26	37	59	82
-3	-94	-71	-49	-38	-27	-16	-9	-5	0	5	17	28	39	61	83
-2	-92	-70	-47	-36	-25	-14	-7	-3	1	7	18	29	41	63	85
-1	-90	-68	-46	-35	-23	-12	-6	-1	2	9	20	31	42	64	87
0	-88	-66	-44	-33	-22	-11	-4	0	4	11	22	33	44	66	88
1	-87	-64	-42	-31	-20	-9	-2	1	6	12	23	35	46	68	90
2	-85	-63	-41	-29	-18	-7	-1	3	7	14	25	36	47	70	92
3	-83	-61	-39	-28	-17	-5	0	5	9	16	27	38	49	71	94
4	-82	-59	-37	-26	-15	-4	2	6	11	18	29	40	51	73	95
5	-80	-58	-35	-24	-13	-2	4	8	13	19	30	41	53	75	97
6	-78	-56	-34	-23	-11	0	5	10	14	21	32	43	54	77	99
7	-76	-54	-32	-21	-10	0	7	12	16	23	34	45	56	78	100
8	-75	-52	-30	-19	-8	2	9	13	18	24	36	47	58	80	102
10	-71	-49	-27	-16	-5	6	12	17	21	28	39	50	61	83	106
12	-68	-46	-23	-12	-1	9	16	20	25	31	42	54	65	87	109
14	-64	-42	-20	-9	1	13	19	24	28	35	46	57	68	90	113
16	-61	-39	-16	-5	5	16	23	27	32	38	49	60	72	94	116
18	-57	-35	-13	-2	8	19	26	31	35	42	53	64	75	97	119
20	-54	-32	-10	1	12	23	30	34	38	45	56	67	78	101	123
25	-45	-23	-1	9	20	31	38	43	47	54	65	76	87	109	131
30	-37	-15	7	18	29	40	47	51	56	62	73	85	96	118	140
40	-20	2	24	35	46	57	64	68	73	80	91	102	113	135	157

Table 5-81

Separation, inch: 9

Leg Distance, inch: 50

SHIM REQUIRED FOR ALIGNMENT

Reverse Indicator Method

DriveN TIR MILS	DriveR TIR, MILS -40	-30	-20	-15	-10	-5	-2	0	2	5	10	15	20	30	40
-40	-202	-174	-146	-132	-118	-105	-96	-91	-85	-77	-63	-49	-35	-7	20
-30	-179	-151	-123	-110	-96	-82	-73	-68	-62	-54	-40	-26	-12	15	42
-25	-168	-140	-112	-98	-84	-70	-62	-56	-51	-43	-29	-15	-1	26	54
-20	-156	-128	-101	-87	-73	-59	-51	-45	-40	-31	-17	-3	10	37	65
-18	-152	-124	-96	-82	-68	-54	-46	-41	-35	-27	-13	0	14	42	70
-16	-147	-119	-92	-78	-64	-50	-42	-36	-30	-22	-8	5	19	46	74
-14	-143	-115	-87	-73	-59	-45	-37	-31	-26	-18	-4	9	23	51	79
-12	-138	-110	-82	-69	-55	-41	-32	-27	-21	-13	0	14	28	56	83
-10	-133	-106	-78	-64	-50	-36	-28	-22	-17	-8	5	18	32	60	88
-8	-129	-101	-73	-59	-46	-32	-23	-18	-12	-4	9	23	37	65	92
-7	-127	-99	-71	-57	-43	-29	-21	-15	-10	-2	11	25	39	67	95
-6	-124	-97	-69	-55	-41	-27	-19	-13	-8	0	14	28	41	69	97
-5	-122	-94	-66	-53	-39	-25	-16	-11	-5	2	16	30	44	71	99
-4	-120	-92	-64	-50	-36	-23	-14	-9	-3	4	18	32	46	74	102
-3	-117	-90	-62	-48	-34	-20	-12	-6	-1	7	20	34	48	76	104
-2	-115	-87	-60	-46	-32	-18	-10	-4	1	9	23	37	51	78	106
-1	-113	-85	-57	-43	-30	-16	-7	-2	3	11	25	39	53	81	108
0	-111	-83	-55	-41	-27	-13	-5	0	5	13	27	41	55	83	111
1	-108	-81	-53	-39	-25	-11	-3	2	7	16	30	43	57	85	113
2	-106	-78	-51	-37	-23	-9	-1	4	10	18	32	46	60	87	115
3	-104	-76	-48	-34	-20	-7	1	6	12	20	34	48	62	90	117
4	-102	-74	-46	-32	-18	-4	3	9	14	23	36	50	64	92	120
5	-99	-71	-44	-30	-16	-2	5	11	16	25	39	53	66	94	122
6	-97	-69	-41	-28	-14	0	8	13	19	27	41	55	69	97	124
7	-95	-67	-39	-25	-11	2	10	15	21	29	43	57	71	99	127
8	-92	-65	-37	-23	-9	4	12	18	23	32	46	59	73	101	129
10	-88	-60	-32	-18	-5	8	17	22	28	36	50	64	78	106	133
12	-83	-56	-28	-14	0	13	21	27	32	41	55	69	82	110	138
14	-79	-51	-23	-9	4	18	26	31	37	45	59	73	87	115	143
16	-74	-46	-19	-5	8	22	30	36	42	50	64	78	92	119	147
18	-70	-42	-14	0	13	27	35	41	46	54	68	82	96	124	152
20	-65	-37	-10	3	17	31	40	45	51	59	73	87	101	128	156
25	-54	-26	1	15	29	43	51	56	62	70	84	98	112	140	168
30	-42	-15	12	26	40	54	62	68	73	82	96	110	123	151	179
40	-20	7	35	49	63	77	85	91	96	105	118	132	146	174	202

Table 5-82

Separation, inch: 9

Leg Distance, inch: 60

SHIM REQUIRED FOR ALIGNMENT

Reverse Indicator Method

DriveN TIR MILS	DriveR TIR, MILS -40	-30	-20	-15	-10	-5	-2	0	2	5	10	15	20	30	40
-40	-246	-213	-180	-163	-146	-130	-120	-113	-106	-96	-80	-63	-46	-13	20
-30	-218	-185	-151	-135	-118	-101	-91	-85	-78	-68	-51	-35	-18	15	48
-25	-204	-170	-137	-120	-104	-87	-77	-70	-64	-54	-37	-20	-4	29	62
-20	-190	-156	-123	-106	-90	-73	-63	-56	-50	-40	-23	-6	10	43	76
-18	-184	-151	-117	-101	-84	-67	-57	-51	-44	-34	-17	-1	15	49	82
-16	-178	-145	-112	-95	-78	-62	-52	-45	-38	-28	-12	4	21	54	88
-14	-173	-139	-106	-89	-73	-56	-46	-39	-33	-23	-6	10	27	60	93
-12	-167	-134	-100	-84	-67	-50	-40	-34	-27	-17	0	16	32	66	99
-10	-161	-128	-95	-78	-61	-45	-35	-28	-21	-11	5	21	38	71	105
-8	-156	-122	-89	-72	-56	-39	-29	-22	-16	-6	10	27	44	77	110
-7	-153	-119	-86	-69	-53	-36	-26	-19	-13	-3	13	30	46	80	113
-6	-150	-117	-83	-67	-50	-33	-23	-17	-10	0	16	33	49	83	116
-5	-147	-114	-80	-64	-47	-30	-20	-14	-7	2	19	35	52	85	119
-4	-144	-111	-78	-61	-44	-28	-18	-11	-4	5	22	38	55	88	122
-3	-141	-108	-75	-58	-41	-25	-15	-8	-1	8	24	41	58	91	124
-2	-139	-105	-72	-55	-39	-22	-12	-5	1	11	27	44	61	94	127
-1	-136	-102	-69	-52	-36	-19	-9	-2	3	13	30	47	63	97	130
0	-133	-100	-66	-50	-33	-16	-6	0	6	16	33	50	66	100	133
1	-130	-97	-63	-47	-30	-13	-3	2	9	19	36	52	69	102	136
2	-127	-94	-61	-44	-27	-11	-1	5	12	22	39	55	72	105	139
3	-124	-91	-58	-41	-24	-8	1	8	15	25	41	58	75	108	141
4	-122	-88	-55	-38	-22	-5	4	11	18	28	44	61	78	111	144
5	-119	-85	-52	-35	-19	-2	7	14	20	30	47	64	80	114	147
6	-116	-83	-49	-33	-16	0	10	17	23	33	50	67	83	117	150
7	-113	-80	-46	-30	-13	3	13	19	26	36	53	69	86	119	153
8	-110	-77	-44	-27	-10	6	16	22	29	39	56	72	89	122	156
10	-105	-71	-38	-21	-5	11	21	28	35	45	61	78	95	128	161
12	-99	-66	-32	-16	0	17	27	34	40	50	67	84	100	134	167
14	-93	-60	-27	-10	6	23	33	39	46	56	73	89	106	139	173
16	-88	-54	-21	-4	12	28	38	45	52	62	78	95	112	145	178
18	-82	-49	-15	1	17	34	44	51	57	67	84	101	117	151	184
20	-76	-43	-10	6	23	40	50	56	63	73	90	106	123	156	190
25	-62	-29	4	20	37	54	64	70	77	87	104	120	137	170	204
30	-48	-15	18	35	51	68	78	85	91	101	118	135	151	185	218
40	-20	13	46	63	80	96	106	113	120	130	146	163	180	213	246

Table 5-83

Separation, inch: 9

Leg Distance, inch: 70

SHIM REQUIRED FOR ALIGNMENT

Reverse Indicator Method

DriveN TIR MILS	DriveR TIR, MILS -40	-30	-20	-15	-10	-5	-2	0	2	5	10	15	20	30	40
-40	-291	-252	-213	-193	-174	-155	-143	-135	-127	-116	-96	-77	-57	-18	20
-30	-257	-218	-179	-160	-140	-121	-109	-101	-93	-82	-62	-43	-23	15	53
-25	-240	-201	-162	-143	-123	-104	-92	-84	-76	-65	-45	-26	-6	31	70
-20	-223	-184	-145	-126	-106	-87	-75	-67	-60	-48	-28	-9	10	48	87
-18	-216	-177	-138	-119	-99	-80	-68	-61	-53	-41	-22	-2	16	55	94
-16	-209	-170	-132	-112	-93	-73	-62	-54	-46	-34	-15	4	23	62	101
-14	-203	-164	-125	-105	-86	-66	-55	-47	-39	-28	-8	10	30	69	108
-12	-196	-157	-118	-99	-79	-60	-48	-40	-32	-21	-1	17	37	76	114
-10	-189	-150	-111	-92	-72	-53	-41	-33	-26	-14	5	24	43	82	121
-8	-182	-143	-104	-85	-66	-46	-34	-27	-19	-7	11	31	50	89	128
-7	-179	-140	-101	-82	-62	-43	-31	-23	-15	-4	15	34	54	92	131
-6	-175	-137	-98	-78	-59	-39	-28	-20	-12	0	18	38	57	96	135
-5	-172	-133	-94	-75	-55	-36	-24	-16	-9	2	21	41	60	99	138
-4	-169	-130	-91	-71	-52	-33	-21	-13	-5	5	25	44	64	103	142
-3	-165	-126	-87	-68	-49	-29	-17	-10	-2	9	28	48	67	106	145
-2	-162	-123	-84	-65	-45	-26	-14	-6	1	12	32	51	71	109	148
-1	-158	-120	-81	-61	-42	-22	-11	-3	4	16	35	54	74	113	152
0	-155	-116	-77	-58	-38	-19	-7	0	7	19	38	58	77	116	155
1	-152	-113	-74	-54	-35	-16	-4	3	11	22	42	61	81	120	158
2	-148	-109	-71	-51	-32	-12	-1	6	14	26	45	65	84	123	162
3	-145	-106	-67	-48	-28	-9	2	10	17	29	49	68	87	126	165
4	-142	-103	-64	-44	-25	-5	5	13	21	33	52	71	91	130	169
5	-138	-99	-60	-41	-21	-2	9	16	24	36	55	75	94	133	172
6	-135	-96	-57	-38	-18	0	12	20	28	39	59	78	98	137	175
7	-131	-92	-54	-34	-15	4	15	23	31	43	62	82	101	140	179
8	-128	-89	-50	-31	-11	7	19	27	34	46	66	85	104	143	182
10	-121	-82	-43	-24	-5	14	26	33	41	53	72	92	111	150	189
12	-114	-76	-37	-17	1	21	32	40	48	60	79	99	118	157	196
14	-108	-69	-30	-10	8	28	39	47	55	66	86	105	125	164	203
16	-101	-62	-23	-4	15	34	46	54	62	73	93	112	132	170	209
18	-94	-55	-16	2	22	41	53	61	68	80	99	119	138	177	216
20	-87	-48	-10	9	28	48	60	67	75	87	106	126	145	184	223
25	-70	-31	6	26	45	65	76	84	92	104	123	143	162	201	240
30	-53	-15	23	43	62	82	93	101	109	121	140	160	179	218	257
40	-20	18	57	77	96	116	127	135	143	155	174	193	213	252	291

Table 5-84

Separation, inch: 9

Leg Distance, inch: 80

SHIM REQUIRED FOR ALIGNMENT

Reverse Indicator Method

DriveN TIR MILS	DriveR TIR, MILS -40	-30	-20	-15	-10	-5	-2	0	2	5	10	15	20	30	40
-40	-335	-291	-246	-224	-202	-180	-166	-157	-148	-135	-113	-91	-68	-24	20
-30	-296	-251	-207	-185	-162	-140	-127	-118	-109	-96	-73	-51	-29	15	59
-25	-276	-231	-187	-165	-143	-120	-107	-98	-89	-76	-54	-31	-9	34	79
-20	-256	-212	-167	-145	-123	-101	-87	-78	-70	-56	-34	-12	10	54	98
-18	-248	-204	-159	-137	-115	-93	-79	-71	-62	-48	-26	-4	17	62	106
-16	-240	-196	-152	-129	-107	-85	-72	-63	-54	-40	-18	3	25	70	114
-14	-233	-188	-144	-121	-99	-77	-64	-55	-46	-33	-10	11	33	78	122
-12	-225	-180	-136	-114	-91	-69	-56	-47	-38	-25	-2	19	41	86	130
-10	-217	-172	-128	-106	-83	-61	-48	-39	-30	-17	5	27	49	93	138
-8	-209	-164	-120	-98	-76	-53	-40	-31	-22	-9	12	35	57	101	146
-7	-205	-160	-116	-94	-72	-49	-36	-27	-18	-5	16	39	61	105	150
-6	-201	-157	-112	-90	-68	-45	-32	-23	-14	-1	20	43	65	109	154
-5	-197	-153	-108	-86	-64	-41	-28	-19	-10	2	24	46	69	113	158
-4	-193	-149	-104	-82	-60	-38	-24	-15	-6	6	28	50	73	117	162
-3	-189	-145	-100	-78	-56	-34	-20	-11	-2	10	32	54	77	121	165
-2	-185	-141	-96	-74	-52	-30	-16	-7	1	14	36	58	81	125	169
-1	-181	-137	-92	-70	-48	-26	-12	-3	4	18	40	62	84	129	173
0	-177	-133	-88	-66	-44	-22	-8	0	8	22	44	66	88	133	177
1	-173	-129	-84	-62	-40	-18	-4	3	12	26	48	70	92	137	181
2	-169	-125	-81	-58	-36	-14	-1	7	16	30	52	74	96	141	185
3	-165	-121	-77	-54	-32	-10	2	11	20	34	56	78	100	145	189
4	-162	-117	-73	-50	-28	-6	6	15	24	38	60	82	104	149	193
5	-158	-113	-69	-46	-24	-2	10	19	28	41	64	86	108	153	197
6	-154	-109	-65	-43	-20	1	14	23	32	45	68	90	112	157	201
7	-150	-105	-61	-39	-16	5	18	27	36	49	72	94	116	160	205
8	-146	-101	-57	-35	-12	9	22	31	40	53	76	98	120	164	209
10	-138	-93	-49	-27	-5	17	30	39	48	61	83	106	128	172	217
12	-130	-86	-41	-19	2	25	38	47	56	69	91	114	136	180	225
14	-122	-78	-33	-11	10	33	46	55	64	77	99	121	144	188	233
16	-114	-70	-25	-3	18	40	54	63	72	85	107	129	152	196	240
18	-106	-62	-17	4	26	48	62	71	79	93	115	137	159	204	248
20	-98	-54	-10	12	34	56	70	78	87	101	123	145	167	212	256
25	-79	-34	9	31	54	76	89	98	107	120	143	165	187	231	276
30	-59	-15	29	51	73	96	109	118	127	140	162	185	207	251	296
40	-20	24	68	91	113	135	148	157	166	180	202	224	246	291	335

Table 5-85

Separation, inch: 9

Leg Distance, inch: 90

SHIM REQUIRED FOR ALIGNMENT

Reverse Indicator Method

DriveR TIR, MILS

riveN IR ILS	-40	-30	-20	-15	-10	-5	-2	0	2	5	10	15	20	30	40
-40	-380	-330	-280	-255	-230	-205	-190	-180	-170	-155	-130	-105	-80	-30	20
-30	-335	-285	-235	-210	-185	-160	-145	-135	-125	-110	-85	-60	-35	15	65
-25	-312	-262	-212	-187	-162	-137	-122	-112	-102	-87	-62	-37	-12	37	87
-20	-290	-240	-190	-165	-140	-115	-100	-90	-80	-65	-40	-15	10	60	110
-18	-281	-231	-181	-156	-131	-106	-91	-81	-71	-56	-31	-6	19	69	119
-16	-272	-222	-172	-147	-122	-97	-82	-72	-62	-47	-22	3	28	78	128
-14	-263	-213	-163	-138	-113	-88	-73	-63	-53	-38	-13	12	37	87	137
-12	-254	-204	-154	-129	-104	-79	-64	-54	-44	-29	-4	21	46	96	146
-10	-245	-195	-145	-120	-95	-70	-55	-45	-35	-20	5	30	55	105	155
-8	-236	-186	-136	-111	-86	-61	-46	-36	-26	-11	14	39	64	114	164
-7	-231	-181	-131	-106	-81	-56	-41	-31	-21	-6	18	43	68	118	168
-6	-227	-177	-127	-102	-77	-52	-37	-27	-17	-2	23	48	73	123	173
-5	-222	-172	-122	-97	-72	-47	-32	-22	-12	2	27	52	77	127	177
-4	-218	-168	-118	-93	-68	-43	-28	-18	-8	7	32	57	82	132	182
-3	-213	-163	-113	-88	-63	-38	-23	-13	-3	11	36	61	86	136	186
-2	-209	-159	-109	-84	-59	-34	-19	-9	1	16	41	66	91	141	191
-1	-204	-154	-104	-79	-54	-29	-14	-4	5	20	45	70	95	145	195
0	-200	-150	-100	-75	-50	-25	-10	0	10	25	50	75	100	150	200
1	-195	-145	-95	-70	-45	-20	-5	4	14	29	54	79	104	154	204
2	-191	-141	-91	-66	-41	-16	-1	9	19	34	59	84	109	159	209
3	-186	-136	-86	-61	-36	-11	3	13	23	38	63	88	113	163	213
4	-182	-132	-82	-57	-32	-7	8	18	28	43	68	93	118	168	218
5	-177	-127	-77	-52	-27	-2	12	22	32	47	72	97	122	172	222
6	-173	-123	-73	-48	-23	2	17	27	37	52	77	102	127	177	227
7	-168	-118	-68	-43	-18	6	21	31	41	56	81	106	131	181	231
8	-164	-114	-64	-39	-14	11	26	36	46	61	86	111	136	186	236
10	-155	-105	-55	-30	-5	20	35	45	55	70	95	120	145	195	245
12	-146	-96	-46	-21	4	29	44	54	64	79	104	129	154	204	254
14	-137	-87	-37	-12	13	38	53	63	73	88	113	138	163	213	263
16	-128	-78	-28	-3	22	47	62	72	82	97	122	147	172	222	272
18	-119	-69	-19	6	31	56	71	81	91	106	131	156	181	231	281
20	-110	-60	-10	15	40	65	80	90	100	115	140	165	190	240	290
25	-87	-37	12	37	62	87	102	112	122	137	162	187	212	262	312
30	-65	-15	35	60	85	110	125	135	145	160	185	210	235	285	335
40	-20	30	80	105	130	155	170	180	190	205	230	255	280	330	380

Table 5-86

Separation, inch: 9

Leg Distance, inch: 100

SHIM REQUIRED FOR ALIGNMENT

Reverse Indicator Method

DriveN TIR MILS	DriveR TIR, MILS -40	-30	-20	-15	-10	-5	-2	0	2	5	10	15	20	30	40
-40	-424	-368	-313	-285	-257	-230	-213	-202	-191	-174	-146	-118	-91	-35	20
-30	-373	-318	-262	-235	-207	-179	-162	-151	-140	-123	-96	-68	-40	15	70
-25	-348	-293	-237	-209	-181	-154	-137	-126	-115	-98	-70	-43	-15	40	95
-20	-323	-267	-212	-184	-156	-128	-112	-101	-90	-73	-45	-17	10	65	121
-18	-313	-257	-202	-174	-146	-118	-102	-91	-79	-63	-35	-7	20	75	131
-16	-303	-247	-192	-164	-136	-108	-92	-80	-69	-53	-25	2	30	85	141
-14	-293	-237	-181	-154	-126	-98	-81	-70	-59	-43	-15	12	40	95	151
-12	-282	-227	-171	-144	-116	-88	-71	-60	-49	-32	-5	22	50	106	161
-10	-272	-217	-161	-133	-106	-78	-61	-50	-39	-22	5	32	60	116	171
-8	-262	-207	-151	-123	-96	-68	-51	-40	-29	-12	15	42	70	126	181
-7	-257	-202	-146	-118	-90	-63	-46	-35	-24	-7	20	47	75	131	186
-6	-252	-197	-141	-113	-85	-58	-41	-30	-19	-2	25	53	80	136	191
-5	-247	-191	-136	-108	-80	-53	-36	-25	-14	2	30	58	85	141	196
-4	-242	-186	-131	-103	-75	-48	-31	-20	-9	7	35	63	90	146	202
-3	-237	-181	-126	-98	-70	-42	-26	-15	-4	12	40	68	95	151	207
-2	-232	-176	-121	-93	-65	-37	-21	-10	1	17	45	73	101	156	212
-1	-227	-171	-116	-88	-60	-32	-16	-5	6	22	50	78	106	161	217
0	-222	-166	-111	-83	-55	-27	-11	0	11	27	55	83	111	166	222
1	-217	-161	-106	-78	-50	-22	-6	5	16	32	60	88	116	171	227
2	-212	-156	-101	-73	-45	-17	-1	10	21	37	65	93	121	176	232
3	-207	-151	-95	-68	-40	-12	4	15	26	42	70	98	126	181	237
4	-202	-146	-90	-63	-35	-7	9	20	31	48	75	103	131	186	242
5	-196	-141	-85	-58	-30	-2	14	25	36	53	80	108	136	191	247
6	-191	-136	-80	-53	-25	2	19	30	41	58	85	113	141	197	252
7	-186	-131	-75	-47	-20	7	24	35	46	63	90	118	146	202	257
8	-181	-126	-70	-42	-15	12	29	40	51	68	96	123	151	207	262
10	-171	-116	-60	-32	-5	22	39	50	61	78	106	133	161	217	272
12	-161	-106	-50	-22	5	32	49	60	71	88	116	144	171	227	282
14	-151	-95	-40	-12	15	43	59	70	81	98	126	154	181	237	293
16	-141	-85	-30	-2	25	53	69	80	92	108	136	164	192	247	303
18	-131	-75	-20	7	35	63	79	91	102	118	146	174	202	257	313
20	-121	-65	-10	17	45	73	90	101	112	128	156	184	212	267	323
25	-95	-40	15	43	70	98	115	126	137	154	181	209	237	293	348
30	-70	-15	40	68	96	123	140	151	162	179	207	235	262	318	373
40	-20	35	91	118	146	174	191	202	213	230	257	285	313	368	424

Table 5-87

Separation, inch: 10

Leg Distance, inch: 25

SHIM REQUIRED FOR ALIGNMENT

Reverse Indicator Method

riveN IR ILS	DriveR TIR, MILS -40	-30	-20	-15	-10	-5	-2	0	2	5	10	15	20	30	40
-40	-80	-67	-55	-48	-42	-36	-32	-30	-27	-23	-17	-11	-5	7	20
-30	-72	-60	-47	-41	-35	-28	-25	-22	-20	-16	-10	-3	2	15	27
-25	-68	-56	-43	-37	-31	-25	-21	-18	-16	-12	-6	0	6	18	31
-20	-65	-52	-40	-33	-27	-21	-17	-15	-12	-8	-2	3	10	22	35
-18	-63	-51	-38	-32	-26	-19	-16	-13	-11	-7	-1	5	11	24	36
-16	-62	-49	-37	-30	-24	-18	-14	-12	-9	-5	0	6	13	25	38
-14	-60	-48	-35	-29	-23	-16	-13	-10	-8	-4	2	8	14	27	39
-12	-59	-46	-34	-27	-21	-15	-11	-9	-6	-2	3	9	16	28	41
-10	-57	-45	-32	-26	-20	-13	-10	-7	-5	-1	5	11	17	30	42
-8	-56	-43	-31	-24	-18	-12	-8	-6	-3	0	6	12	19	31	44
-7	-55	-42	-30	-24	-17	-11	-7	-5	-2	1	7	13	19	32	44
-6	-54	-42	-29	-23	-17	-10	-7	-4	-2	1	8	14	20	33	45
-5	-53	-41	-28	-22	-16	-10	-6	-3	-1	2	8	15	21	33	46
-4	-53	-40	-28	-21	-15	-9	-5	-3	0	3	9	15	22	34	47
-3	-52	-39	-27	-21	-14	-8	-4	-2	0	4	10	16	22	35	47
-2	-51	-39	-26	-20	-14	-7	-4	-1	1	4	11	17	23	36	48
-1	-50	-38	-25	-19	-13	-7	-3	0	1	5	11	18	24	36	49
0	-50	-37	-25	-18	-12	-6	-2	0	2	6	12	18	25	37	50
1	-49	-36	-24	-18	-11	-5	-1	0	3	7	13	19	25	38	50
2	-48	-36	-23	-17	-11	-4	-1	1	4	7	14	20	26	39	51
3	-47	-35	-22	-16	-10	-4	0	2	4	8	14	21	27	39	52
4	-47	-34	-22	-15	-9	-3	0	3	5	9	15	21	28	40	53
5	-46	-33	-21	-15	-8	-2	1	3	6	10	16	22	28	41	53
6	-45	-33	-20	-14	-8	-1	2	4	7	10	17	23	29	42	54
7	-44	-32	-19	-13	-7	-1	2	5	7	11	17	24	30	42	55
8	-44	-31	-19	-12	-6	0	3	6	8	12	18	24	31	43	56
10	-42	-30	-17	-11	-5	1	5	7	10	13	20	26	32	45	57
12	-41	-28	-16	-9	-3	2	6	9	11	15	21	27	34	46	59
14	-39	-27	-14	-8	-2	4	8	10	13	16	23	29	35	48	60
16	-38	-25	-13	-6	0	5	9	12	14	18	24	30	37	49	62
18	-36	-24	-11	-5	1	7	11	13	16	19	26	32	38	51	63
20	-35	-22	-10	-3	2	8	12	15	17	21	27	33	40	52	65
25	-31	-18	-6	0	6	12	16	18	21	25	31	37	43	56	68
30	-27	-15	-2	3	10	16	20	22	25	28	35	41	47	60	72
40	-20	-7	5	11	17	23	27	30	32	36	42	48	55	67	80

Table 5-88

Separation, inch: 10

Leg Distance, inch: 30

SHIM REQUIRED FOR ALIGNMENT

Reverse Indicator Method

DriveN TIR MILS	DriveR TIR, MILS -40	-30	-20	-15	-10	-5	-2	0	2	5	10	15	20	30	40
-40	-100	-85	-70	-62	-55	-47	-43	-40	-37	-32	-25	-17	-10	5	20
-30	-90	-75	-60	-52	-45	-37	-33	-30	-27	-22	-15	-7	0	15	30
-25	-85	-70	-55	-47	-40	-32	-28	-25	-22	-17	-10	-2	5	20	35
-20	-80	-65	-50	-42	-35	-27	-23	-20	-17	-12	-5	2	10	25	40
-18	-78	-63	-48	-40	-33	-25	-21	-18	-15	-10	-3	4	12	27	42
-16	-76	-61	-46	-38	-31	-23	-19	-16	-13	-8	-1	6	14	29	44
-14	-74	-59	-44	-36	-29	-21	-17	-14	-11	-6	1	8	16	31	46
-12	-72	-57	-42	-34	-27	-19	-15	-12	-9	-4	3	10	18	33	48
-10	-70	-55	-40	-32	-25	-17	-13	-10	-7	-2	5	12	20	35	50
-8	-68	-53	-38	-30	-23	-15	-11	-8	-5	0	7	14	22	37	52
-7	-67	-52	-37	-29	-22	-14	-10	-7	-4	0	8	15	23	38	53
-6	-66	-51	-36	-28	-21	-13	-9	-6	-3	1	9	16	24	39	54
-5	-65	-50	-35	-27	-20	-12	-8	-5	-2	2	10	17	25	40	55
-4	-64	-49	-34	-26	-19	-11	-7	-4	-1	3	11	18	26	41	56
-3	-63	-48	-33	-25	-18	-10	-6	-3	0	4	12	19	27	42	57
-2	-62	-47	-32	-24	-17	-9	-5	-2	1	5	13	20	28	43	58
-1	-61	-46	-31	-23	-16	-8	-4	-1	2	6	14	21	29	44	59
0	-60	-45	-30	-22	-15	-7	-3	0	3	7	15	22	30	45	60
1	-59	-44	-29	-21	-14	-6	-2	1	4	8	16	23	31	46	61
2	-58	-43	-28	-20	-13	-5	-1	2	5	9	17	24	32	47	62
3	-57	-42	-27	-19	-12	-4	0	3	6	10	18	25	33	48	63
4	-56	-41	-26	-18	-11	-3	1	4	7	11	19	26	34	49	64
5	-55	-40	-25	-17	-10	-2	2	5	8	12	20	27	35	50	65
6	-54	-39	-24	-16	-9	-1	3	6	9	13	21	28	36	51	66
7	-53	-38	-23	-15	-8	0	4	7	10	14	22	29	37	52	67
8	-52	-37	-22	-14	-7	0	5	8	11	15	23	30	38	53	68
10	-50	-35	-20	-12	-5	2	7	10	13	17	25	32	40	55	70
12	-48	-33	-18	-10	-3	4	9	12	15	19	27	34	42	57	72
14	-46	-31	-16	-8	-1	6	11	14	17	21	29	36	44	59	74
16	-44	-29	-14	-6	1	8	13	16	19	23	31	38	46	61	76
18	-42	-27	-12	-4	3	10	15	18	21	25	33	40	48	63	78
20	-40	-25	-10	-2	5	12	17	20	23	27	35	42	50	65	80
25	-35	-20	-5	2	10	17	22	25	28	32	40	47	55	70	85
30	-30	-15	0	7	15	22	27	30	33	37	45	52	60	75	90
40	-20	-5	10	17	25	32	37	40	43	47	55	62	70	85	100

Table 5-89

Separation, inch: 10

Leg Distance, inch: 35

SHIM REQUIRED FOR ALIGNMENT

Reverse Indicator Method

riveN IR ILS	DriveR TIR, MILS -40	-30	-20	-15	-10	-5	-2	0	2	5	10	15	20	30	40
-40	-120	-102	-85	-76	-67	-58	-53	-50	-46	-41	-32	-23	-15	2	20
-30	-107	-90	-72	-63	-55	-46	-41	-37	-34	-28	-20	-11	-2	15	32
-25	-101	-83	-66	-57	-48	-40	-34	-31	-27	-22	-13	-5	3	21	38
-20	-95	-77	-60	-51	-42	-33	-28	-25	-21	-16	-7	1	10	27	45
-18	-92	-75	-57	-48	-40	-31	-26	-22	-19	-13	-5	3	12	30	47
-16	-90	-72	-55	-46	-37	-28	-23	-20	-16	-11	-2	6	15	32	50
-14	-87	-70	-52	-43	-35	-26	-21	-17	-14	-8	0	8	17	35	52
-12	-85	-67	-50	-41	-32	-23	-18	-15	-11	-6	2	11	20	37	55
-10	-82	-65	-47	-38	-30	-21	-16	-12	-9	-3	5	13	22	40	57
-8	-80	-62	-45	-36	-27	-18	-13	-10	-6	-1	7	16	25	42	60
-7	-78	-61	-43	-35	-26	-17	-12	-8	-5	0	8	17	26	43	61
-6	-77	-60	-42	-33	-25	-16	-11	-7	-4	1	10	18	27	45	62
-5	-76	-58	-41	-32	-23	-15	-9	-6	-2	2	11	20	28	46	63
-4	-75	-57	-40	-31	-22	-13	-8	-5	-1	3	12	21	30	47	65
-3	-73	-56	-38	-30	-21	-12	-7	-3	0	5	13	22	31	48	66
-2	-72	-55	-37	-28	-20	-11	-6	-2	1	6	15	23	32	50	67
-1	-71	-53	-36	-27	-18	-10	-4	-1	2	7	16	25	33	51	68
0	-70	-52	-35	-26	-17	-8	-3	0	3	8	17	26	35	52	70
1	-68	-51	-33	-25	-16	-7	-2	1	4	10	18	27	36	53	71
2	-67	-50	-32	-23	-15	-6	-1	2	6	11	20	28	37	55	72
3	-66	-48	-31	-22	-13	-5	0	3	7	12	21	30	38	56	73
4	-65	-47	-30	-21	-12	-3	1	5	8	13	22	31	40	57	75
5	-63	-46	-28	-20	-11	-2	2	6	9	15	23	32	41	58	76
6	-62	-45	-27	-18	-10	-1	4	7	11	16	25	33	42	60	77
7	-61	-43	-26	-17	-8	0	5	8	12	17	26	35	43	61	78
8	-60	-42	-25	-16	-7	1	6	10	13	18	27	36	45	62	80
10	-57	-40	-22	-13	-5	3	9	12	16	21	30	38	47	65	82
12	-55	-37	-20	-11	-2	6	11	15	18	23	32	41	50	67	85
14	-52	-35	-17	-8	0	8	14	17	21	26	35	43	52	70	87
16	-50	-32	-15	-6	2	11	16	20	23	28	37	46	55	72	90
18	-47	-30	-12	-3	5	13	19	22	26	31	40	48	57	75	92
20	-45	-27	-10	-1	7	16	21	25	28	33	42	51	60	77	95
25	-38	-21	-3	5	13	22	27	31	34	40	48	57	66	83	101
30	-32	-15	2	11	20	28	34	37	41	46	55	63	72	90	107
40	-20	-2	15	23	32	41	46	50	53	58	67	76	85	102	120

Table 5-90

Separation, inch: 10

Leg Distance, inch: 40

SHIM REQUIRED FOR ALIGNMENT

Reverse Indicator Method

DriveN TIR MILS	DriveR TIR, MILS -40	-30	-20	-15	-10	-5	-2	0	2	5	10	15	20	30	40
-40	-140	-120	-100	-90	-80	-70	-64	-60	-56	-50	-40	-30	-20	0	20
-30	-125	-105	-85	-75	-65	-55	-49	-45	-41	-35	-25	-15	-5	15	35
-25	-117	-97	-77	-67	-57	-47	-41	-37	-33	-27	-17	-7	2	22	42
-20	-110	-90	-70	-60	-50	-40	-34	-30	-26	-20	-10	0	10	30	50
-18	-107	-87	-67	-57	-47	-37	-31	-27	-23	-17	-7	3	13	33	53
-16	-104	-84	-64	-54	-44	-34	-28	-24	-20	-14	-4	6	16	36	56
-14	-101	-81	-61	-51	-41	-31	-25	-21	-17	-11	-1	9	19	39	59
-12	-98	-78	-58	-48	-38	-28	-22	-18	-14	-8	2	12	22	42	62
-10	-95	-75	-55	-45	-35	-25	-19	-15	-11	-5	5	15	25	45	65
-8	-92	-72	-52	-42	-32	-22	-16	-12	-8	-2	8	18	28	48	68
-7	-90	-70	-50	-40	-30	-20	-14	-10	-6	0	9	19	29	49	69
-6	-89	-69	-49	-39	-29	-19	-13	-9	-5	1	11	21	31	51	71
-5	-87	-67	-47	-37	-27	-17	-11	-7	-3	2	12	22	32	52	72
-4	-86	-66	-46	-36	-26	-16	-10	-6	-2	4	14	24	34	54	74
-3	-84	-64	-44	-34	-24	-14	-8	-4	0	5	15	25	35	55	75
-2	-83	-63	-43	-33	-23	-13	-7	-3	1	7	17	27	37	57	77
-1	-81	-61	-41	-31	-21	-11	-5	-1	2	8	18	28	38	58	78
0	-80	-60	-40	-30	-20	-10	-4	0	4	10	20	30	40	60	80
1	-78	-58	-38	-28	-18	-8	-2	1	5	11	21	31	41	61	81
2	-77	-57	-37	-27	-17	-7	-1	3	7	13	23	33	43	63	83
3	-75	-55	-35	-25	-15	-5	0	4	8	14	24	34	44	64	84
4	-74	-54	-34	-24	-14	-4	2	6	10	16	26	36	46	66	86
5	-72	-52	-32	-22	-12	-2	3	7	11	17	27	37	47	67	87
6	-71	-51	-31	-21	-11	-1	5	9	13	19	29	39	49	69	89
7	-69	-49	-29	-19	-9	0	6	10	14	20	30	40	50	70	90
8	-68	-48	-28	-18	-8	2	8	12	16	22	32	42	52	72	92
10	-65	-45	-25	-15	-5	5	11	15	19	25	35	45	55	75	95
12	-62	-42	-22	-12	-2	8	14	18	22	28	38	48	58	78	98
14	-59	-39	-19	-9	1	11	17	21	25	31	41	51	61	81	101
16	-56	-36	-16	-6	4	14	20	24	28	34	44	54	64	84	104
18	-53	-33	-13	-3	7	17	23	27	31	37	47	57	67	87	107
20	-50	-30	-10	0	10	20	26	30	34	40	50	60	70	90	110
25	-42	-22	-2	7	17	27	33	37	41	47	57	67	77	97	117
30	-35	-15	5	15	25	35	41	45	49	55	65	75	85	105	125
40	-20	0	20	30	40	50	56	60	64	70	80	90	100	120	140

Table 5-91

Separation, inch: 10

Leg Distance, inch: 50

SHIM REQUIRED FOR ALIGNMENT

Reverse Indicator Method

riveN IR ILS	DriveR TIR, MILS -40	-30	-20	-15	-10	-5	-2	0	2	5	10	15	20	30	40
-40	-180	-155	-130	-117	-105	-92	-85	-80	-75	-67	-55	-42	-30	-5	20
-30	-160	-135	-110	-97	-85	-72	-65	-60	-55	-47	-35	-22	-10	15	40
-25	-150	-125	-100	-87	-75	-62	-55	-50	-45	-37	-25	-12	0	25	50
-20	-140	-115	-90	-77	-65	-52	-45	-40	-35	-27	-15	-2	10	35	60
-18	-136	-111	-86	-73	-61	-48	-41	-36	-31	-23	-11	1	14	39	64
-16	-132	-107	-82	-69	-57	-44	-37	-32	-27	-19	-7	5	18	43	68
-14	-128	-103	-78	-65	-53	-40	-33	-28	-23	-15	-3	9	22	47	72
-12	-124	-99	-74	-61	-49	-36	-29	-24	-19	-11	1	13	26	51	76
-10	-120	-95	-70	-57	-45	-32	-25	-20	-15	-7	5	17	30	55	80
-8	-116	-91	-66	-53	-41	-28	-21	-16	-11	-3	9	21	34	59	84
-7	-114	-89	-64	-51	-39	-26	-19	-14	-9	-1	11	23	36	61	86
-6	-112	-87	-62	-49	-37	-24	-17	-12	-7	0	13	25	38	63	88
-5	-110	-85	-60	-47	-35	-22	-15	-10	-5	2	15	27	40	65	90
-4	-108	-83	-58	-45	-33	-20	-13	-8	-3	4	17	29	42	67	92
-3	-106	-81	-56	-43	-31	-18	-11	-6	-1	6	19	31	44	69	94
-2	-104	-79	-54	-41	-29	-16	-9	-4	1	8	21	33	46	71	96
-1	-102	-77	-52	-39	-27	-14	-7	-2	3	10	23	35	48	73	98
0	-100	-75	-50	-37	-25	-12	-5	0	5	12	25	37	50	75	100
1	-98	-73	-48	-35	-23	-10	-3	2	7	14	27	39	52	77	102
2	-96	-71	-46	-33	-21	-8	-1	4	9	16	29	41	54	79	104
3	-94	-69	-44	-31	-19	-6	1	6	11	18	31	43	56	81	106
4	-92	-67	-42	-29	-17	-4	3	8	13	20	33	45	58	83	108
5	-90	-65	-40	-27	-15	-2	5	10	15	22	35	47	60	85	110
6	-88	-63	-38	-25	-13	0	7	12	17	24	37	49	62	87	112
7	-86	-61	-36	-23	-11	1	9	14	19	26	39	51	64	89	114
8	-84	-59	-34	-21	-9	3	11	16	21	28	41	53	66	91	116
10	-80	-55	-30	-17	-5	7	15	20	25	32	45	57	70	95	120
12	-76	-51	-26	-13	-1	11	19	24	29	36	49	61	74	99	124
14	-72	-47	-22	-9	3	15	23	28	33	40	53	65	78	103	128
16	-68	-43	-18	-5	7	19	27	32	37	44	57	69	82	107	132
18	-64	-39	-14	-1	11	23	31	36	41	48	61	73	86	111	136
20	-60	-35	-10	2	15	27	35	40	45	52	65	77	90	115	140
25	-50	-25	0	12	25	37	45	50	55	62	75	87	100	125	150
30	-40	-15	10	22	35	47	55	60	65	72	85	97	110	135	160
40	-20	5	30	42	55	67	75	80	85	92	105	117	130	155	180

Table 5-92

Separation, inch: 10

Leg Distance, inch: 60

SHIM REQUIRED FOR ALIGNMENT

Reverse Indicator Method

DriveN TIR MILS	DriveR TIR, MILS -40	-30	-20	-15	-10	-5	-2	0	2	5	10	15	20	30	40
-40	-220	-190	-160	-145	-130	-115	-106	-100	-94	-85	-70	-55	-40	-10	20
-30	-195	-165	-135	-120	-105	-90	-81	-75	-69	-60	-45	-30	-15	15	45
-25	-182	-152	-122	-107	-92	-77	-68	-62	-56	-47	-32	-17	-2	27	57
-20	-170	-140	-110	-95	-80	-65	-56	-50	-44	-35	-20	-5	10	40	70
-18	-165	-135	-105	-90	-75	-60	-51	-45	-39	-30	-15	0	15	45	75
-16	-160	-130	-100	-85	-70	-55	-46	-40	-34	-25	-10	5	20	50	80
-14	-155	-125	-95	-80	-65	-50	-41	-35	-29	-20	-5	10	25	55	85
-12	-150	-120	-90	-75	-60	-45	-36	-30	-24	-15	0	15	30	60	90
-10	-145	-115	-85	-70	-55	-40	-31	-25	-19	-10	5	20	35	65	95
-8	-140	-110	-80	-65	-50	-35	-26	-20	-14	-5	10	25	40	70	100
-7	-137	-107	-77	-62	-47	-32	-23	-17	-11	-2	12	27	42	72	102
-6	-135	-105	-75	-60	-45	-30	-21	-15	-9	0	15	30	45	75	105
-5	-132	-102	-72	-57	-42	-27	-18	-12	-6	2	17	32	47	77	107
-4	-130	-100	-70	-55	-40	-25	-16	-10	-4	5	20	35	50	80	110
-3	-127	-97	-67	-52	-37	-22	-13	-7	-1	7	22	37	52	82	112
-2	-125	-95	-65	-50	-35	-20	-11	-5	1	10	25	40	55	85	115
-1	-122	-92	-62	-47	-32	-17	-8	-2	3	12	27	42	57	87	117
0	-120	-90	-60	-45	-30	-15	-6	0	6	15	30	45	60	90	120
1	-117	-87	-57	-42	-27	-12	-3	2	8	17	32	47	62	92	122
2	-115	-85	-55	-40	-25	-10	-1	5	11	20	35	50	65	95	125
3	-112	-82	-52	-37	-22	-7	1	7	13	22	37	52	67	97	127
4	-110	-80	-50	-35	-20	-5	4	10	16	25	40	55	70	100	130
5	-107	-77	-47	-32	-17	-2	6	12	18	27	42	57	72	102	132
6	-105	-75	-45	-30	-15	0	9	15	21	30	45	60	75	105	135
7	-102	-72	-42	-27	-12	2	11	17	23	32	47	62	77	107	137
8	-100	-70	-40	-25	-10	5	14	20	26	35	50	65	80	110	140
10	-95	-65	-35	-20	-5	10	19	25	31	40	55	70	85	115	145
12	-90	-60	-30	-15	0	15	24	30	36	45	60	75	90	120	150
14	-85	-55	-25	-10	5	20	29	35	41	50	65	80	95	125	155
16	-80	-50	-20	-5	10	25	34	40	46	55	70	85	100	130	160
18	-75	-45	-15	0	15	30	39	45	51	60	75	90	105	135	165
20	-70	-40	-10	5	20	35	44	50	56	65	80	95	110	140	170
25	-57	-27	2	17	32	47	56	62	68	77	92	107	122	152	182
30	-45	-15	15	30	45	60	69	75	81	90	105	120	135	165	195
40	-20	10	40	55	70	85	94	100	106	115	130	145	160	190	220

Table 5-93

Separation, inch: 10

Leg Distance, inch: 70

SHIM REQUIRED FOR ALIGNMENT

Reverse Indicator Method

DriveN TIR MILS	DriveR TIR, MILS														
	-40	-30	-20	-15	-10	-5	-2	0	2	5	10	15	20	30	40
-40	-260	-225	-190	-172	-155	-137	-127	-120	-113	-102	-85	-67	-50	-15	20
-30	-230	-195	-160	-142	-125	-107	-97	-90	-83	-72	-55	-37	-20	15	50
-25	-215	-180	-145	-127	-110	-92	-82	-75	-68	-57	-40	-22	-5	30	65
-20	-200	-165	-130	-112	-95	-77	-67	-60	-53	-42	-25	-7	10	45	80
-18	-194	-159	-124	-106	-89	-71	-61	-54	-47	-36	-19	-1	16	51	86
-16	-188	-153	-118	-100	-83	-65	-55	-48	-41	-30	-13	4	22	57	92
-14	-182	-147	-112	-94	-77	-59	-49	-42	-35	-24	-7	10	28	63	98
-12	-176	-141	-106	-88	-71	-53	-43	-36	-29	-18	-1	16	34	69	104
-10	-170	-135	-100	-82	-65	-47	-37	-30	-23	-12	5	22	40	75	110
-8	-164	-129	-94	-76	-59	-41	-31	-24	-17	-6	11	28	46	81	116
-7	-161	-126	-91	-73	-56	-38	-28	-21	-14	-3	14	31	49	84	119
-6	-158	-123	-88	-70	-53	-35	-25	-18	-11	0	17	34	52	87	122
-5	-155	-120	-85	-67	-50	-32	-22	-15	-8	2	20	37	55	90	125
-4	-152	-117	-82	-64	-47	-29	-19	-12	-5	5	23	40	58	93	128
-3	-149	-114	-79	-61	-44	-26	-16	-9	-2	8	26	43	61	96	131
-2	-146	-111	-76	-58	-41	-23	-13	-6	1	11	29	46	64	99	134
-1	-143	-108	-73	-55	-38	-20	-10	-3	4	14	32	49	67	102	137
0	-140	-105	-70	-52	-35	-17	-7	0	7	17	35	52	70	105	140
1	-137	-102	-67	-49	-32	-14	-4	3	10	20	38	55	73	108	143
2	-134	-99	-64	-46	-29	-11	-1	6	13	23	41	58	76	111	146
3	-131	-96	-61	-43	-26	-8	2	9	16	26	44	61	79	114	149
4	-128	-93	-58	-40	-23	-5	5	12	19	29	47	64	82	117	152
5	-125	-90	-55	-37	-20	-2	8	15	22	32	50	67	85	120	155
6	-122	-87	-52	-34	-17	0	11	18	25	35	53	70	88	123	158
7	-119	-84	-49	-31	-14	3	14	21	28	38	56	73	91	126	161
8	-116	-81	-46	-28	-11	6	17	24	31	41	59	76	94	129	164
10	-110	-75	-40	-22	-5	12	23	30	37	47	65	82	100	135	170
12	-104	-69	-34	-16	1	18	29	36	43	53	71	88	106	141	176
14	-98	-63	-28	-10	7	24	35	42	49	59	77	94	112	147	182
16	-92	-57	-22	-4	13	30	41	48	55	65	83	100	118	153	188
18	-86	-51	-16	1	19	36	47	54	61	71	89	106	124	159	194
20	-80	-45	-10	7	25	42	53	60	67	77	95	112	130	165	200
25	-65	-30	5	22	40	57	68	75	82	92	110	127	145	180	215
30	-50	-15	20	37	55	72	83	90	97	107	125	142	160	195	230
40	-20	15	50	67	85	102	113	120	127	137	155	172	190	225	260

Table 5-94

Separation, inch: 10

Leg Distance, inch: 80

SHIM REQUIRED FOR ALIGNMENT

Reverse Indicator Method

DriveN TIR MILS	DriveR TIR, MILS -40	-30	-20	-15	-10	-5	-2	0	2	5	10	15	20	30	40
-40	-300	-260	-220	-200	-180	-160	-148	-140	-132	-120	-100	-80	-60	-20	20
-30	-265	-225	-185	-165	-145	-125	-113	-105	-97	-85	-65	-45	-25	15	55
-25	-247	-207	-167	-147	-127	-107	-95	-87	-79	-67	-47	-27	-7	32	72
-20	-230	-190	-150	-130	-110	-90	-78	-70	-62	-50	-30	-10	10	50	90
-18	-223	-183	-143	-123	-103	-83	-71	-63	-55	-43	-23	-3	17	57	97
-16	-216	-176	-136	-116	-96	-76	-64	-56	-48	-36	-16	4	24	64	104
-14	-209	-169	-129	-109	-89	-69	-57	-49	-41	-29	-9	11	31	71	111
-12	-202	-162	-122	-102	-82	-62	-50	-42	-34	-22	-2	18	38	78	118
-10	-195	-155	-115	-95	-75	-55	-43	-35	-27	-15	5	25	45	85	125
-8	-188	-148	-108	-88	-68	-48	-36	-28	-20	-8	12	32	52	92	132
-7	-184	-144	-104	-84	-64	-44	-32	-24	-16	-4	15	35	55	95	135
-6	-181	-141	-101	-81	-61	-41	-29	-21	-13	-1	19	39	59	99	139
-5	-177	-137	-97	-77	-57	-37	-25	-17	-9	2	22	42	62	102	142
-4	-174	-134	-94	-74	-54	-34	-22	-14	-6	6	26	46	66	106	146
-3	-170	-130	-90	-70	-50	-30	-18	-10	-2	9	29	49	69	109	149
-2	-167	-127	-87	-67	-47	-27	-15	-7	1	13	33	53	73	113	153
-1	-163	-123	-83	-63	-43	-23	-11	-3	4	16	36	56	76	116	156
0	-160	-120	-80	-60	-40	-20	-8	0	8	20	40	60	80	120	160
1	-156	-116	-76	-56	-36	-16	-4	3	11	23	43	63	83	123	163
2	-153	-113	-73	-53	-33	-13	-1	7	15	27	47	67	87	127	167
3	-149	-109	-69	-49	-29	-9	2	10	18	30	50	70	90	130	170
4	-146	-106	-66	-46	-26	-6	6	14	22	34	54	74	94	134	174
5	-142	-102	-62	-42	-22	-2	9	17	25	37	57	77	97	137	177
6	-139	-99	-59	-39	-19	1	13	21	29	41	61	81	101	141	181
7	-135	-95	-55	-35	-15	4	16	24	32	44	64	84	104	144	184
8	-132	-92	-52	-32	-12	8	20	28	36	48	68	88	108	148	188
10	-125	-85	-45	-25	-5	15	27	35	43	55	75	95	115	155	195
12	-118	-78	-38	-18	2	22	34	42	50	62	82	102	122	162	202
14	-111	-71	-31	-11	9	29	41	49	57	69	89	109	129	169	209
16	-104	-64	-24	-4	16	36	48	56	64	76	96	116	136	176	216
18	-97	-57	-17	3	23	43	55	63	71	83	103	123	143	183	223
20	-90	-50	-10	10	30	50	62	70	78	90	110	130	150	190	230
25	-72	-32	7	27	47	67	79	87	95	107	127	147	167	207	247
30	-55	-15	25	45	65	85	97	105	113	125	145	165	185	225	265
40	-20	20	60	80	100	120	132	140	148	160	180	200	220	260	300

Table 5-95

Separation, inch: 10

Leg Distance, inch: 90

SHIM REQUIRED FOR ALIGNMENT

Reverse Indicator Method

DriveN TIR MILS	DriveR TIR, MILS -40	-30	-20	-15	-10	-5	-2	0	2	5	10	15	20	30	40
-40	-340	-295	-250	-227	-205	-182	-169	-160	-151	-137	-115	-92	-70	-25	20
-30	-300	-255	-210	-187	-165	-142	-129	-120	-111	-97	-75	-52	-30	15	60
-25	-280	-235	-190	-167	-145	-122	-109	-100	-91	-77	-55	-32	-10	35	80
-20	-260	-215	-170	-147	-125	-102	-89	-80	-71	-57	-35	-12	10	55	100
-18	-252	-207	-162	-139	-117	-94	-81	-72	-63	-49	-27	-4	18	63	108
-16	-244	-199	-154	-131	-109	-86	-73	-64	-55	-41	-19	3	26	71	116
-14	-236	-191	-146	-123	-101	-78	-65	-56	-47	-33	-11	11	34	79	124
-12	-228	-183	-138	-115	-93	-70	-57	-48	-39	-25	-3	19	42	87	132
-10	-220	-175	-130	-107	-85	-62	-49	-40	-31	-17	5	27	50	95	140
-8	-212	-167	-122	-99	-77	-54	-41	-32	-23	-9	13	35	58	103	148
-7	-208	-163	-118	-95	-73	-50	-37	-28	-19	-5	17	39	62	107	152
-6	-204	-159	-114	-91	-69	-46	-33	-24	-15	-1	21	43	66	111	156
-5	-200	-155	-110	-87	-65	-42	-29	-20	-11	2	25	47	70	115	160
-4	-196	-151	-106	-83	-61	-38	-25	-16	-7	6	29	51	74	119	164
-3	-192	-147	-102	-79	-57	-34	-21	-12	-3	10	33	55	78	123	168
-2	-188	-143	-98	-75	-53	-30	-17	-8	1	14	37	59	82	127	172
-1	-184	-139	-94	-71	-49	-26	-13	-4	5	18	41	63	86	131	176
0	-180	-135	-90	-67	-45	-22	-9	0	9	22	45	67	90	135	180
1	-176	-131	-86	-63	-41	-18	-5	4	13	26	49	71	94	139	184
2	-172	-127	-82	-59	-37	-14	-1	8	17	30	53	75	98	143	188
3	-168	-123	-78	-55	-33	-10	3	12	21	34	57	79	102	147	192
4	-164	-119	-74	-51	-29	-6	7	16	25	38	61	83	106	151	196
5	-160	-115	-70	-47	-25	-2	11	20	29	42	65	87	110	155	200
6	-156	-111	-66	-43	-21	1	15	24	33	46	69	91	114	159	204
7	-152	-107	-62	-39	-17	5	19	28	37	50	73	95	118	163	208
8	-148	-103	-58	-35	-13	9	23	32	41	54	77	99	122	167	212
10	-140	-95	-50	-27	-5	17	31	40	49	62	85	107	130	175	220
12	-132	-87	-42	-19	3	25	39	48	57	70	93	115	138	183	228
14	-124	-79	-34	-11	11	33	47	56	65	78	101	123	146	191	236
16	-116	-71	-26	-3	19	41	55	64	73	86	109	131	154	199	244
18	-108	-63	-18	4	27	49	63	72	81	94	117	139	162	207	252
20	-100	-55	-10	12	35	57	71	80	89	102	125	147	170	215	260
25	-80	-35	10	32	55	77	91	100	109	122	145	167	190	235	280
30	-60	-15	30	52	75	97	111	120	129	142	165	187	210	255	300
40	-20	25	70	92	115	137	151	160	169	182	205	227	250	295	340

Table 5-96

Separation, inch: 10

Leg Distance, inch: 100

SHIM REQUIRED FOR ALIGNMENT

Reverse Indicator Method

DriveN TIR MILS	DriveR TIR, MILS -40	-30	-20	-15	-10	-5	-2	0	2	5	10	15	20	30	40
-40	-380	-330	-280	-255	-230	-205	-190	-180	-170	-155	-130	-105	-80	-30	20
-30	-335	-285	-235	-210	-185	-160	-145	-135	-125	-110	-85	-60	-35	15	65
-25	-312	-262	-212	-187	-162	-137	-122	-112	-102	-87	-62	-37	-12	37	87
-20	-290	-240	-190	-165	-140	-115	-100	-90	-80	-65	-40	-15	10	60	110
-18	-281	-231	-181	-156	-131	-106	-91	-81	-71	-56	-31	-6	19	69	119
-16	-272	-222	-172	-147	-122	-97	-82	-72	-62	-47	-22	3	28	78	128
-14	-263	-213	-163	-138	-113	-88	-73	-63	-53	-38	-13	12	37	87	137
-12	-254	-204	-154	-129	-104	-79	-64	-54	-44	-29	-4	21	46	96	146
-10	-245	-195	-145	-120	-95	-70	-55	-45	-35	-20	5	30	55	105	155
-8	-236	-186	-136	-111	-86	-61	-46	-36	-26	-11	14	39	64	114	164
-7	-231	-181	-131	-106	-81	-56	-41	-31	-21	-6	18	43	68	118	168
-6	-227	-177	-127	-102	-77	-52	-37	-27	-17	-2	23	48	73	123	173
-5	-222	-172	-122	-97	-72	-47	-32	-22	-12	2	27	52	77	127	177
-4	-218	-168	-118	-93	-68	-43	-28	-18	-8	7	32	57	82	132	182
-3	-213	-163	-113	-88	-63	-38	-23	-13	-3	11	36	61	86	136	186
-2	-209	-159	-109	-84	-59	-34	-19	-9	1	16	41	66	91	141	191
-1	-204	-154	-104	-79	-54	-29	-14	-4	5	20	45	70	95	145	195
0	-200	-150	-100	-75	-50	-25	-10	0	10	25	50	75	100	150	200
1	-195	-145	-95	-70	-45	-20	-5	4	14	29	54	79	104	154	204
2	-191	-141	-91	-66	-41	-16	-1	9	19	34	59	84	109	159	209
3	-186	-136	-86	-61	-36	-11	3	13	23	38	63	88	113	163	213
4	-182	-132	-82	-57	-32	-7	8	18	28	43	68	93	118	168	218
5	-177	-127	-77	-52	-27	-2	12	22	32	47	72	97	122	172	222
6	-173	-123	-73	-48	-23	2	17	27	37	52	77	102	127	177	227
7	-168	-118	-68	-43	-18	6	21	31	41	56	81	106	131	181	231
8	-164	-114	-64	-39	-14	11	26	36	46	61	86	111	136	186	236
10	-155	-105	-55	-30	-5	20	35	45	55	70	95	120	145	195	245
12	-146	-96	-46	-21	4	29	44	54	64	79	104	129	154	204	254
14	-137	-87	-37	-12	13	38	53	63	73	88	113	138	163	213	263
16	-128	-78	-28	-3	22	47	62	72	82	97	122	147	172	222	272
18	-119	-69	-19	6	31	56	71	81	91	106	131	156	181	231	281
20	-110	-60	-10	15	40	65	80	90	100	115	140	165	190	240	290
25	-87	-37	12	37	62	87	102	112	122	137	162	187	212	262	312
30	-65	-15	35	60	85	110	125	135	145	160	185	210	235	285	335
40	-20	30	80	105	130	155	170	180	190	205	230	255	280	330	380

Table 5-97

Separation, inch: 12

Leg Distance, inch: 25

SHIM REQUIRED FOR ALIGNMENT

Reverse Indicator Method

DriveN TIR MILS	DriveR TIR, MILS -50	-40	-30	-20	-15	-10	-5	0	5	10	15	20	30	40	50
-70	-90	-79	-69	-58	-53	-48	-43	-37	-32	-27	-22	-17	-6	3	14
-60	-84	-74	-63	-53	-48	-42	-37	-32	-27	-22	-16	-11	-1	9	19
-55	-81	-71	-61	-50	-45	-40	-35	-29	-24	-19	-14	-8	1	11	22
-50	-79	-68	-58	-47	-42	-37	-32	-27	-21	-16	-11	-6	4	14	25
-45	-76	-66	-55	-45	-40	-34	-29	-24	-19	-13	-8	-3	6	17	27
-40	-73	-63	-52	-42	-37	-32	-26	-21	-16	-11	-6	0	9	20	30
-35	-71	-60	-50	-39	-34	-29	-24	-18	-13	-8	-3	1	12	22	33
-30	-68	-57	-47	-37	-31	-26	-21	-16	-11	-5	0	4	15	25	35
-25	-65	-55	-44	-34	-29	-23	-18	-13	-8	-3	2	7	17	28	38
-20	-62	-52	-42	-31	-26	-21	-16	-10	-5	0	4	10	20	30	41
-15	-60	-49	-39	-28	-23	-18	-13	-8	-2	2	7	12	23	33	43
-12	-58	-48	-37	-27	-22	-16	-11	-6	-1	3	9	14	24	35	45
-10	-57	-47	-36	-26	-21	-15	-10	-5	0	5	10	15	25	36	46
-8	-56	-46	-35	-25	-19	-14	-9	-4	0	6	11	16	26	37	47
-6	-55	-44	-34	-24	-18	-13	-8	-3	1	7	12	17	28	38	48
-4	-54	-43	-33	-23	-17	-12	-7	-2	3	8	13	18	29	39	49
-2	-53	-42	-32	-21	-16	-11	-6	-1	4	9	14	19	30	40	51
0	-52	-41	-31	-20	-15	-10	-5	0	5	10	15	20	31	41	52
2	-51	-40	-30	-19	-14	-9	-4	1	6	11	16	21	32	42	53
4	-49	-39	-29	-18	-13	-8	-3	2	7	12	17	23	33	43	54
6	-48	-38	-28	-17	-12	-7	-1	3	8	13	18	24	34	44	55
8	-47	-37	-26	-16	-11	-6	0	4	9	14	19	25	35	46	56
10	-46	-36	-25	-15	-10	-5	0	5	10	15	21	26	36	47	57
12	-45	-35	-24	-14	-9	-3	1	6	11	16	22	27	37	48	58
15	-43	-33	-23	-12	-7	-2	2	8	13	18	23	28	39	49	60
20	-41	-30	-20	-10	-4	0	5	10	16	21	26	31	42	52	62
25	-38	-28	-17	-7	-2	3	8	13	18	23	29	34	44	55	65
30	-35	-25	-15	-4	0	5	11	16	21	26	31	37	47	57	68
35	-33	-22	-12	-1	3	8	13	18	24	29	34	39	50	60	71
40	-30	-20	-9	0	6	11	16	21	26	32	37	42	52	63	73
45	-27	-17	-6	3	8	13	19	24	29	34	40	45	55	66	76
50	-25	-14	-4	6	11	16	21	27	32	37	42	47	58	68	79
55	-22	-11	-1	8	14	19	24	29	35	40	45	50	61	71	81
60	-19	-9	1	11	16	22	27	32	37	42	48	53	63	74	84
70	-14	-3	6	17	22	27	32	37	43	48	53	58	69	79	90

Table 5-98

Separation, inch: 12

Leg Distance, inch: 30

SHIM REQUIRED FOR ALIGNMENT

Reverse Indicator Method

DriveN TIR MILS	DriveR TIR, MILS -50	-40	-30	-20	-15	-10	-5	0	5	10	15	20	30	40	50
-70	-115	-102	-90	-77	-71	-65	-58	-52	-46	-40	-33	-27	-15	-2	10
-60	-107	-95	-82	-70	-63	-57	-51	-45	-38	-32	-26	-20	-7	5	17
-55	-103	-91	-78	-66	-60	-53	-47	-41	-35	-28	-22	-16	-3	8	21
-50	-100	-87	-75	-62	-56	-50	-43	-37	-31	-25	-18	-12	0	12	25
-45	-96	-83	-71	-58	-52	-46	-40	-33	-27	-21	-15	-8	3	16	28
-40	-92	-80	-67	-55	-48	-42	-36	-30	-23	-17	-11	-5	7	20	32
-35	-88	-76	-63	-51	-45	-38	-32	-26	-20	-13	-7	-1	11	23	36
-30	-85	-72	-60	-47	-41	-35	-28	-22	-16	-10	-3	2	15	27	40
-25	-81	-68	-56	-43	-37	-31	-25	-18	-12	-6	0	6	18	31	43
-20	-77	-65	-52	-40	-33	-27	-21	-15	-8	-2	3	10	22	35	47
-15	-73	-61	-48	-36	-30	-23	-17	-11	-5	1	7	13	26	38	51
-12	-71	-59	-46	-34	-27	-21	-15	-9	-2	3	9	16	28	41	53
-10	-70	-57	-45	-32	-26	-20	-13	-7	-1	5	11	17	30	42	55
-8	-68	-56	-43	-31	-24	-18	-12	-6	0	6	12	19	31	44	56
-6	-67	-54	-42	-29	-23	-17	-10	-4	1	8	14	20	33	45	58
-4	-65	-53	-40	-28	-21	-15	-9	-3	3	9	15	22	34	47	59
-2	-64	-51	-39	-26	-20	-14	-7	-1	4	11	17	23	36	48	61
0	-62	-50	-37	-25	-18	-12	-6	0	6	12	18	25	37	50	62
2	-61	-48	-36	-23	-17	-11	-4	1	7	14	20	26	39	51	64
4	-59	-47	-34	-22	-15	-9	-3	3	9	15	21	28	40	53	65
6	-58	-45	-33	-20	-14	-8	-1	4	10	17	23	29	42	54	67
8	-56	-44	-31	-19	-12	-6	0	6	12	18	24	31	43	56	68
10	-55	-42	-30	-17	-11	-5	1	7	13	20	26	32	45	57	70
12	-53	-41	-28	-16	-9	-3	2	9	15	21	27	34	46	59	71
15	-51	-38	-26	-13	-7	-1	5	11	17	23	30	36	48	61	73
20	-47	-35	-22	-10	-3	2	8	15	21	27	33	40	52	65	77
25	-43	-31	-18	-6	0	6	12	18	25	31	37	43	56	68	81
30	-40	-27	-15	-2	3	10	16	22	28	35	41	47	60	72	85
35	-36	-23	-11	1	7	13	20	26	32	38	45	51	63	76	88
40	-32	-20	-7	5	11	17	23	30	36	42	48	55	67	80	92
45	-28	-16	-3	8	15	21	27	33	40	46	52	58	71	83	96
50	-25	-12	0	12	18	25	31	37	43	50	56	62	75	87	100
55	-21	-8	3	16	22	28	35	41	47	53	60	66	78	91	103
60	-17	-5	7	20	26	32	38	45	51	57	63	70	82	95	107
70	-10	2	15	27	33	40	46	52	58	65	71	77	90	102	115

Table 5-99

Separation, inch: 12

Leg Distance, inch: 35

SHIM REQUIRED FOR ALIGNMENT

Reverse Indicator Method

riveN IR ILS	DriveR TIR, MILS														
	-50	-40	-30	-20	-15	-10	-5	0	5	10	15	20	30	40	50
-70	-140	-125	-110	-96	-88	-81	-74	-67	-59	-52	-45	-37	-23	-8	5
-60	-130	-115	-101	-86	-79	-72	-64	-57	-50	-42	-35	-28	-13	0	15
-55	-125	-111	-96	-81	-74	-67	-60	-52	-45	-38	-30	-23	-8	5	20
-50	-120	-106	-91	-77	-69	-62	-55	-47	-40	-33	-26	-18	-4	10	25
-45	-116	-101	-86	-72	-65	-57	-50	-43	-35	-28	-21	-13	0	15	29
-40	-111	-96	-82	-67	-60	-52	-45	-38	-31	-23	-16	-9	5	20	34
-35	-106	-91	-77	-62	-55	-48	-40	-33	-26	-18	-11	-4	10	24	39
-30	-101	-87	-72	-57	-50	-43	-36	-28	-21	-14	-6	0	15	29	44
-25	-96	-82	-67	-53	-45	-38	-31	-23	-16	-9	-2	5	19	34	48
-20	-92	-77	-62	-48	-41	-33	-26	-19	-11	-4	2	10	24	39	53
-15	-87	-72	-58	-43	-36	-28	-21	-14	-7	0	7	14	29	43	58
-12	-84	-69	-55	-40	-33	-26	-18	-11	-4	3	10	17	32	46	61
-10	-82	-67	-53	-38	-31	-24	-16	-9	-2	5	12	19	34	48	63
-8	-80	-66	-51	-36	-29	-22	-14	-7	0	6	14	21	36	50	65
-6	-78	-64	-49	-34	-27	-20	-13	-5	1	8	16	23	38	52	67
-4	-76	-62	-47	-33	-25	-18	-11	-3	3	10	18	25	39	54	69
-2	-74	-60	-45	-31	-23	-16	-9	-1	5	12	19	27	41	56	71
0	-72	-58	-43	-29	-21	-14	-7	0	7	14	21	29	43	58	72
2	-71	-56	-41	-27	-19	-12	-5	1	9	16	23	31	45	60	74
4	-69	-54	-39	-25	-18	-10	-3	3	11	18	25	33	47	62	76
6	-67	-52	-38	-23	-16	-8	-1	5	13	20	27	34	49	64	78
8	-65	-50	-36	-21	-14	-6	0	7	14	22	29	36	51	66	80
10	-63	-48	-34	-19	-12	-5	2	9	16	24	31	38	53	67	82
12	-61	-46	-32	-17	-10	-3	4	11	18	26	33	40	55	69	84
15	-58	-43	-29	-14	-7	0	7	14	21	28	36	43	58	72	87
20	-53	-39	-24	-10	-2	4	11	19	26	33	41	48	62	77	92
25	-48	-34	-19	-5	2	9	16	23	31	38	45	53	67	82	96
30	-44	-29	-15	0	6	14	21	28	36	43	50	57	72	87	101
35	-39	-24	-10	4	11	18	26	33	40	48	55	62	77	91	106
40	-34	-20	-5	9	16	23	31	38	45	52	60	67	82	96	111
45	-29	-15	0	13	21	28	35	43	50	57	65	72	86	101	116
50	-25	-10	4	18	26	33	40	47	55	62	69	77	91	106	120
55	-20	-5	8	23	30	38	45	52	60	67	74	81	96	111	125
60	-15	0	13	28	35	42	50	57	64	72	79	86	101	115	130
70	-5	8	23	37	45	52	59	67	74	81	88	96	110	125	140

Table 5-100

Separation, inch: 12

Leg Distance, inch: 40

SHIM REQUIRED FOR ALIGNMENT

Reverse Indicator Method

DriveN TIR MILS	DriveR TIR, MILS -50	-40	-30	-20	-15	-10	-5	0	5	10	15	20	30	40	50
-70	-165	-148	-131	-115	-106	-98	-90	-81	-73	-65	-56	-48	-31	-15	1
-60	-153	-136	-120	-103	-95	-86	-78	-70	-61	-53	-45	-36	-20	-3	13
-55	-147	-130	-114	-97	-89	-80	-72	-64	-55	-47	-39	-30	-14	2	19
-50	-141	-125	-108	-91	-83	-75	-66	-58	-50	-41	-33	-25	-8	8	25
-45	-135	-119	-102	-85	-77	-69	-60	-52	-44	-35	-27	-19	-2	14	30
-40	-130	-113	-96	-80	-71	-63	-55	-46	-38	-30	-21	-13	3	20	36
-35	-124	-107	-90	-74	-65	-57	-49	-40	-32	-24	-15	-7	9	25	42
-30	-118	-101	-85	-68	-60	-51	-43	-35	-26	-18	-10	-1	15	31	48
-25	-112	-95	-79	-62	-54	-45	-37	-29	-20	-12	-4	4	20	37	54
-20	-106	-90	-73	-56	-48	-40	-31	-23	-15	-6	1	10	26	43	60
-15	-100	-84	-67	-50	-42	-34	-25	-17	-9	0	7	15	32	49	65
-12	-97	-80	-64	-47	-39	-30	-22	-14	-5	2	11	19	36	52	69
-10	-95	-78	-61	-45	-36	-28	-20	-11	-3	5	13	21	38	55	71
-8	-92	-76	-59	-42	-34	-26	-17	-9	-1	7	15	24	40	57	74
-6	-90	-73	-57	-40	-32	-23	-15	-7	1	9	18	26	43	59	76
-4	-88	-71	-54	-38	-29	-21	-13	-4	3	12	20	28	45	62	78
-2	-85	-69	-52	-35	-27	-19	-10	-2	6	14	22	31	47	64	81
0	-83	-66	-50	-33	-25	-16	-8	0	8	16	25	33	50	66	83
2	-81	-64	-47	-31	-22	-14	-6	2	10	19	27	35	52	69	85
4	-78	-62	-45	-28	-20	-12	-3	4	13	21	29	38	54	71	88
6	-76	-59	-43	-26	-18	-9	-1	7	15	23	32	40	57	73	90
8	-74	-57	-40	-24	-15	-7	1	9	17	26	34	42	59	76	92
10	-71	-55	-38	-21	-13	-5	3	11	20	28	36	45	61	78	95
12	-69	-52	-36	-19	-11	-2	5	14	22	30	39	47	64	80	97
15	-65	-49	-32	-15	-7	0	9	17	25	34	42	50	67	84	100
20	-60	-43	-26	-10	-1	6	15	23	31	40	48	56	73	90	106
25	-54	-37	-20	-4	4	12	20	29	37	45	54	62	79	95	112
30	-48	-31	-15	1	10	18	26	35	43	51	60	68	85	101	118
35	-42	-25	-9	7	15	24	32	40	49	57	65	74	90	107	124
40	-36	-20	-3	13	21	30	38	46	55	63	71	80	96	113	130
45	-30	-14	2	19	27	35	44	52	60	69	77	85	102	119	135
50	-25	-8	8	25	33	41	50	58	66	75	83	91	108	125	141
55	-19	-2	14	30	39	47	55	64	72	80	89	97	114	130	147
60	-13	3	20	36	45	53	61	70	78	86	95	103	120	136	153
70	-1	15	31	48	56	65	73	81	90	98	106	115	131	148	165

Table 5-101

Separation, inch: 12

Leg Distance, inch: 50

SHIM REQUIRED FOR ALIGNMENT

Reverse Indicator Method

riveN IR ILS	DriveR TIR, MILS														
	-50	-40	-30	-20	-15	-10	-5	0	5	10	15	20	30	40	50
-70	-215	-194	-173	-152	-142	-131	-121	-110	-100	-90	-79	-69	-48	-27	-6
-60	-199	-178	-157	-136	-126	-115	-105	-95	-84	-74	-63	-53	-32	-11	9
-55	-191	-170	-149	-128	-118	-107	-97	-87	-76	-66	-55	-45	-24	-3	17
-50	-183	-162	-141	-120	-110	-100	-89	-79	-68	-58	-47	-37	-16	4	25
-45	-175	-154	-133	-112	-102	-92	-81	-71	-60	-50	-40	-29	-8	12	32
-40	-167	-146	-125	-105	-94	-84	-73	-63	-52	-42	-32	-21	0	20	40
-35	-159	-138	-117	-97	-86	-76	-65	-55	-45	-34	-24	-13	7	27	48
-30	-151	-130	-110	-89	-78	-68	-57	-47	-37	-26	-16	-5	15	35	56
-25	-143	-122	-102	-81	-70	-60	-50	-39	-29	-18	-8	2	22	43	64
-20	-135	-115	-94	-73	-62	-52	-42	-31	-21	-10	0	10	30	51	72
-15	-127	-107	-86	-65	-55	-44	-34	-23	-13	-2	7	17	38	59	80
-12	-123	-102	-81	-60	-50	-39	-29	-19	-8	1	12	22	43	64	85
-10	-120	-99	-78	-57	-47	-36	-26	-15	-5	5	15	25	46	67	88
-8	-116	-96	-75	-54	-43	-33	-23	-12	-2	8	18	29	49	70	91
-6	-113	-92	-72	-51	-40	-30	-19	-9	0	11	21	32	53	73	94
-4	-110	-89	-68	-48	-37	-27	-16	-6	4	14	24	35	56	77	97
-2	-107	-86	-65	-44	-34	-24	-13	-3	7	17	28	38	59	80	101
0	-104	-83	-62	-41	-31	-20	-10	0	10	20	31	41	62	83	104
2	-101	-80	-59	-38	-28	-17	-7	3	13	24	34	44	65	86	107
4	-97	-77	-56	-35	-24	-14	-4	6	16	27	37	48	68	89	110
6	-94	-73	-53	-32	-21	-11	0	9	19	30	40	51	72	92	113
8	-91	-70	-49	-29	-18	-8	2	12	23	33	43	54	75	96	116
10	-88	-67	-46	-25	-15	-5	5	15	26	36	47	57	78	99	120
12	-85	-64	-43	-22	-12	-1	8	19	29	39	50	60	81	102	123
15	-80	-59	-38	-17	-7	2	13	23	34	44	55	65	86	107	127
20	-72	-51	-30	-10	0	10	21	31	42	52	62	73	94	115	135
25	-64	-43	-22	-2	8	18	29	39	50	60	70	81	102	122	143
30	-56	-35	-15	5	16	26	37	47	57	68	78	89	110	130	151
35	-48	-27	-7	13	24	34	45	55	65	76	86	97	117	138	159
40	-40	-20	0	21	32	42	52	63	73	84	94	105	125	146	167
45	-32	-12	8	29	40	50	60	71	81	92	102	112	133	154	175
50	-25	-4	16	37	47	58	68	79	89	100	110	120	141	162	183
55	-17	3	24	45	55	66	76	87	97	107	118	128	149	170	191
60	-9	11	32	53	63	74	84	95	105	115	126	136	157	178	199
70	6	27	48	69	79	90	100	110	121	131	142	152	173	194	215

Table 5-102

Separation, inch: 12

Leg Distance, inch: 60

SHIM REQUIRED FOR ALIGNMENT

Reverse Indicator Method

DriveN TIR MILS	DriveR TIR, MILS -50	-40	-30	-20	-15	-10	-5	0	5	10	15	20	30	40	50
-70	-265	-240	-215	-190	-177	-165	-152	-140	-127	-115	-102	-90	-65	-40	-15
-60	-245	-220	-195	-170	-157	-145	-132	-120	-107	-95	-82	-70	-45	-20	5
-55	-235	-210	-185	-160	-147	-135	-122	-110	-97	-85	-72	-60	-35	-10	15
-50	-225	-200	-175	-150	-137	-125	-112	-100	-87	-75	-62	-50	-25	0	25
-45	-215	-190	-165	-140	-127	-115	-102	-90	-77	-65	-52	-40	-15	10	35
-40	-205	-180	-155	-130	-117	-105	-92	-80	-67	-55	-42	-30	-5	20	45
-35	-195	-170	-145	-120	-107	-95	-82	-70	-57	-45	-32	-20	5	30	55
-30	-185	-160	-135	-110	-97	-85	-72	-60	-47	-35	-22	-10	15	40	65
-25	-175	-150	-125	-100	-87	-75	-62	-50	-37	-25	-12	0	25	50	75
-20	-165	-140	-115	-90	-77	-65	-52	-40	-27	-15	-2	10	35	60	85
-15	-155	-130	-105	-80	-67	-55	-42	-30	-17	-5	7	20	45	70	95
-12	-149	-124	-99	-74	-61	-49	-36	-24	-11	1	13	26	51	76	101
-10	-145	-120	-95	-70	-57	-45	-32	-20	-7	5	17	30	55	80	105
-8	-141	-116	-91	-66	-53	-41	-28	-16	-3	9	21	34	59	84	109
-6	-137	-112	-87	-62	-49	-37	-24	-12	0	13	25	38	63	88	113
-4	-133	-108	-83	-58	-45	-33	-20	-8	4	17	29	42	67	92	117
-2	-129	-104	-79	-54	-41	-29	-16	-4	8	21	33	46	71	96	121
0	-125	-100	-75	-50	-37	-25	-12	0	12	25	37	50	75	100	125
2	-121	-96	-71	-46	-33	-21	-8	4	16	29	41	54	79	104	129
4	-117	-92	-67	-42	-29	-17	-4	8	20	33	45	58	83	108	133
6	-113	-88	-63	-38	-25	-13	0	12	24	37	49	62	87	112	137
8	-109	-84	-59	-34	-21	-9	3	16	28	41	53	66	91	116	141
10	-105	-80	-55	-30	-17	-5	7	20	32	45	57	70	95	120	145
12	-101	-76	-51	-26	-13	-1	11	24	36	49	61	74	99	124	149
15	-95	-70	-45	-20	-7	5	17	30	42	55	67	80	105	130	155
20	-85	-60	-35	-10	2	15	27	40	52	65	77	90	115	140	165
25	-75	-50	-25	0	12	25	37	50	62	75	87	100	125	150	175
30	-65	-40	-15	10	22	35	47	60	72	85	97	110	135	160	185
35	-55	-30	-5	20	32	45	57	70	82	95	107	120	145	170	195
40	-45	-20	5	30	42	55	67	80	92	105	117	130	155	180	205
45	-35	-10	15	40	52	65	77	90	102	115	127	140	165	190	215
50	-25	0	25	50	62	75	87	100	112	125	137	150	175	200	225
55	-15	10	35	60	72	85	97	110	122	135	147	160	185	210	235
60	-5	20	45	70	82	95	107	120	132	145	157	170	195	220	245
70	15	40	65	90	102	115	127	140	152	165	177	190	215	240	265

Table 5-103

Separation, inch: 12

Leg Distance, inch: 70

SHIM REQUIRED FOR ALIGNMENT

Reverse Indicator Method

riveN IR ILS	DriveR TIR, MILS -50	-40	-30	-20	-15	-10	-5	0	5	10	15	20	30	40	50
-70	-315	-285	-256	-227	-212	-198	-183	-169	-154	-140	-125	-110	-81	-52	-23
-60	-290	-261	-232	-203	-188	-174	-159	-145	-130	-115	-101	-86	-57	-28	0
-55	-278	-249	-220	-191	-176	-162	-147	-132	-118	-103	-89	-74	-45	-16	12
-50	-266	-237	-208	-179	-164	-150	-135	-120	-106	-91	-77	-62	-33	-4	25
-45	-254	-225	-196	-167	-152	-137	-123	-108	-94	-79	-65	-50	-21	7	37
-40	-242	-213	-184	-155	-140	-125	-111	-96	-82	-67	-52	-38	-9	20	49
-35	-230	-201	-172	-142	-128	-113	-99	-84	-70	-55	-40	-26	2	32	61
-30	-218	-189	-160	-130	-116	-101	-87	-72	-57	-43	-28	-14	15	44	73
-25	-206	-177	-147	-118	-104	-89	-75	-60	-45	-31	-16	-2	27	56	85
-20	-194	-165	-135	-106	-92	-77	-62	-48	-33	-19	-4	10	39	68	97
-15	-182	-152	-123	-94	-80	-65	-50	-36	-21	-7	7	22	51	80	109
-12	-174	-145	-116	-87	-72	-58	-43	-29	-14	0	14	29	58	87	116
-10	-170	-140	-111	-82	-67	-53	-38	-24	-9	5	19	34	63	92	121
-8	-165	-136	-106	-77	-63	-48	-33	-19	-4	9	24	39	68	97	126
-6	-160	-131	-102	-72	-58	-43	-29	-14	0	14	29	43	73	102	131
-4	-155	-126	-97	-68	-53	-38	-24	-9	4	19	34	48	77	107	136
-2	-150	-121	-92	-63	-48	-34	-19	-4	9	24	38	53	82	111	141
0	-145	-116	-87	-58	-43	-29	-14	0	14	29	43	58	87	116	145
2	-141	-111	-82	-53	-38	-24	-9	4	19	34	48	63	92	121	150
4	-136	-107	-77	-48	-34	-19	-4	9	24	38	53	68	97	126	155
6	-131	-102	-73	-43	-29	-14	0	14	29	43	58	72	102	131	160
8	-126	-97	-68	-39	-24	-9	4	19	33	48	63	77	106	136	165
10	-121	-92	-63	-34	-19	-5	9	24	38	53	67	82	111	140	170
12	-116	-87	-58	-29	-14	0	14	29	43	58	72	87	116	145	174
15	-109	-80	-51	-22	-7	7	21	36	50	65	80	94	123	152	182
20	-97	-68	-39	-10	4	19	33	48	62	77	92	106	135	165	194
25	-85	-56	-27	2	16	31	45	60	75	89	104	118	147	177	206
30	-73	-44	-15	14	28	43	57	72	87	101	116	130	160	189	218
35	-61	-32	-2	26	40	55	70	84	99	113	128	142	172	201	230
40	-49	-20	9	38	52	67	82	96	111	125	140	155	184	213	242
45	-37	-7	21	50	65	79	94	108	123	137	152	167	196	225	254
50	-25	4	33	62	77	91	106	120	135	150	164	179	208	237	266
55	-12	16	45	74	89	103	118	132	147	162	176	191	220	249	278
60	0	28	57	86	101	115	130	145	159	174	188	203	232	261	290
70	23	52	81	110	125	140	154	169	183	198	212	227	256	285	315

Table 5-104

Separation, inch: 12

Leg Distance, inch: 80

SHIM REQUIRED FOR ALIGNMENT

Reverse Indicator Method

DriveN TIR MILS	DriveR TIR, MILS -50	-40	-30	-20	-15	-10	-5	0	5	10	15	20	30	40	50
-70	-365	-331	-298	-265	-248	-231	-215	-198	-181	-165	-148	-131	-98	-65	-31
-60	-336	-303	-270	-236	-220	-203	-186	-170	-153	-136	-120	-103	-70	-36	-3
-55	-322	-289	-255	-222	-205	-189	-172	-155	-139	-122	-105	-89	-55	-22	10
-50	-308	-275	-241	-208	-191	-175	-158	-141	-125	-108	-91	-75	-41	-8	25
-45	-294	-260	-227	-194	-177	-160	-144	-127	-110	-94	-77	-60	-27	5	39
-40	-280	-246	-213	-180	-163	-146	-130	-113	-96	-80	-63	-46	-13	20	53
-35	-265	-232	-199	-165	-149	-132	-115	-99	-82	-65	-49	-32	0	34	67
-30	-251	-218	-185	-151	-135	-118	-101	-85	-68	-51	-35	-18	15	48	81
-25	-237	-204	-170	-137	-120	-104	-87	-70	-54	-37	-20	-4	29	62	95
-20	-223	-190	-156	-123	-106	-90	-73	-56	-40	-23	-6	10	43	76	110
-15	-209	-175	-142	-109	-92	-75	-59	-42	-25	-9	7	24	57	90	124
-12	-200	-167	-134	-100	-84	-67	-50	-34	-17	0	16	32	66	99	132
-10	-195	-161	-128	-95	-78	-61	-45	-28	-11	5	21	38	71	105	138
-8	-189	-156	-122	-89	-72	-56	-39	-22	-6	10	27	44	77	110	144
-6	-183	-150	-117	-83	-67	-50	-33	-17	0	16	33	49	83	116	149
-4	-178	-144	-111	-78	-61	-44	-28	-11	5	22	38	55	88	122	155
-2	-172	-139	-105	-72	-55	-39	-22	-5	11	27	44	61	94	127	161
0	-166	-133	-100	-66	-50	-33	-16	0	16	33	50	66	100	133	166
2	-161	-127	-94	-61	-44	-27	-11	5	22	39	55	72	105	139	172
4	-155	-122	-88	-55	-38	-22	-5	11	28	44	61	78	111	144	178
6	-149	-116	-83	-49	-33	-16	0	17	33	50	67	83	117	150	183
8	-144	-110	-77	-44	-27	-10	6	22	39	56	72	89	122	156	189
10	-138	-105	-71	-38	-21	-5	11	28	45	61	78	95	128	161	195
12	-132	-99	-66	-32	-16	0	17	34	50	67	84	100	134	167	200
15	-124	-90	-57	-24	-7	9	25	42	59	75	92	109	142	175	209
20	-110	-76	-43	-10	6	23	40	56	73	90	106	123	156	190	223
25	-95	-62	-29	4	20	37	54	70	87	104	120	137	170	204	237
30	-81	-48	-15	18	35	51	68	85	101	118	135	151	185	218	251
35	-67	-34	0	32	49	65	82	99	115	132	149	165	199	232	265
40	-53	-20	13	46	63	80	96	113	130	146	163	180	213	246	280
45	-39	-5	27	60	77	94	110	127	144	160	177	194	227	260	294
50	-25	8	41	75	91	108	125	141	158	175	191	208	241	275	308
55	-10	22	55	89	105	122	139	155	172	189	205	222	255	289	322
60	3	36	70	103	120	136	153	170	186	203	220	236	270	303	336
70	31	65	98	131	148	165	181	198	215	231	248	265	298	331	365

Table 5-105

Separation, inch: 12

Leg Distance, inch: 90

SHIM REQUIRED FOR ALIGNMENT

Reverse Indicator Method

riveN IR ILS	DriveR TIR, MILS -50	-40	-30	-20	-15	-10	-5	0	5	10	15	20	30	40	50
-70	-415	-377	-340	-302	-283	-265	-246	-227	-208	-190	-171	-152	-115	-77	-40
-60	-382	-345	-307	-270	-251	-232	-213	-195	-176	-157	-138	-120	-82	-45	-7
-55	-366	-328	-291	-253	-235	-216	-197	-178	-160	-141	-122	-103	-66	-28	8
-50	-350	-312	-275	-237	-218	-200	-181	-162	-143	-125	-106	-87	-50	-12	25
-45	-333	-296	-258	-221	-202	-183	-165	-146	-127	-108	-90	-71	-33	3	41
-40	-317	-280	-242	-205	-186	-167	-148	-130	-111	-92	-73	-55	-17	20	57
-35	-301	-263	-226	-188	-170	-151	-132	-113	-95	-76	-57	-38	-1	36	73
-30	-285	-247	-210	-172	-153	-135	-116	-97	-78	-60	-41	-22	15	52	90
-25	-268	-231	-193	-156	-137	-118	-100	-81	-62	-43	-25	-6	31	68	106
-20	-252	-215	-177	-140	-121	-102	-83	-65	-46	-27	-8	10	47	85	122
-15	-236	-198	-161	-123	-105	-86	-67	-48	-30	-11	7	26	63	101	138
-12	-226	-189	-151	-114	-95	-76	-57	-39	-20	-1	17	36	73	111	148
-10	-220	-182	-145	-107	-88	-70	-51	-32	-13	5	23	42	80	117	155
-8	-213	-176	-138	-101	-82	-63	-44	-26	-7	11	30	49	86	124	161
-6	-207	-169	-132	-94	-75	-57	-38	-19	0	18	36	55	93	130	168
-4	-200	-163	-125	-88	-69	-50	-31	-13	5	24	43	62	99	137	174
-2	-194	-156	-119	-81	-62	-44	-25	-6	12	31	49	68	106	143	181
0	-187	-150	-112	-75	-56	-37	-18	0	18	37	56	75	112	150	187
2	-181	-143	-106	-68	-49	-31	-12	6	25	44	62	81	119	156	194
4	-174	-137	-99	-62	-43	-24	-5	13	31	50	69	88	125	163	200
6	-168	-130	-93	-55	-36	-18	0	19	38	57	75	94	132	169	207
8	-161	-124	-86	-49	-30	-11	7	26	44	63	82	101	138	176	213
10	-155	-117	-80	-42	-23	-5	13	32	51	70	88	107	145	182	220
12	-148	-111	-73	-36	-17	1	20	39	57	76	95	114	151	189	226
15	-138	-101	-63	-26	-7	11	30	48	67	86	105	123	161	198	236
20	-122	-85	-47	-10	8	27	46	65	83	102	121	140	177	215	252
25	-106	-68	-31	6	25	43	62	81	100	118	137	156	193	231	268
30	-90	-52	-15	22	41	60	78	97	116	135	153	172	210	247	285
35	-73	-36	1	38	57	76	95	113	132	151	170	188	226	263	301
40	-57	-20	17	55	73	92	111	130	148	167	186	205	242	280	317
45	-41	-3	33	71	90	108	127	146	165	183	202	221	258	296	333
50	-25	12	50	87	106	125	143	162	181	200	218	237	275	312	350
55	-8	28	66	103	122	141	160	178	197	216	235	253	291	328	366
60	7	45	82	120	138	157	176	195	213	232	251	270	307	345	382
70	40	77	115	152	171	190	208	227	246	265	283	302	340	377	415

Table 5-106

Separation, inch: 12

Leg Distance, inch: 100

SHIM REQUIRED FOR ALIGNMENT

Reverse Indicator Method

DriveN TIR MILS	DriveR TIR, MILS -50	-40	-30	-20	-15	-10	-5	0	5	10	15	20	30	40	50
-70	-465	-423	-381	-340	-319	-298	-277	-256	-235	-215	-194	-173	-131	-90	-48
-60	-428	-386	-345	-303	-282	-261	-240	-220	-199	-178	-157	-136	-95	-53	-11
-55	-410	-368	-326	-285	-264	-243	-222	-201	-180	-160	-139	-118	-76	-35	6
-50	-391	-350	-308	-266	-245	-225	-204	-183	-162	-141	-120	-100	-58	-16	25
-45	-373	-331	-290	-248	-227	-206	-185	-165	-144	-123	-102	-81	-40	1	43
-40	-355	-313	-271	-230	-209	-188	-167	-146	-125	-105	-84	-63	-21	20	61
-35	-336	-295	-253	-211	-190	-170	-149	-128	-107	-86	-65	-45	-3	38	80
-30	-318	-276	-235	-193	-172	-151	-130	-110	-89	-68	-47	-26	15	56	98
-25	-300	-258	-216	-175	-154	-133	-112	-91	-70	-50	-29	-8	33	75	116
-20	-281	-240	-198	-156	-135	-115	-94	-73	-52	-31	-10	10	51	93	135
-15	-263	-221	-180	-138	-117	-96	-75	-55	-34	-13	7	28	70	111	153
-12	-252	-210	-169	-127	-106	-85	-64	-44	-23	-2	18	39	81	122	164
-10	-245	-203	-161	-120	-99	-78	-57	-36	-15	5	25	46	88	130	171
-8	-237	-196	-154	-112	-91	-71	-50	-29	-8	12	33	54	95	137	179
-6	-230	-188	-147	-105	-84	-63	-42	-22	-1	19	40	61	103	144	186
-4	-223	-181	-139	-98	-77	-56	-35	-14	6	27	47	68	110	152	193
-2	-215	-174	-132	-90	-69	-49	-28	-7	13	34	55	76	117	159	201
0	-208	-166	-125	-83	-62	-41	-20	0	20	41	62	83	125	166	208
2	-201	-159	-117	-76	-55	-34	-13	7	28	49	69	90	132	174	215
4	-193	-152	-110	-68	-47	-27	-6	14	35	56	77	98	139	181	223
6	-186	-144	-103	-61	-40	-19	1	22	42	63	84	105	147	188	230
8	-179	-137	-95	-54	-33	-12	8	29	50	71	91	112	154	196	237
10	-171	-130	-88	-46	-25	-5	15	36	57	78	99	120	161	203	245
12	-164	-122	-81	-39	-18	2	23	44	64	85	106	127	169	210	252
15	-153	-111	-70	-28	-7	13	34	55	75	96	117	138	180	221	263
20	-135	-93	-51	-10	10	31	52	73	94	115	135	156	198	240	281
25	-116	-75	-33	8	29	50	70	91	112	133	154	175	216	258	300
30	-98	-56	-15	26	47	68	89	110	130	151	172	193	235	276	318
35	-80	-38	3	45	65	86	107	128	149	170	190	211	253	295	336
40	-61	-20	21	63	84	105	125	146	167	188	209	230	271	313	355
45	-43	-1	40	81	102	123	144	165	185	206	227	248	290	331	373
50	-25	16	58	100	120	141	162	183	204	225	245	266	308	350	391
55	-6	35	76	118	139	160	180	201	222	243	264	285	326	368	410
60	11	53	95	136	157	178	199	220	240	261	282	303	345	386	428
70	48	90	131	173	194	215	235	256	277	298	319	340	381	423	465

Table 5-107

Separation, inch: 14

Leg Distance, inch: 25

SHIM REQUIRED FOR ALIGNMENT

Reverse Indicator Method

DriveN TIR MILS	DriveR TIR, MILS -50	-40	-30	-20	-15	-10	-5	0	5	10	15	20	30	40	50
-70	-72	-63	-54	-45	-40	-36	-31	-27	-23	-18	-14	-9	0	8	17
-60	-68	-59	-50	-41	-36	-32	-28	-23	-19	-14	-10	-5	3	12	21
-55	-66	-57	-48	-39	-35	-30	-26	-21	-17	-12	-8	-3	5	14	23
-50	-64	-55	-46	-37	-33	-28	-24	-19	-15	-10	-6	-1	7	16	25
-45	-62	-53	-44	-35	-31	-26	-22	-17	-13	-8	-4	0	9	18	26
-40	-60	-51	-42	-33	-29	-24	-20	-15	-11	-6	-2	2	11	20	28
-35	-58	-49	-40	-31	-27	-22	-18	-13	-9	-4	0	4	13	21	30
-30	-56	-47	-38	-29	-25	-20	-16	-11	-7	-2	1	6	15	23	32
-25	-54	-45	-36	-27	-23	-18	-14	-9	-5	0	3	8	16	25	34
-20	-52	-43	-34	-25	-21	-16	-12	-7	-3	1	5	10	18	27	36
-15	-50	-41	-32	-23	-19	-14	-10	-5	-1	3	7	11	20	29	38
-12	-49	-40	-31	-22	-18	-13	-9	-4	0	4	8	13	22	31	39
-10	-48	-39	-30	-21	-17	-12	-8	-3	0	5	9	13	22	31	40
-8	-47	-38	-29	-21	-16	-12	-7	-3	1	5	10	14	23	32	41
-6	-47	-38	-29	-20	-15	-11	-6	-2	2	6	11	15	24	33	42
-4	-46	-37	-28	-19	-14	-10	-6	-1	2	7	11	16	25	34	43
-2	-45	-36	-27	-18	-14	-9	-5	0	3	8	12	17	26	34	43
0	-44	-35	-26	-17	-13	-8	-4	0	4	8	13	17	26	35	44
2	-43	-34	-26	-17	-12	-8	-3	0	5	9	14	18	27	36	45
4	-43	-34	-25	-16	-11	-7	-2	1	6	10	14	19	28	37	46
6	-42	-33	-24	-15	-11	-6	-2	2	6	11	15	20	29	38	47
8	-41	-32	-23	-14	-10	-5	-1	3	7	12	16	21	29	38	47
10	-40	-31	-22	-13	-9	-5	0	3	8	12	17	21	30	39	48
12	-39	-31	-22	-13	-8	-4	0	4	9	13	18	22	31	40	49
15	-38	-29	-20	-11	-7	-3	1	5	10	14	19	23	32	41	50
20	-36	-27	-18	-10	-5	-1	3	7	12	16	21	25	34	43	52
25	-34	-25	-16	-8	-3	0	5	9	14	18	23	27	36	45	54
30	-32	-23	-15	-6	-1	2	7	11	16	20	25	29	38	47	56
35	-30	-21	-13	-4	0	4	9	13	18	22	27	31	40	49	58
40	-28	-20	-11	-2	2	6	11	15	20	24	29	33	42	51	60
45	-26	-18	-9	0	4	8	13	17	22	26	31	35	44	53	62
50	-25	-16	-7	1	6	10	15	19	24	28	33	37	46	55	64
55	-23	-14	-5	3	8	12	17	21	26	30	35	39	48	57	66
60	-21	-12	-3	5	10	14	19	23	28	32	36	41	50	59	68
70	-17	-8	0	9	14	18	23	27	31	36	40	45	54	63	72

Table 5-108

Separation, inch: 14

Leg Distance, inch: 30

SHIM REQUIRED FOR ALIGNMENT

Reverse Indicator Method

DriveN TIR MILS	DriveR TIR, MILS -50	-40	-30	-20	-15	-10	-5	0	5	10	15	20	30	40	50
-70	-93	-82	-72	-61	-56	-50	-45	-40	-34	-29	-23	-18	-7	2	13
-60	-87	-77	-66	-55	-50	-45	-39	-34	-28	-23	-18	-12	-2	8	19
-55	-85	-74	-63	-52	-47	-42	-36	-31	-26	-20	-15	-10	0	11	22
-50	-82	-71	-60	-50	-44	-39	-33	-28	-23	-17	-12	-7	3	14	25
-45	-79	-68	-57	-47	-41	-36	-31	-25	-20	-15	-9	-4	6	17	27
-40	-76	-65	-55	-44	-38	-33	-28	-22	-17	-12	-6	-1	9	20	30
-35	-73	-62	-52	-41	-36	-30	-25	-20	-14	-9	-3	1	12	22	33
-30	-70	-60	-49	-38	-33	-27	-22	-17	-11	-6	-1	4	15	25	36
-25	-67	-57	-46	-35	-30	-25	-19	-14	-8	-3	1	7	17	28	39
-20	-65	-54	-43	-32	-27	-22	-16	-11	-6	0	4	10	20	31	42
-15	-62	-51	-40	-30	-24	-19	-13	-8	-3	2	7	12	23	34	45
-12	-60	-49	-39	-28	-22	-17	-12	-6	-1	3	9	14	25	36	46
-10	-59	-48	-37	-27	-21	-16	-11	-5	0	5	10	15	26	37	47
-8	-58	-47	-36	-26	-20	-15	-9	-4	0	6	11	16	27	38	49
-6	-57	-46	-35	-24	-19	-14	-8	-3	1	7	12	18	28	39	50
-4	-55	-45	-34	-23	-18	-13	-7	-2	3	8	13	19	29	40	51
-2	-54	-44	-33	-22	-17	-11	-6	-1	4	9	14	20	31	41	52
0	-53	-42	-32	-21	-16	-10	-5	0	5	10	16	21	32	42	53
2	-52	-41	-31	-20	-14	-9	-4	1	6	11	17	22	33	44	54
4	-51	-40	-29	-19	-13	-8	-3	2	7	13	18	23	34	45	55
6	-50	-39	-28	-18	-12	-7	-1	3	8	14	19	24	35	46	57
8	-49	-38	-27	-16	-11	-6	0	4	9	15	20	26	36	47	58
10	-47	-37	-26	-15	-10	-5	0	5	11	16	21	27	37	48	59
12	-46	-36	-25	-14	-9	-3	1	6	12	17	22	28	39	49	60
15	-45	-34	-23	-12	-7	-2	3	8	13	19	24	30	40	51	62
20	-42	-31	-20	-10	-4	0	6	11	16	22	27	32	43	54	65
25	-39	-28	-17	-7	-1	3	8	14	19	25	30	35	46	57	67
30	-36	-25	-15	-4	1	6	11	17	22	27	33	38	49	60	70
35	-33	-22	-12	-1	3	9	14	20	25	30	36	41	52	62	73
40	-30	-20	-9	1	6	12	17	22	28	33	38	44	55	65	76
45	-27	-17	-6	4	9	15	20	25	31	36	41	47	57	68	79
50	-25	-14	-3	7	12	17	23	28	33	39	44	50	60	71	82
55	-22	-11	0	10	15	20	26	31	36	42	47	52	63	74	85
60	-19	-8	2	12	18	23	28	34	39	45	50	55	66	77	87
70	-13	-2	7	18	23	29	34	40	45	50	56	61	72	82	93

Table 5-109

Separation, inch: 14

Leg Distance, inch: 35

SHIM REQUIRED FOR ALIGNMENT

Reverse Indicator Method

riveN IR ILS	DriveR TIR, MILS -50	-40	-30	-20	-15	-10	-5	0	5	10	15	20	30	40	50
-70	-115	-102	-90	-77	-71	-65	-58	-52	-46	-40	-33	-27	-15	-2	10
-60	-107	-95	-82	-70	-63	-57	-51	-45	-38	-32	-26	-20	-7	5	17
-55	-103	-91	-78	-66	-60	-53	-47	-41	-35	-28	-22	-16	-3	8	21
-50	-100	-87	-75	-62	-56	-50	-43	-37	-31	-25	-18	-12	0	12	25
-45	-96	-83	-71	-58	-52	-46	-40	-33	-27	-21	-15	-8	3	16	28
-40	-92	-80	-67	-55	-48	-42	-36	-30	-23	-17	-11	-5	7	20	32
-35	-88	-76	-63	-51	-45	-38	-32	-26	-20	-13	-7	-1	11	23	36
-30	-85	-72	-60	-47	-41	-35	-28	-22	-16	-10	-3	2	15	27	40
-25	-81	-68	-56	-43	-37	-31	-25	-18	-12	-6	0	6	18	31	43
-20	-77	-65	-52	-40	-33	-27	-21	-15	-8	-2	3	10	22	35	47
-15	-73	-61	-48	-36	-30	-23	-17	-11	-5	1	7	13	26	38	51
-12	-71	-59	-46	-34	-27	-21	-15	-9	-2	3	9	16	28	41	53
-10	-70	-57	-45	-32	-26	-20	-13	-7	-1	5	11	17	30	42	55
-8	-68	-56	-43	-31	-24	-18	-12	-6	0	6	12	19	31	44	56
-6	-67	-54	-42	-29	-23	-17	-10	-4	1	8	14	20	33	45	58
-4	-65	-53	-40	-28	-21	-15	-9	-3	3	9	15	22	34	47	59
-2	-64	-51	-39	-26	-20	-14	-7	-1	4	11	17	23	36	48	61
0	-62	-50	-37	-25	-18	-12	-6	0	6	12	18	25	37	50	62
2	-61	-48	-36	-23	-17	-11	-4	1	7	14	20	26	39	51	64
4	-59	-47	-34	-22	-15	-9	-3	3	9	15	21	28	40	53	65
6	-58	-45	-33	-20	-14	-8	-1	4	10	17	23	29	42	54	67
8	-56	-44	-31	-19	-12	-6	0	6	12	18	24	31	43	56	68
10	-55	-42	-30	-17	-11	-5	1	7	13	20	26	32	45	57	70
12	-53	-41	-28	-16	-9	-3	2	9	15	21	27	34	46	59	71
15	-51	-38	-26	-13	-7	-1	5	11	17	23	30	36	48	61	73
20	-47	-35	-22	-10	-3	2	8	15	21	27	33	40	52	65	77
25	-43	-31	-18	-6	0	6	12	18	25	31	37	43	56	68	81
30	-40	-27	-15	-2	3	10	16	22	28	35	41	47	60	72	85
35	-36	-23	-11	1	7	13	20	26	32	38	45	51	63	76	88
40	-32	-20	-7	5	11	17	23	30	36	42	48	55	67	80	92
45	-28	-16	-3	8	15	21	27	33	40	46	52	58	71	83	96
50	-25	-12	0	12	18	25	31	37	43	50	56	62	75	87	100
55	-21	-8	3	16	22	28	35	41	47	53	60	66	78	91	103
60	-17	-5	7	20	26	32	38	45	51	57	63	70	82	95	107
70	-10	2	15	27	33	40	46	52	58	65	71	77	90	102	115

Table 5-110

Separation, inch: 14

Leg Distance, inch: 40

SHIM REQUIRED FOR ALIGNMENT

Reverse Indicator Method

DriveN TIR MILS	DriveR TIR, MILS -50	-40	-30	-20	-15	-10	-5	0	5	10	15	20	30	40	50
-70	-136	-122	-107	-93	-86	-79	-72	-65	-57	-50	-43	-36	-22	-7	6
-60	-127	-112	-98	-84	-77	-70	-62	-55	-48	-41	-34	-27	-12	1	15
-55	-122	-108	-93	-79	-72	-65	-58	-51	-43	-36	-29	-22	-8	6	20
-50	-117	-103	-89	-75	-67	-60	-53	-46	-39	-32	-25	-17	-3	10	25
-45	-113	-98	-84	-70	-63	-56	-48	-41	-34	-27	-20	-13	1	15	29
-40	-108	-94	-80	-65	-58	-51	-44	-37	-30	-22	-15	-8	5	20	34
-35	-103	-89	-75	-61	-53	-46	-39	-32	-25	-18	-11	-3	10	24	38
-30	-99	-85	-70	-56	-49	-42	-35	-27	-20	-13	-6	0	15	29	43
-25	-94	-80	-66	-51	-44	-37	-30	-23	-16	-8	-1	5	19	33	48
-20	-90	-75	-61	-47	-40	-32	-25	-18	-11	-4	2	10	24	38	52
-15	-85	-71	-56	-42	-35	-28	-21	-13	-6	0	7	14	28	43	57
-12	-82	-68	-54	-39	-32	-25	-18	-11	-4	3	10	17	31	46	60
-10	-80	-66	-52	-37	-30	-23	-16	-9	-2	5	12	19	33	47	62
-8	-78	-64	-50	-36	-28	-21	-14	-7	0	6	14	21	35	49	64
-6	-77	-62	-48	-34	-27	-19	-12	-5	1	8	15	23	37	51	65
-4	-75	-60	-46	-32	-25	-18	-10	-3	3	10	17	24	39	53	67
-2	-73	-59	-44	-30	-23	-16	-9	-1	5	12	19	26	41	55	69
0	-71	-57	-42	-28	-21	-14	-7	0	7	14	21	28	42	57	71
2	-69	-55	-41	-26	-19	-12	-5	1	9	16	23	30	44	59	73
4	-67	-53	-39	-24	-17	-10	-3	3	10	18	25	32	46	60	75
6	-65	-51	-37	-23	-15	-8	-1	5	12	19	27	34	48	62	77
8	-64	-49	-35	-21	-14	-6	0	7	14	21	28	36	50	64	78
10	-62	-47	-33	-19	-12	-5	2	9	16	23	30	37	52	66	80
12	-60	-46	-31	-17	-10	-3	4	11	18	25	32	39	54	68	82
15	-57	-43	-28	-14	-7	0	6	13	21	28	35	42	56	71	85
20	-52	-38	-24	-10	-2	4	11	18	25	32	40	47	61	75	90
25	-48	-33	-19	-5	1	8	16	23	30	37	44	51	66	80	94
30	-43	-29	-15	0	6	13	20	27	35	42	49	56	70	85	99
35	-38	-24	-10	3	11	18	25	32	39	46	53	61	75	89	103
40	-34	-20	-5	8	15	22	30	37	44	51	58	65	80	94	108
45	-29	-15	-1	13	20	27	34	41	48	56	63	70	84	98	113
50	-25	-10	3	17	25	32	39	46	53	60	67	75	89	103	117
55	-20	-6	8	22	29	36	43	51	58	65	72	79	93	108	122
60	-15	-1	12	27	34	41	48	55	62	70	77	84	98	112	127
70	-6	7	22	36	43	50	57	65	72	79	86	93	107	122	136

Table 5-111

Separation, inch: 14

Leg Distance, inch: 50

SHIM REQUIRED FOR ALIGNMENT

Reverse Indicator Method

riveN IR — DriveR TIR, MILS

ILS	-50	-40	-30	-20	-15	-10	-5	0	5	10	15	20	30	40	50
-70	-179	-161	-143	-125	-116	-107	-98	-90	-81	-72	-63	-54	-36	-18	0
-60	-166	-148	-130	-112	-103	-95	-86	-77	-68	-59	-50	-41	-23	-5	12
-55	-160	-142	-124	-106	-97	-88	-79	-70	-61	-52	-43	-35	-17	0	18
-50	-153	-135	-117	-100	-91	-82	-73	-64	-55	-46	-37	-28	-10	7	25
-45	-147	-129	-111	-93	-84	-75	-66	-57	-48	-40	-31	-22	-4	13	31
-40	-140	-122	-105	-87	-78	-69	-60	-51	-42	-33	-24	-15	2	20	37
-35	-134	-116	-98	-80	-71	-62	-53	-45	-36	-27	-18	-9	8	26	44
-30	-127	-110	-92	-74	-65	-56	-47	-38	-29	-20	-11	-2	15	32	50
-25	-121	-103	-85	-67	-58	-50	-41	-32	-23	-14	-5	3	21	39	57
-20	-115	-97	-79	-61	-52	-43	-34	-25	-16	-7	1	10	27	45	63
-15	-108	-90	-72	-55	-46	-37	-28	-19	-10	-1	7	16	34	52	70
-12	-104	-86	-69	-51	-42	-33	-24	-15	-6	2	11	20	38	56	73
-10	-102	-84	-66	-48	-39	-30	-21	-12	-3	5	13	22	40	58	76
-8	-99	-81	-63	-46	-37	-28	-19	-10	-1	7	16	25	43	61	79
-6	-97	-79	-61	-43	-34	-25	-16	-7	1	10	19	28	45	63	81
-4	-94	-76	-58	-40	-31	-23	-14	-5	3	12	21	30	48	66	84
-2	-91	-74	-56	-38	-29	-20	-11	-2	6	15	24	33	51	68	86
0	-89	-71	-53	-35	-26	-17	-8	0	8	17	26	35	53	71	89
2	-86	-68	-51	-33	-24	-15	-6	2	11	20	29	38	56	74	91
4	-84	-66	-48	-30	-21	-12	-3	5	14	23	31	40	58	76	94
6	-81	-63	-45	-28	-19	-10	-1	7	16	25	34	43	61	79	97
8	-79	-61	-43	-25	-16	-7	1	10	19	28	37	46	63	81	99
10	-76	-58	-40	-22	-13	-5	3	12	21	30	39	48	66	84	102
12	-73	-56	-38	-20	-11	-2	6	15	24	33	42	51	69	86	104
15	-70	-52	-34	-16	-7	1	10	19	28	37	46	55	72	90	108
20	-63	-45	-27	-10	-1	7	16	25	34	43	52	61	79	97	115
25	-57	-39	-21	-3	5	14	23	32	41	50	58	67	85	103	121
30	-50	-32	-15	2	11	20	29	38	47	56	65	74	92	110	127
35	-44	-26	-8	9	18	27	36	45	53	62	71	80	98	116	134
40	-37	-20	-2	15	24	33	42	51	60	69	78	87	105	122	140
45	-31	-13	4	22	31	40	48	57	66	75	84	93	111	129	147
50	-25	-7	10	28	37	46	55	64	73	82	91	100	117	135	153
55	-18	0	17	35	43	52	61	70	79	88	97	106	124	142	160
60	-12	5	23	41	50	59	68	77	86	95	103	112	130	148	166
70	0	18	36	54	63	72	81	90	98	107	116	125	143	161	179

Table 5-112

Separation, inch: 14

Leg Distance, inch: 60

SHIM REQUIRED FOR ALIGNMENT

Reverse Indicator Method

DriveN TIR MILS	DriveR TIR, MILS -50	-40	-30	-20	-15	-10	-5	0	5	10	15	20	30	40	50
-70	-222	-200	-179	-157	-147	-136	-125	-115	-104	-93	-82	-72	-50	-29	-7
-60	-205	-184	-162	-141	-130	-120	-109	-98	-87	-77	-66	-55	-34	-12	8
-55	-197	-176	-154	-133	-122	-111	-101	-90	-79	-68	-58	-47	-26	-4	16
-50	-189	-167	-146	-125	-114	-103	-92	-82	-71	-60	-50	-39	-17	3	25
-45	-181	-159	-138	-116	-106	-95	-84	-73	-63	-52	-41	-31	-9	11	33
-40	-172	-151	-130	-108	-97	-87	-76	-65	-55	-44	-33	-22	-1	20	41
-35	-164	-143	-121	-100	-89	-78	-68	-57	-46	-36	-25	-14	6	28	49
-30	-156	-135	-113	-92	-81	-70	-60	-49	-38	-27	-17	-6	15	36	57
-25	-148	-126	-105	-83	-73	-62	-51	-41	-30	-19	-8	1	23	44	66
-20	-140	-118	-97	-75	-65	-54	-43	-32	-22	-11	0	10	31	52	74
-15	-131	-110	-88	-67	-56	-46	-35	-24	-13	-3	7	18	39	61	82
-12	-126	-105	-84	-62	-51	-41	-30	-19	-9	1	12	23	44	66	87
-10	-123	-102	-80	-59	-48	-37	-27	-16	-5	5	15	26	47	69	90
-8	-120	-98	-77	-56	-45	-34	-23	-13	-2	8	19	29	51	72	94
-6	-117	-95	-74	-52	-42	-31	-20	-9	0	11	22	33	54	75	97
-4	-113	-92	-70	-49	-38	-28	-17	-6	4	14	25	36	57	79	100
-2	-110	-89	-67	-46	-35	-24	-14	-3	7	18	28	39	61	82	103
0	-107	-85	-64	-42	-32	-21	-10	0	10	21	32	42	64	85	107
2	-103	-82	-61	-39	-28	-18	-7	3	14	24	35	46	67	89	110
4	-100	-79	-57	-36	-25	-14	-4	6	17	28	38	49	70	92	113
6	-97	-75	-54	-33	-22	-11	0	9	20	31	42	52	74	95	117
8	-94	-72	-51	-29	-19	-8	2	13	23	34	45	56	77	98	120
10	-90	-69	-47	-26	-15	-5	5	16	27	37	48	59	80	102	123
12	-87	-66	-44	-23	-12	-1	9	19	30	41	51	62	84	105	126
15	-82	-61	-39	-18	-7	3	13	24	35	46	56	67	88	110	131
20	-74	-52	-31	-10	0	11	22	32	43	54	65	75	97	118	140
25	-66	-44	-23	-1	8	19	30	41	51	62	73	83	105	126	148
30	-57	-36	-15	6	17	27	38	49	60	70	81	92	113	135	156
35	-49	-28	-6	14	25	36	46	57	68	78	89	100	121	143	164
40	-41	-20	1	22	33	44	55	65	76	87	97	108	130	151	172
45	-33	-11	9	31	41	52	63	73	84	95	106	116	138	159	181
50	-25	-3	17	39	50	60	71	82	92	103	114	125	146	167	189
55	-16	4	26	47	58	68	79	90	101	111	122	133	154	176	197
60	-8	12	34	55	66	77	87	98	109	120	130	141	162	184	205
70	7	29	50	72	82	93	104	115	125	136	147	157	179	200	222

Table 5-113

Separation, inch: 14

Leg Distance, inch: 70

SHIM REQUIRED FOR ALIGNMENT

Reverse Indicator Method

riveN IR ILS	DriveR TIR, MILS -50	-40	-30	-20	-15	-10	-5	0	5	10	15	20	30	40	50
-70	-265	-240	-215	-190	-177	-165	-152	-140	-127	-115	-102	-90	-65	-40	-15
-60	-245	-220	-195	-170	-157	-145	-132	-120	-107	-95	-82	-70	-45	-20	5
-55	-235	-210	-185	-160	-147	-135	-122	-110	-97	-85	-72	-60	-35	-10	15
-50	-225	-200	-175	-150	-137	-125	-112	-100	-87	-75	-62	-50	-25	0	25
-45	-215	-190	-165	-140	-127	-115	-102	-90	-77	-65	-52	-40	-15	10	35
-40	-205	-180	-155	-130	-117	-105	-92	-80	-67	-55	-42	-30	-5	20	45
-35	-195	-170	-145	-120	-107	-95	-82	-70	-57	-45	-32	-20	5	30	55
-30	-185	-160	-135	-110	-97	-85	-72	-60	-47	-35	-22	-10	15	40	65
-25	-175	-150	-125	-100	-87	-75	-62	-50	-37	-25	-12	0	25	50	75
-20	-165	-140	-115	-90	-77	-65	-52	-40	-27	-15	-2	10	35	60	85
-15	-155	-130	-105	-80	-67	-55	-42	-30	-17	-5	7	20	45	70	95
-12	-149	-124	-99	-74	-61	-49	-36	-24	-11	1	13	26	51	76	101
-10	-145	-120	-95	-70	-57	-45	-32	-20	-7	5	17	30	55	80	105
-8	-141	-116	-91	-66	-53	-41	-28	-16	-3	9	21	34	59	84	109
-6	-137	-112	-87	-62	-49	-37	-24	-12	0	13	25	38	63	88	113
-4	-133	-108	-83	-58	-45	-33	-20	-8	4	17	29	42	67	92	117
-2	-129	-104	-79	-54	-41	-29	-16	-4	8	21	33	46	71	96	121
0	-125	-100	-75	-50	-37	-25	-12	0	12	25	37	50	75	100	125
2	-121	-96	-71	-46	-33	-21	-8	4	16	29	41	54	79	104	129
4	-117	-92	-67	-42	-29	-17	-4	8	20	33	45	58	83	108	133
6	-113	-88	-63	-38	-25	-13	0	12	24	37	49	62	87	112	137
8	-109	-84	-59	-34	-21	-9	3	16	28	41	53	66	91	116	141
10	-105	-80	-55	-30	-17	-5	7	20	32	45	57	70	95	120	145
12	-101	-76	-51	-26	-13	-1	11	24	36	49	61	74	99	124	149
15	-95	-70	-45	-20	-7	5	17	30	42	55	67	80	105	130	155
20	-85	-60	-35	-10	2	15	27	40	52	65	77	90	115	140	165
25	-75	-50	-25	0	12	25	37	50	62	75	87	100	125	150	175
30	-65	-40	-15	10	22	35	47	60	72	85	97	110	135	160	185
35	-55	-30	-5	20	32	45	57	70	82	95	107	120	145	170	195
40	-45	-20	5	30	42	55	67	80	92	105	117	130	155	180	205
45	-35	-10	15	40	52	65	77	90	102	115	127	140	165	190	215
50	-25	0	25	50	62	75	87	100	112	125	137	150	175	200	225
55	-15	10	35	60	72	85	97	110	122	135	147	160	185	210	235
60	-5	20	45	70	82	95	107	120	132	145	157	170	195	220	245
70	15	40	65	90	102	115	127	140	152	165	177	190	215	240	265

Table 5-114

Separation, inch: 14

Leg Distance, inch: 80

SHIM REQUIRED FOR ALIGNMENT

Reverse Indicator Method

DriveN TIR MILS	DriveR TIR, MILS -50	-40	-30	-20	-15	-10	-5	0	5	10	15	20	30	40	50
-70	-307	-279	-250	-222	-207	-193	-179	-165	-150	-136	-122	-107	-79	-50	-22
-60	-284	-255	-227	-198	-184	-170	-155	-141	-127	-112	-98	-84	-55	-27	1
-55	-272	-243	-215	-186	-172	-158	-143	-129	-115	-101	-86	-72	-43	-15	13
-50	-260	-232	-203	-175	-160	-146	-132	-117	-103	-89	-75	-60	-32	-3	25
-45	-248	-220	-191	-163	-148	-134	-120	-106	-91	-77	-63	-48	-20	8	36
-40	-237	-208	-180	-151	-137	-122	-108	-94	-80	-65	-51	-37	-8	20	48
-35	-225	-196	-168	-139	-125	-111	-96	-82	-68	-53	-39	-25	3	31	60
-30	-213	-185	-156	-127	-113	-99	-85	-70	-56	-42	-27	-13	15	43	72
-25	-201	-173	-144	-116	-101	-87	-73	-58	-44	-30	-16	-1	26	55	83
-20	-190	-161	-132	-104	-90	-75	-61	-47	-32	-18	-4	10	38	67	95
-15	-178	-149	-121	-92	-78	-63	-49	-35	-21	-6	7	21	50	78	107
-12	-171	-142	-114	-85	-71	-56	-42	-28	-14	0	14	28	57	86	114
-10	-166	-137	-109	-80	-66	-52	-37	-23	-9	5	19	33	62	90	119
-8	-161	-133	-104	-76	-61	-47	-33	-18	-4	9	24	38	66	95	124
-6	-157	-128	-99	-71	-57	-42	-28	-14	0	14	28	43	71	100	128
-4	-152	-123	-95	-66	-52	-38	-23	-9	4	19	33	47	76	104	133
-2	-147	-119	-90	-61	-47	-33	-19	-4	9	23	38	52	81	109	138
0	-142	-114	-85	-57	-42	-28	-14	0	14	28	42	57	85	114	142
2	-138	-109	-81	-52	-38	-23	-9	4	19	33	47	61	90	119	147
4	-133	-104	-76	-47	-33	-19	-4	9	23	38	52	66	95	123	152
6	-128	-100	-71	-43	-28	-14	0	14	28	42	57	71	99	128	157
8	-124	-95	-66	-38	-24	-9	4	18	33	47	61	76	104	133	161
10	-119	-90	-62	-33	-19	-5	9	23	37	52	66	80	109	137	166
12	-114	-86	-57	-28	-14	0	14	28	42	56	71	85	114	142	171
15	-107	-78	-50	-21	-7	6	21	35	49	63	78	92	121	149	178
20	-95	-67	-38	-10	4	18	32	47	61	75	90	104	132	161	190
25	-83	-55	-26	1	16	30	44	58	73	87	101	116	144	173	201
30	-72	-43	-15	13	27	42	56	70	85	99	113	127	156	185	213
35	-60	-31	-3	25	39	53	68	82	96	111	125	139	168	196	225
40	-48	-20	8	37	51	65	80	94	108	122	137	151	180	208	237
45	-36	-8	20	48	63	77	91	106	120	134	148	163	191	220	248
50	-25	3	32	60	75	89	103	117	132	146	160	175	203	232	260
55	-13	15	43	72	86	101	115	129	143	158	172	186	215	243	272
60	-1	27	55	84	98	112	127	141	155	170	184	198	227	255	284
70	22	50	79	107	122	136	150	165	179	193	207	222	250	279	307

Table 5-115

Separation, inch: 14

Leg Distance, inch: 90

SHIM REQUIRED FOR ALIGNMENT

Reverse Indicator Method

riveN IR ILS	DriveR TIR, MILS														
	-50	-40	-30	-20	-15	-10	-5	0	5	10	15	20	30	40	50
-70	-350	-318	-286	-254	-238	-222	-206	-190	-173	-157	-141	-125	-93	-61	-29
-60	-323	-291	-259	-227	-211	-195	-178	-162	-146	-130	-114	-98	-66	-34	-2
-55	-310	-277	-245	-213	-197	-181	-165	-149	-133	-117	-101	-85	-52	-20	11
-50	-296	-264	-232	-200	-183	-167	-151	-135	-119	-103	-87	-71	-39	-7	25
-45	-282	-250	-218	-186	-170	-154	-138	-122	-106	-90	-73	-57	-25	6	38
-40	-269	-237	-205	-172	-156	-140	-124	-108	-92	-76	-60	-44	-12	20	52
-35	-255	-223	-191	-159	-143	-127	-111	-95	-78	-62	-46	-30	1	33	65
-30	-242	-210	-177	-145	-129	-113	-97	-81	-65	-49	-33	-17	15	47	79
-25	-228	-196	-164	-132	-116	-100	-83	-67	-51	-35	-19	-3	28	60	92
-20	-215	-182	-150	-118	-102	-86	-70	-54	-38	-22	-6	10	42	74	106
-15	-201	-169	-137	-105	-88	-72	-56	-40	-24	-8	7	23	55	87	120
-12	-193	-161	-129	-96	-80	-64	-48	-32	-16	0	15	31	63	96	128
-10	-187	-155	-123	-91	-75	-59	-43	-27	-11	5	21	37	69	101	133
-8	-182	-150	-118	-86	-69	-53	-37	-21	-5	10	26	42	74	106	139
-6	-177	-144	-112	-80	-64	-48	-32	-16	0	15	31	48	80	112	144
-4	-171	-139	-107	-75	-59	-43	-26	-10	5	21	37	53	85	117	149
-2	-166	-134	-101	-69	-53	-37	-21	-5	10	26	42	58	91	123	155
0	-160	-128	-96	-64	-48	-32	-16	0	16	32	48	64	96	128	160
2	-155	-123	-91	-58	-42	-26	-10	5	21	37	53	69	101	134	166
4	-149	-117	-85	-53	-37	-21	-5	10	26	43	59	75	107	139	171
6	-144	-112	-80	-48	-31	-15	0	16	32	48	64	80	112	144	177
8	-139	-106	-74	-42	-26	-10	5	21	37	53	69	86	118	150	182
10	-133	-101	-69	-37	-21	-5	11	27	43	59	75	91	123	155	187
12	-128	-96	-63	-31	-15	0	16	32	48	64	80	96	129	161	193
15	-120	-87	-55	-23	-7	8	24	40	56	72	88	105	137	169	201
20	-106	-74	-42	-10	6	22	38	54	70	86	102	118	150	182	215
25	-92	-60	-28	3	19	35	51	67	83	100	116	132	164	196	228
30	-79	-47	-15	17	33	49	65	81	97	113	129	145	177	210	242
35	-65	-33	-1	30	46	62	78	95	111	127	143	159	191	223	255
40	-52	-20	12	44	60	76	92	108	124	140	156	172	205	237	269
45	-38	-6	25	57	73	90	106	122	138	154	170	186	218	250	282
50	-25	7	39	71	87	103	119	135	151	167	183	200	232	264	296
55	-11	20	52	85	101	117	133	149	165	181	197	213	245	277	310
60	2	34	66	98	114	130	146	162	178	195	211	227	259	291	323
70	29	61	93	125	141	157	173	190	206	222	238	254	286	318	350

Table 5-116

Separation, inch: 14

Leg Distance, inch: 100

SHIM REQUIRED FOR ALIGNMENT

Reverse Indicator Method

DriveN TIR MILS	DriveR TIR, MILS -50	-40	-30	-20	-15	-10	-5	0	5	10	15	20	30	40	50
-70	-393	-357	-322	-286	-268	-250	-232	-215	-197	-179	-161	-143	-107	-72	-36
-60	-362	-327	-291	-255	-237	-220	-202	-184	-166	-148	-130	-112	-77	-41	-5
-55	-347	-311	-276	-240	-222	-204	-186	-168	-151	-133	-115	-97	-61	-26	9
-50	-332	-296	-260	-225	-207	-189	-171	-153	-135	-117	-100	-82	-46	-10	25
-45	-316	-281	-245	-209	-191	-173	-156	-138	-120	-102	-84	-66	-31	4	40
-40	-301	-265	-230	-194	-176	-158	-140	-122	-105	-87	-69	-51	-15	20	55
-35	-286	-250	-214	-178	-161	-143	-125	-107	-89	-71	-53	-36	0	35	71
-30	-270	-235	-199	-163	-145	-127	-110	-92	-74	-56	-38	-20	15	50	86
-25	-255	-219	-183	-148	-130	-112	-94	-76	-58	-41	-23	-5	30	66	101
-20	-240	-204	-168	-132	-115	-97	-79	-61	-43	-25	-7	10	45	81	117
-15	-224	-188	-153	-117	-99	-81	-63	-46	-28	-10	7	25	61	96	132
-12	-215	-179	-144	-108	-90	-72	-54	-36	-19	-1	16	34	70	106	141
-10	-209	-173	-137	-102	-84	-66	-48	-30	-12	5	22	40	76	112	147
-8	-203	-167	-131	-96	-78	-60	-42	-24	-6	11	29	46	82	118	154
-6	-197	-161	-125	-89	-72	-54	-36	-18	0	17	35	53	88	124	160
-4	-190	-155	-119	-83	-65	-48	-30	-12	5	23	41	59	94	130	166
-2	-184	-149	-113	-77	-59	-41	-24	-6	11	29	47	65	101	136	172
0	-178	-142	-107	-71	-53	-35	-17	0	17	35	53	71	107	142	178
2	-172	-136	-101	-65	-47	-29	-11	6	24	41	59	77	113	149	184
4	-166	-130	-94	-59	-41	-23	-5	12	30	48	65	83	119	155	190
6	-160	-124	-88	-53	-35	-17	0	18	36	54	72	89	125	161	197
8	-154	-118	-82	-46	-29	-11	6	24	42	60	78	96	131	167	203
10	-147	-112	-76	-40	-22	-5	12	30	48	66	84	102	137	173	209
12	-141	-106	-70	-34	-16	1	19	36	54	72	90	108	144	179	215
15	-132	-96	-61	-25	-7	10	28	46	63	81	99	117	153	188	224
20	-117	-81	-45	-10	7	25	43	61	79	97	115	132	168	204	240
25	-101	-66	-30	5	23	41	58	76	94	112	130	148	183	219	255
30	-86	-50	-15	20	38	56	74	92	110	127	145	163	199	235	270
35	-71	-35	0	36	53	71	89	107	125	143	161	178	214	250	286
40	-55	-20	15	51	69	87	105	122	140	158	176	194	230	265	301
45	-40	-4	31	66	84	102	120	138	156	173	191	209	245	281	316
50	-25	10	46	82	100	117	135	153	171	189	207	225	260	296	332
55	-9	26	61	97	115	133	151	168	186	204	222	240	276	311	347
60	5	41	77	112	130	148	166	184	202	220	237	255	291	327	362
70	36	72	107	143	161	179	197	215	232	250	268	286	322	357	393

Table 5-117

Separation, inch: 16

Leg Distance, inch: 25

SHIM REQUIRED FOR ALIGNMENT

Reverse Indicator Method

DriveN TIR MILS	DriveR TIR, MILS -50	-40	-30	-20	-15	-10	-5	0	5	10	15	20	30	40	50
-70	-58	-50	-43	-35	-31	-27	-23	-19	-15	-11	-7	-4	3	11	19
-60	-55	-48	-40	-32	-28	-24	-20	-16	-12	-9	-5	-1	6	14	22
-55	-54	-46	-38	-31	-27	-23	-19	-15	-11	-7	-3	0	7	15	23
-50	-53	-45	-37	-29	-25	-21	-17	-14	-10	-6	-2	1	9	17	25
-45	-51	-43	-36	-28	-24	-20	-16	-12	-8	-4	0	2	10	18	26
-40	-50	-42	-34	-26	-22	-19	-15	-11	-7	-3	0	4	12	20	27
-35	-48	-41	-33	-25	-21	-17	-13	-9	-5	-2	1	5	13	21	29
-30	-47	-39	-31	-24	-20	-16	-12	-8	-4	0	3	7	15	22	30
-25	-46	-38	-30	-22	-18	-14	-10	-7	-3	0	4	8	16	24	32
-20	-44	-36	-29	-21	-17	-13	-9	-5	-1	2	6	10	17	25	33
-15	-43	-35	-27	-19	-15	-12	-8	-4	0	3	7	11	19	27	34
-12	-42	-34	-26	-19	-15	-11	-7	-3	0	4	8	12	20	27	35
-10	-41	-34	-26	-18	-14	-10	-6	-2	1	5	8	12	20	28	36
-8	-41	-33	-25	-17	-13	-10	-6	-2	1	5	9	13	21	29	36
-6	-40	-32	-25	-17	-13	-9	-5	-1	2	6	10	13	21	29	37
-4	-40	-32	-24	-16	-12	-8	-5	-1	2	6	10	14	22	30	37
-2	-39	-31	-24	-16	-12	-8	-4	0	3	7	11	15	22	30	38
0	-39	-31	-23	-15	-11	-7	-3	0	3	7	11	15	23	31	39
2	-38	-30	-22	-15	-11	-7	-3	0	4	8	12	16	24	31	39
4	-37	-30	-22	-14	-10	-6	-2	1	5	8	12	16	24	32	40
6	-37	-29	-21	-13	-10	-6	-2	1	5	9	13	17	25	32	40
8	-36	-29	-21	-13	-9	-5	-1	2	6	10	13	17	25	33	41
10	-36	-28	-20	-12	-8	-5	-1	2	6	10	14	18	26	34	41
12	-35	-27	-20	-12	-8	-4	0	3	7	11	15	19	26	34	42
15	-34	-27	-19	-11	-7	-3	0	4	8	12	15	19	27	35	43
20	-33	-25	-17	-10	-6	-2	1	5	9	13	17	21	29	36	44
25	-32	-24	-16	-8	-4	0	3	7	10	14	18	22	30	38	46
30	-30	-22	-15	-7	-3	0	4	8	12	16	20	24	31	39	47
35	-29	-21	-13	-5	-1	2	5	9	13	17	21	25	33	41	48
40	-27	-20	-12	-4	0	3	7	11	15	19	22	26	34	42	50
45	-26	-18	-10	-2	0	4	8	12	16	20	24	28	36	43	51
50	-25	-17	-9	-1	2	6	10	14	17	21	25	29	37	45	53
55	-23	-15	-7	0	3	7	11	15	19	23	27	31	38	46	54
60	-22	-14	-6	1	5	9	12	16	20	24	28	32	40	48	55
70	-19	-11	-3	4	7	11	15	19	23	27	31	35	43	50	58

Table 5-118

Separation, inch: 16

Leg Distance, inch: 30

SHIM REQUIRED FOR ALIGNMENT

Reverse Indicator Method

DriveN TIR MILS	DriveR TIR, MILS -50	-40	-30	-20	-15	-10	-5	0	5	10	15	20	30	40	50
-70	-77	-68	-58	-49	-44	-40	-35	-30	-25	-21	-16	-11	-2	6	16
-60	-73	-63	-54	-45	-40	-35	-30	-26	-21	-16	-12	-7	1	11	20
-55	-70	-61	-52	-42	-38	-33	-28	-24	-19	-14	-10	-5	4	13	22
-50	-68	-59	-50	-40	-35	-31	-26	-21	-17	-12	-7	-3	6	15	25
-45	-66	-57	-47	-38	-33	-29	-24	-19	-15	-10	-5	0	8	17	27
-40	-64	-55	-45	-36	-31	-26	-22	-17	-12	-8	-3	1	10	20	29
-35	-62	-52	-43	-34	-29	-24	-20	-15	-10	-5	-1	3	12	22	31
-30	-60	-50	-41	-31	-27	-22	-17	-13	-8	-3	0	5	15	24	33
-25	-57	-48	-39	-29	-25	-20	-15	-10	-6	-1	3	7	17	26	35
-20	-55	-46	-36	-27	-22	-18	-13	-8	-4	0	5	10	19	28	38
-15	-53	-44	-34	-25	-20	-15	-11	-6	-1	2	7	12	21	30	40
-12	-52	-42	-33	-24	-19	-14	-9	-5	0	4	8	13	22	32	41
-10	-51	-41	-32	-23	-18	-13	-9	-4	0	5	9	14	23	33	42
-8	-50	-41	-31	-22	-17	-12	-8	-3	1	5	10	15	24	34	43
-6	-49	-40	-30	-21	-16	-12	-7	-2	2	6	11	16	25	34	44
-4	-48	-39	-29	-20	-15	-11	-6	-1	2	7	12	17	26	35	45
-2	-47	-38	-29	-19	-14	-10	-5	0	3	8	13	17	27	36	46
0	-46	-37	-28	-18	-14	-9	-4	0	4	9	14	18	28	37	46
2	-46	-36	-27	-17	-13	-8	-3	0	5	10	14	19	29	38	47
4	-45	-35	-26	-17	-12	-7	-2	1	6	11	15	20	29	39	48
6	-44	-34	-25	-16	-11	-6	-2	2	7	12	16	21	30	40	49
8	-43	-34	-24	-15	-10	-5	-1	3	8	12	17	22	31	41	50
10	-42	-33	-23	-14	-9	-5	0	4	9	13	18	23	32	41	51
12	-41	-32	-22	-13	-8	-4	0	5	9	14	19	24	33	42	52
15	-40	-30	-21	-12	-7	-2	1	6	11	15	20	25	34	44	53
20	-38	-28	-19	-10	-5	0	4	8	13	18	22	27	36	46	55
25	-35	-26	-17	-7	-3	1	6	10	15	20	25	29	39	48	57
30	-33	-24	-15	-5	0	3	8	13	17	22	27	31	41	50	60
35	-31	-22	-12	-3	1	5	10	15	20	24	29	34	43	52	62
40	-29	-20	-10	-1	3	8	12	17	22	26	31	36	45	55	64
45	-27	-17	-8	0	5	10	15	19	24	29	33	38	47	57	66
50	-25	-15	-6	3	7	12	17	21	26	31	35	40	50	59	68
55	-22	-13	-4	5	10	14	19	24	28	33	38	42	52	61	70
60	-20	-11	-1	7	12	16	21	26	30	35	40	45	54	63	73
70	-16	-6	2	11	16	21	25	30	35	40	44	49	58	68	77

Table 5-119

Separation, inch: 16

Leg Distance, inch: 35

SHIM REQUIRED FOR ALIGNMENT

Reverse Indicator Method

DriveN TIR MILS	DriveR TIR, MILS -50	-40	-30	-20	-15	-10	-5	0	5	10	15	20	30	40	50
-70	-96	-85	-74	-63	-57	-52	-47	-41	-36	-30	-25	-19	-8	2	13
-60	-90	-79	-68	-57	-52	-46	-41	-35	-30	-24	-19	-13	-2	8	19
-55	-87	-76	-65	-54	-49	-43	-38	-32	-27	-21	-16	-10	0	11	22
-50	-84	-73	-62	-51	-46	-40	-35	-29	-24	-18	-13	-7	3	14	25
-45	-81	-70	-59	-48	-43	-37	-32	-26	-21	-15	-10	-4	6	17	27
-40	-78	-67	-56	-45	-40	-34	-29	-23	-18	-12	-7	-1	9	20	30
-35	-75	-64	-53	-42	-37	-31	-26	-20	-15	-9	-4	1	12	22	33
-30	-72	-61	-50	-39	-34	-28	-23	-17	-12	-6	-1	4	15	25	36
-25	-69	-58	-47	-36	-31	-25	-20	-14	-9	-3	1	7	17	28	39
-20	-66	-55	-44	-33	-28	-22	-17	-11	-6	0	4	10	20	31	42
-15	-63	-52	-41	-30	-25	-19	-14	-8	-3	2	7	12	23	34	45
-12	-61	-50	-39	-29	-23	-18	-12	-7	-1	3	9	14	25	36	47
-10	-60	-49	-38	-27	-22	-16	-11	-5	0	5	10	15	26	37	48
-8	-59	-48	-37	-26	-21	-15	-10	-4	0	6	11	17	28	39	49
-6	-58	-47	-36	-25	-19	-14	-9	-3	1	7	12	18	29	40	51
-4	-57	-46	-35	-24	-18	-13	-7	-2	3	8	14	19	30	41	52
-2	-55	-44	-34	-23	-17	-12	-6	-1	4	9	15	20	31	42	53
0	-54	-43	-32	-21	-16	-10	-5	0	5	10	16	21	32	43	54
2	-53	-42	-31	-20	-15	-9	-4	1	6	12	17	23	34	44	55
4	-52	-41	-30	-19	-14	-8	-3	2	7	13	18	24	35	46	57
6	-51	-40	-29	-18	-12	-7	-1	3	9	14	19	25	36	47	58
8	-49	-39	-28	-17	-11	-6	0	4	10	15	21	26	37	48	59
10	-48	-37	-26	-15	-10	-5	0	5	11	16	22	27	38	49	60
12	-47	-36	-25	-14	-9	-3	1	7	12	18	23	29	39	50	61
15	-45	-34	-23	-12	-7	-2	3	8	14	19	25	30	41	52	63
20	-42	-31	-20	-10	-4	0	6	11	17	22	28	33	44	55	66
25	-39	-28	-17	-7	-1	3	9	14	20	25	31	36	47	58	69
30	-36	-25	-15	-4	1	6	12	17	23	28	34	39	50	61	72
35	-33	-22	-12	-1	4	9	15	20	26	31	37	42	53	64	75
40	-30	-20	-9	1	7	12	18	23	29	34	40	45	56	67	78
45	-27	-17	-6	4	10	15	21	26	32	37	43	48	59	70	81
50	-25	-14	-3	7	13	18	24	29	35	40	46	51	62	73	84
55	-22	-11	0	10	16	21	27	32	38	43	49	54	65	76	87
60	-19	-8	2	13	19	24	30	35	41	46	52	57	68	79	90
70	-13	-2	8	19	25	30	36	41	47	52	57	63	74	85	96

Table 5-120

Separation, inch: 16

Leg Distance, inch: 40

SHIM REQUIRED FOR ALIGNMENT

Reverse Indicator Method

DriveN TIR MILS	DriveR TIR, MILS -50	-40	-30	-20	-15	-10	-5	0	5	10	15	20	30	40	50
-70	-115	-102	-90	-77	-71	-65	-58	-52	-46	-40	-33	-27	-15	-2	10
-60	-107	-95	-82	-70	-63	-57	-51	-45	-38	-32	-26	-20	-7	5	17
-55	-103	-91	-78	-66	-60	-53	-47	-41	-35	-28	-22	-16	-3	8	21
-50	-100	-87	-75	-62	-56	-50	-43	-37	-31	-25	-18	-12	0	12	25
-45	-96	-83	-71	-58	-52	-46	-40	-33	-27	-21	-15	-8	3	16	28
-40	-92	-80	-67	-55	-48	-42	-36	-30	-23	-17	-11	-5	7	20	32
-35	-88	-76	-63	-51	-45	-38	-32	-26	-20	-13	-7	-1	11	23	36
-30	-85	-72	-60	-47	-41	-35	-28	-22	-16	-10	-3	2	15	27	40
-25	-81	-68	-56	-43	-37	-31	-25	-18	-12	-6	0	6	18	31	43
-20	-77	-65	-52	-40	-33	-27	-21	-15	-8	-2	3	10	22	35	47
-15	-73	-61	-48	-36	-30	-23	-17	-11	-5	1	7	13	26	38	51
-12	-71	-59	-46	-34	-27	-21	-15	-9	-2	3	9	16	28	41	53
-10	-70	-57	-45	-32	-26	-20	-13	-7	-1	5	11	17	30	42	55
-8	-68	-56	-43	-31	-24	-18	-12	-6	0	6	12	19	31	44	56
-6	-67	-54	-42	-29	-23	-17	-10	-4	1	8	14	20	33	45	58
-4	-65	-53	-40	-28	-21	-15	-9	-3	3	9	15	22	34	47	59
-2	-64	-51	-39	-26	-20	-14	-7	-1	4	11	17	23	36	48	61
0	-62	-50	-37	-25	-18	-12	-6	0	6	12	18	25	37	50	62
2	-61	-48	-36	-23	-17	-11	-4	1	7	14	20	26	39	51	64
4	-59	-47	-34	-22	-15	-9	-3	3	9	15	21	28	40	53	65
6	-58	-45	-33	-20	-14	-8	-1	4	10	17	23	29	42	54	67
8	-56	-44	-31	-19	-12	-6	0	6	12	18	24	31	43	56	68
10	-55	-42	-30	-17	-11	-5	1	7	13	20	26	32	45	57	70
12	-53	-41	-28	-16	-9	-3	2	9	15	21	27	34	46	59	71
15	-51	-38	-26	-13	-7	-1	5	11	17	23	30	36	48	61	73
20	-47	-35	-22	-10	-3	2	8	15	21	27	33	40	52	65	77
25	-43	-31	-18	-6	0	6	12	18	25	31	37	43	56	68	81
30	-40	-27	-15	-2	3	10	16	22	28	35	41	47	60	72	85
35	-36	-23	-11	1	7	13	20	26	32	38	45	51	63	76	88
40	-32	-20	-7	5	11	17	23	30	36	42	48	55	67	80	92
45	-28	-16	-3	8	15	21	27	33	40	46	52	58	71	83	96
50	-25	-12	0	12	18	25	31	37	43	50	56	62	75	87	100
55	-21	-8	3	16	22	28	35	41	47	53	60	66	78	91	103
60	-17	-5	7	20	26	32	38	45	51	57	63	70	82	95	107
70	-10	2	15	27	33	40	46	52	58	65	71	77	90	102	115

Table 5-121

Separation, inch: 16

Leg Distance, inch: 50

SHIM REQUIRED FOR ALIGNMENT

Reverse Indicator Method

DriveN TIR MILS	DriveR TIR, MILS -50	-40	-30	-20	-15	-10	-5	0	5	10	15	20	30	40	50
-70	-152	-136	-121	-105	-97	-90	-82	-74	-66	-58	-50	-43	-27	-11	3
-60	-141	-126	-110	-95	-87	-79	-71	-63	-55	-48	-40	-32	-16	-1	14
-55	-136	-120	-105	-89	-81	-74	-66	-58	-50	-42	-35	-27	-11	4	19
-50	-131	-115	-100	-84	-76	-68	-60	-53	-45	-37	-29	-21	-6	9	25
-45	-125	-110	-94	-79	-71	-63	-55	-47	-40	-32	-24	-16	0	14	30
-40	-120	-105	-89	-73	-65	-58	-50	-42	-34	-26	-19	-11	4	20	35
-35	-115	-99	-84	-68	-60	-52	-45	-37	-29	-21	-13	-5	9	25	40
-30	-110	-94	-78	-63	-55	-47	-39	-31	-24	-16	-8	0	15	30	46
-25	-104	-89	-73	-57	-50	-42	-34	-26	-18	-10	-3	4	20	35	51
-20	-99	-83	-68	-52	-44	-36	-29	-21	-13	-5	2	10	25	41	56
-15	-94	-78	-62	-47	-39	-31	-23	-15	-8	0	7	15	30	46	62
-12	-90	-75	-59	-44	-36	-28	-20	-12	-4	2	10	18	34	49	65
-10	-88	-73	-57	-41	-34	-26	-18	-10	-2	5	12	20	36	51	67
-8	-86	-71	-55	-39	-31	-24	-16	-8	0	7	14	22	38	54	69
-6	-84	-68	-53	-37	-29	-22	-14	-6	1	9	17	24	40	56	71
-4	-82	-66	-51	-35	-27	-19	-12	-4	3	11	19	27	42	58	73
-2	-80	-64	-49	-33	-25	-17	-9	-2	5	13	21	29	44	60	76
0	-78	-62	-46	-31	-23	-15	-7	0	7	15	23	31	46	62	78
2	-76	-60	-44	-29	-21	-13	-5	2	9	17	25	33	49	64	80
4	-73	-58	-42	-27	-19	-11	-3	4	12	19	27	35	51	66	82
6	-71	-56	-40	-24	-17	-9	-1	6	14	22	29	37	53	68	84
8	-69	-54	-38	-22	-14	-7	0	8	16	24	31	39	55	71	86
10	-67	-51	-36	-20	-12	-5	2	10	18	26	34	41	57	73	88
12	-65	-49	-34	-18	-10	-2	4	12	20	28	36	44	59	75	90
15	-62	-46	-30	-15	-7	0	8	15	23	31	39	47	62	78	94
20	-56	-41	-25	-10	-2	5	13	21	29	36	44	52	68	83	99
25	-51	-35	-20	-4	3	10	18	26	34	42	50	57	73	89	104
30	-46	-30	-15	0	8	16	24	31	39	47	55	63	78	94	110
35	-40	-25	-9	5	13	21	29	37	45	52	60	68	84	99	115
40	-35	-20	-4	11	19	26	34	42	50	58	65	73	89	105	120
45	-30	-14	0	16	24	32	40	47	55	63	71	79	94	110	125
50	-25	-9	6	21	29	37	45	53	60	68	76	84	100	115	131
55	-19	-4	11	27	35	42	50	58	66	74	81	89	105	120	136
60	-14	1	16	32	40	48	55	63	71	79	87	95	110	126	141
70	-3	11	27	43	50	58	66	74	82	90	97	105	121	136	152

Table 5-122

Separation, inch: 16

Leg Distance, inch: 60

SHIM REQUIRED FOR ALIGNMENT

Reverse Indicator Method

DriveN TIR MILS	DriveR TIR, MILS -50	-40	-30	-20	-15	-10	-5	0	5	10	15	20	30	40	50
-70	-190	-171	-152	-133	-124	-115	-105	-96	-86	-77	-68	-58	-40	-21	-2
-60	-176	-157	-138	-120	-110	-101	-91	-82	-73	-63	-54	-45	-26	-7	11
-55	-169	-150	-131	-113	-103	-94	-85	-75	-66	-56	-47	-38	-19	0	18
-50	-162	-143	-125	-106	-96	-87	-78	-68	-59	-50	-40	-31	-12	6	25
-45	-155	-136	-118	-99	-90	-80	-71	-61	-52	-43	-33	-24	-5	13	31
-40	-148	-130	-111	-92	-83	-73	-64	-55	-45	-36	-26	-17	1	20	38
-35	-141	-123	-104	-85	-76	-66	-57	-48	-38	-29	-20	-10	8	26	45
-30	-135	-116	-97	-78	-69	-60	-50	-41	-31	-22	-13	-3	15	33	52
-25	-128	-109	-90	-71	-62	-53	-43	-34	-25	-15	-6	3	21	40	59
-20	-121	-102	-83	-65	-55	-46	-36	-27	-18	-8	0	10	28	47	66
-15	-114	-95	-76	-58	-48	-39	-30	-20	-11	-1	7	16	35	54	73
-12	-110	-91	-72	-54	-44	-35	-25	-16	-7	2	11	21	39	58	77
-10	-107	-88	-70	-51	-41	-32	-23	-13	-4	5	14	23	42	61	80
-8	-104	-86	-67	-48	-39	-29	-20	-11	-1	7	17	26	45	64	82
-6	-102	-83	-64	-45	-36	-27	-17	-8	1	10	19	29	48	66	85
-4	-99	-80	-61	-43	-33	-24	-14	-5	3	13	22	32	50	69	88
-2	-96	-77	-59	-40	-30	-21	-12	-2	6	16	25	34	53	72	91
0	-93	-75	-56	-37	-28	-18	-9	0	9	18	28	37	56	75	93
2	-91	-72	-53	-34	-25	-16	-6	2	12	21	30	40	59	77	96
4	-88	-69	-50	-32	-22	-13	-3	5	14	24	33	43	61	80	99
6	-85	-66	-48	-29	-19	-10	-1	8	17	27	36	45	64	83	102
8	-82	-64	-45	-26	-17	-7	1	11	20	29	39	48	67	86	104
10	-80	-61	-42	-23	-14	-5	4	13	23	32	41	51	70	88	107
12	-77	-58	-39	-21	-11	-2	7	16	25	35	44	54	72	91	110
15	-73	-54	-35	-16	-7	1	11	20	30	39	48	58	76	95	114
20	-66	-47	-28	-10	0	8	18	27	36	46	55	65	83	102	121
25	-59	-40	-21	-3	6	15	25	34	43	53	62	71	90	109	128
30	-52	-33	-15	3	13	22	31	41	50	60	69	78	97	116	135
35	-45	-26	-8	10	20	29	38	48	57	66	76	85	104	123	141
40	-38	-20	-1	17	26	36	45	55	64	73	83	92	111	130	148
45	-31	-13	5	24	33	43	52	61	71	80	90	99	118	136	155
50	-25	-6	12	31	40	50	59	68	78	87	96	106	125	143	162
55	-18	0	19	38	47	56	66	75	85	94	103	113	131	150	169
60	-11	7	26	45	54	63	73	82	91	101	110	120	138	157	176
70	2	21	40	58	68	77	86	96	105	115	124	133	152	171	190

Table 5-123

Separation, inch: 16

Leg Distance, inch: 70

SHIM REQUIRED FOR ALIGNMENT

Reverse Indicator Method

DriveN TIR MILS	DriveR TIR, MILS -50	-40	-30	-20	-15	-10	-5	0	5	10	15	20	30	40	50
-70	-227	-205	-183	-161	-150	-140	-129	-118	-107	-96	-85	-74	-52	-30	-8
-60	-210	-188	-166	-145	-134	-123	-112	-101	-90	-79	-68	-57	-35	-13	8
-55	-202	-180	-158	-136	-125	-114	-103	-92	-81	-70	-60	-49	-27	-5	16
-50	-193	-171	-150	-128	-117	-106	-95	-84	-73	-62	-51	-40	-18	3	25
-45	-185	-163	-141	-119	-108	-97	-86	-75	-65	-54	-43	-32	-10	11	33
-40	-176	-155	-133	-111	-100	-89	-78	-67	-56	-45	-34	-23	-1	20	41
-35	-168	-146	-124	-102	-91	-80	-70	-59	-48	-37	-26	-15	6	28	50
-30	-160	-138	-116	-94	-83	-72	-61	-50	-39	-28	-17	-6	15	36	58
-25	-151	-129	-107	-85	-75	-64	-53	-42	-31	-20	-9	1	23	45	67
-20	-143	-121	-99	-77	-66	-55	-44	-33	-22	-11	0	10	31	53	75
-15	-134	-112	-90	-69	-58	-47	-36	-25	-14	-3	7	18	40	62	84
-12	-129	-107	-85	-64	-53	-42	-31	-20	-9	1	12	23	45	67	89
-10	-126	-104	-82	-60	-49	-38	-27	-16	-5	5	15	26	48	70	92
-8	-122	-101	-79	-57	-46	-35	-24	-13	-2	8	19	30	52	74	95
-6	-119	-97	-75	-53	-42	-32	-21	-10	0	11	22	33	55	77	99
-4	-116	-94	-72	-50	-39	-28	-17	-6	4	15	26	37	58	80	102
-2	-112	-90	-69	-47	-36	-25	-14	-3	7	18	29	40	62	84	106
0	-109	-87	-65	-43	-32	-21	-10	0	10	21	32	43	65	87	109
2	-106	-84	-62	-40	-29	-18	-7	3	14	25	36	47	69	90	112
4	-102	-80	-58	-37	-26	-15	-4	6	17	28	39	50	72	94	116
6	-99	-77	-55	-33	-22	-11	0	10	21	32	42	53	75	97	119
8	-95	-74	-52	-30	-19	-8	2	13	24	35	46	57	79	101	122
10	-92	-70	-48	-26	-15	-5	5	16	27	38	49	60	82	104	126
12	-89	-67	-45	-23	-12	-1	9	20	31	42	53	64	85	107	129
15	-84	-62	-40	-18	-7	3	14	25	36	47	58	69	90	112	134
20	-75	-53	-31	-10	0	11	22	33	44	55	66	77	99	121	143
25	-67	-45	-23	-1	9	20	31	42	53	64	75	85	107	129	151
30	-58	-36	-15	6	17	28	39	50	61	72	83	94	116	138	160
35	-50	-28	-6	15	26	37	48	59	70	80	91	102	124	146	168
40	-41	-20	1	23	34	45	56	67	78	89	100	111	133	155	176
45	-33	-11	10	32	43	54	65	75	86	97	108	119	141	163	185
50	-25	-3	18	40	51	62	73	84	95	106	117	128	150	171	193
55	-16	5	27	49	60	70	81	92	103	114	125	136	158	180	202
60	-8	13	35	57	68	79	90	101	112	123	134	145	166	188	210
70	8	30	52	74	85	96	107	118	129	140	150	161	183	205	227

Table 5-124

Separation, inch: 16

Leg Distance, inch: 80

SHIM REQUIRED FOR ALIGNMENT

Reverse Indicator Method

DriveN TIR MILS	DriveR TIR, MILS -50	-40	-30	-20	-15	-10	-5	0	5	10	15	20	30	40	50
-70	-265	-240	-215	-190	-177	-165	-152	-140	-127	-115	-102	-90	-65	-40	-15
-60	-245	-220	-195	-170	-157	-145	-132	-120	-107	-95	-82	-70	-45	-20	5
-55	-235	-210	-185	-160	-147	-135	-122	-110	-97	-85	-72	-60	-35	-10	15
-50	-225	-200	-175	-150	-137	-125	-112	-100	-87	-75	-62	-50	-25	0	25
-45	-215	-190	-165	-140	-127	-115	-102	-90	-77	-65	-52	-40	-15	10	35
-40	-205	-180	-155	-130	-117	-105	-92	-80	-67	-55	-42	-30	-5	20	45
-35	-195	-170	-145	-120	-107	-95	-82	-70	-57	-45	-32	-20	5	30	55
-30	-185	-160	-135	-110	-97	-85	-72	-60	-47	-35	-22	-10	15	40	65
-25	-175	-150	-125	-100	-87	-75	-62	-50	-37	-25	-12	0	25	50	75
-20	-165	-140	-115	-90	-77	-65	-52	-40	-27	-15	-2	10	35	60	85
-15	-155	-130	-105	-80	-67	-55	-42	-30	-17	-5	7	20	45	70	95
-12	-149	-124	-99	-74	-61	-49	-36	-24	-11	1	13	26	51	76	101
-10	-145	-120	-95	-70	-57	-45	-32	-20	-7	5	17	30	55	80	105
-8	-141	-116	-91	-66	-53	-41	-28	-16	-3	9	21	34	59	84	109
-6	-137	-112	-87	-62	-49	-37	-24	-12	0	13	25	38	63	88	113
-4	-133	-108	-83	-58	-45	-33	-20	-8	4	17	29	42	67	92	117
-2	-129	-104	-79	-54	-41	-29	-16	-4	8	21	33	46	71	96	121
0	-125	-100	-75	-50	-37	-25	-12	0	12	25	37	50	75	100	125
2	-121	-96	-71	-46	-33	-21	-8	4	16	29	41	54	79	104	129
4	-117	-92	-67	-42	-29	-17	-4	8	20	33	45	58	83	108	133
6	-113	-88	-63	-38	-25	-13	0	12	24	37	49	62	87	112	137
8	-109	-84	-59	-34	-21	-9	3	16	28	41	53	66	91	116	141
10	-105	-80	-55	-30	-17	-5	7	20	32	45	57	70	95	120	145
12	-101	-76	-51	-26	-13	-1	11	24	36	49	61	74	99	124	149
15	-95	-70	-45	-20	-7	5	17	30	42	55	67	80	105	130	155
20	-85	-60	-35	-10	2	15	27	40	52	65	77	90	115	140	165
25	-75	-50	-25	0	12	25	37	50	62	75	87	100	125	150	175
30	-65	-40	-15	10	22	35	47	60	72	85	97	110	135	160	185
35	-55	-30	-5	20	32	45	57	70	82	95	107	120	145	170	195
40	-45	-20	5	30	42	55	67	80	92	105	117	130	155	180	205
45	-35	-10	15	40	52	65	77	90	102	115	127	140	165	190	215
50	-25	0	25	50	62	75	87	100	112	125	137	150	175	200	225
55	-15	10	35	60	72	85	97	110	122	135	147	160	185	210	235
60	-5	20	45	70	82	95	107	120	132	145	157	170	195	220	245
70	15	40	65	90	102	115	127	140	152	165	177	190	215	240	265

Table 5-125

Separation, inch: 16

Leg Distance, inch: 90

SHIM REQUIRED FOR ALIGNMENT

Reverse Indicator Method

DriveN TIR MILS	DriveR TIR, MILS -50	-40	-30	-20	-15	-10	-5	0	5	10	15	20	30	40	50
-70	-302	-274	-246	-218	-204	-190	-175	-161	-147	-133	-119	-105	-77	-49	-21
-60	-279	-251	-223	-195	-180	-166	-152	-138	-124	-110	-96	-82	-54	-26	1
-55	-267	-239	-211	-183	-169	-155	-141	-127	-113	-99	-85	-70	-42	-14	13
-50	-256	-228	-200	-171	-157	-143	-129	-115	-101	-87	-73	-59	-31	-3	25
-45	-244	-216	-188	-160	-146	-132	-118	-104	-90	-75	-61	-47	-19	8	36
-40	-233	-205	-176	-148	-134	-120	-106	-92	-78	-64	-50	-36	-8	20	48
-35	-221	-193	-165	-137	-123	-109	-95	-80	-66	-52	-38	-24	3	31	59
-30	-210	-181	-153	-125	-111	-97	-83	-69	-55	-41	-27	-13	15	43	71
-25	-198	-170	-142	-114	-100	-85	-71	-57	-43	-29	-15	-1	26	54	82
-20	-186	-158	-130	-102	-88	-74	-60	-46	-32	-18	-4	10	38	66	94
-15	-175	-147	-119	-90	-76	-62	-48	-34	-20	-6	7	21	49	77	105
-12	-168	-140	-112	-84	-69	-55	-41	-27	-13	0	14	28	56	84	112
-10	-163	-135	-107	-79	-65	-51	-37	-23	-9	5	19	33	61	89	117
-8	-159	-131	-102	-74	-60	-46	-32	-18	-4	9	23	37	65	94	122
-6	-154	-126	-98	-70	-56	-42	-27	-13	0	14	28	42	70	98	126
-4	-149	-121	-93	-65	-51	-37	-23	-9	4	18	32	47	75	103	131
-2	-145	-117	-89	-60	-46	-32	-18	-4	9	23	37	51	79	107	136
0	-140	-112	-84	-56	-42	-28	-14	0	14	28	42	56	84	112	140
2	-136	-107	-79	-51	-37	-23	-9	4	18	32	46	60	89	117	145
4	-131	-103	-75	-47	-32	-18	-4	9	23	37	51	65	93	121	149
6	-126	-98	-70	-42	-28	-14	0	13	27	42	56	70	98	126	154
8	-122	-94	-65	-37	-23	-9	4	18	32	46	60	74	102	131	159
10	-117	-89	-61	-33	-19	-5	9	23	37	51	65	79	107	135	163
12	-112	-84	-56	-28	-14	0	13	27	41	55	69	84	112	140	168
15	-105	-77	-49	-21	-7	6	20	34	48	62	76	90	119	147	175
20	-94	-66	-38	-10	4	18	32	46	60	74	88	102	130	158	186
25	-82	-54	-26	1	15	29	43	57	71	85	100	114	142	170	198
30	-71	-43	-15	13	27	41	55	69	83	97	111	125	153	181	210
35	-59	-31	-3	24	38	52	66	80	95	109	123	137	165	193	221
40	-48	-20	8	36	50	64	78	92	106	120	134	148	176	205	233
45	-36	-8	19	47	61	75	90	104	118	132	146	160	188	216	244
50	-25	3	31	59	73	87	101	115	129	143	157	171	200	228	256
55	-13	14	42	70	85	99	113	127	141	155	169	183	211	239	267
60	-1	26	54	82	96	110	124	138	152	166	180	195	223	251	279
70	21	49	77	105	119	133	147	161	175	190	204	218	246	274	302

Table 5-126

Separation, inch: 16

Leg Distance, inch: 100

SHIM REQUIRED FOR ALIGNMENT

Reverse Indicator Method

DriveN TIR MILS	DriveR TIR, MILS -50	-40	-30	-20	-15	-10	-5	0	5	10	15	20	30	40	50
-70	-340	-308	-277	-246	-230	-215	-199	-183	-168	-152	-136	-121	-90	-58	-27
-60	-313	-282	-251	-220	-204	-188	-173	-157	-141	-126	-110	-95	-63	-32	-1
-55	-300	-269	-238	-206	-191	-175	-160	-144	-128	-113	-97	-81	-50	-19	11
-50	-287	-256	-225	-193	-178	-162	-146	-131	-115	-100	-84	-68	-37	-6	25
-45	-274	-243	-211	-180	-165	-149	-133	-118	-102	-86	-71	-55	-24	6	38
-40	-261	-230	-198	-167	-151	-136	-120	-105	-89	-73	-58	-42	-11	20	51
-35	-248	-216	-185	-154	-138	-123	-107	-91	-76	-60	-45	-29	1	33	64
-30	-235	-203	-172	-141	-125	-110	-94	-78	-63	-47	-31	-16	15	46	77
-25	-221	-190	-159	-128	-112	-96	-81	-65	-50	-34	-18	-3	28	59	90
-20	-208	-177	-146	-115	-99	-83	-68	-52	-36	-21	-5	10	41	72	103
-15	-195	-164	-133	-101	-86	-70	-55	-39	-23	-8	7	23	54	85	116
-12	-187	-156	-125	-94	-78	-62	-47	-31	-15	0	15	31	62	93	124
-10	-182	-151	-120	-88	-73	-57	-41	-26	-10	5	20	36	67	98	130
-8	-177	-146	-114	-83	-67	-52	-36	-21	-5	10	25	41	72	104	135
-6	-172	-140	-109	-78	-62	-47	-31	-15	0	15	31	46	78	109	140
-4	-166	-135	-104	-73	-57	-41	-26	-10	5	20	36	52	83	114	145
-2	-161	-130	-99	-67	-52	-36	-20	-5	10	26	41	57	88	119	151
0	-156	-125	-93	-62	-46	-31	-15	0	15	31	46	62	93	125	156
2	-151	-119	-88	-57	-41	-26	-10	5	20	36	52	67	99	130	161
4	-145	-114	-83	-52	-36	-20	-5	10	26	41	57	73	104	135	166
6	-140	-109	-78	-46	-31	-15	0	15	31	47	62	78	109	140	172
8	-135	-104	-72	-41	-25	-10	5	21	36	52	67	83	114	146	177
10	-130	-98	-67	-36	-20	-5	10	26	41	57	73	88	120	151	182
12	-124	-93	-62	-31	-15	0	15	31	47	62	78	94	125	156	187
15	-116	-85	-54	-23	-7	8	23	39	55	70	86	101	133	164	195
20	-103	-72	-41	-10	5	21	36	52	68	83	99	115	146	177	208
25	-90	-59	-28	3	18	34	50	65	81	96	112	128	159	190	221
30	-77	-46	-15	16	31	47	63	78	94	110	125	141	172	203	235
35	-64	-33	-1	29	45	60	76	91	107	123	138	154	185	216	248
40	-51	-20	11	42	58	73	89	105	120	136	151	167	198	230	261
45	-38	-6	24	55	71	86	102	118	133	149	165	180	211	243	274
50	-25	6	37	68	84	100	115	131	146	162	178	193	225	256	287
55	-11	19	50	81	97	113	128	144	160	175	191	206	238	269	300
60	1	32	63	95	110	126	141	157	173	188	204	220	251	282	313
70	27	58	90	121	136	152	168	183	199	215	230	246	277	308	340

Table 5-127

Separation, inch: 18

Leg Distance, inch: 25

SHIM REQUIRED FOR ALIGNMENT

Reverse Indicator Method

riveN IR ILS	DriveR TIR, MILS -50	-40	-30	-20	-15	-10	-5	0	5	10	15	20	30	40	50
-70	-48	-41	-34	-27	-24	-20	-17	-13	-10	-6	-3	0	7	14	21
-60	-46	-39	-32	-25	-22	-18	-15	-11	-8	-4	-1	2	9	16	23
-55	-45	-38	-31	-24	-21	-17	-14	-10	-7	-3	0	3	10	17	24
-50	-44	-37	-30	-23	-20	-16	-13	-9	-6	-2	0	4	11	18	25
-45	-43	-36	-29	-22	-19	-15	-12	-8	-5	-1	1	5	12	19	25
-40	-42	-35	-28	-21	-18	-14	-11	-7	-4	0	2	6	13	20	26
-35	-41	-34	-27	-20	-17	-13	-10	-6	-3	0	3	7	14	20	27
-30	-40	-33	-26	-19	-16	-12	-9	-5	-2	1	4	8	15	21	28
-25	-39	-32	-25	-18	-15	-11	-8	-4	-1	2	5	9	15	22	29
-20	-38	-31	-24	-17	-14	-10	-7	-3	0	3	6	10	16	23	30
-15	-37	-30	-23	-16	-13	-9	-6	-2	0	4	7	10	17	24	31
-12	-37	-30	-23	-16	-12	-9	-5	-2	1	4	8	11	18	25	32
-10	-36	-29	-22	-15	-12	-8	-5	-1	1	5	8	11	18	25	32
-8	-36	-29	-22	-15	-11	-8	-5	-1	1	5	8	12	19	26	33
-6	-35	-28	-22	-15	-11	-8	-4	-1	2	5	9	12	19	26	33
-4	-35	-28	-21	-14	-11	-7	-4	0	2	6	9	13	20	27	33
-2	-35	-28	-21	-14	-10	-7	-3	0	3	6	10	13	20	27	34
0	-34	-27	-20	-13	-10	-6	-3	0	3	6	10	13	20	27	34
2	-34	-27	-20	-13	-10	-6	-3	0	3	7	10	14	21	28	35
4	-33	-27	-20	-13	-9	-6	-2	0	4	7	11	14	21	28	35
6	-33	-26	-19	-12	-9	-5	-2	1	4	8	11	15	22	28	35
8	-33	-26	-19	-12	-8	-5	-1	1	5	8	11	15	22	29	36
10	-32	-25	-18	-11	-8	-5	-1	1	5	8	12	15	22	29	36
12	-32	-25	-18	-11	-8	-4	-1	2	5	9	12	16	23	30	37
15	-31	-24	-17	-10	-7	-4	0	2	6	9	13	16	23	30	37
20	-30	-23	-16	-10	-6	-3	0	3	7	10	14	17	24	31	38
25	-29	-22	-15	-9	-5	-2	1	4	8	11	15	18	25	32	39
30	-28	-21	-15	-8	-4	-1	2	5	9	12	16	19	26	33	40
35	-27	-20	-14	-7	-3	0	3	6	10	13	17	20	27	34	41
40	-26	-20	-13	-6	-2	0	4	7	11	14	18	21	28	35	42
45	-25	-19	-12	-5	-1	1	5	8	12	15	19	22	29	36	43
50	-25	-18	-11	-4	0	2	6	9	13	16	20	23	30	37	44
55	-24	-17	-10	-3	0	3	7	10	14	17	21	24	31	38	45
60	-23	-16	-9	-2	1	4	8	11	15	18	22	25	32	39	46
70	-21	-14	-7	0	3	6	10	13	17	20	24	27	34	41	48

Table 5-128

Separation, inch: 18

Leg Distance, inch: 30

SHIM REQUIRED FOR ALIGNMENT

Reverse Indicator Method

DriveN TIR MILS	DriveR TIR, MILS -50	-40	-30	-20	-15	-10	-5	0	5	10	15	20	30	40	50
-70	-65	-56	-48	-40	-35	-31	-27	-23	-19	-15	-10	-6	1	10	18
-60	-61	-53	-45	-36	-32	-28	-24	-20	-15	-11	-7	-3	5	13	21
-55	-60	-51	-43	-35	-30	-26	-22	-18	-14	-10	-5	-1	6	15	23
-50	-58	-50	-41	-33	-29	-25	-20	-16	-12	-8	-4	0	8	16	25
-45	-56	-48	-40	-31	-27	-23	-19	-15	-10	-6	-2	1	10	18	26
-40	-55	-46	-38	-30	-25	-21	-17	-13	-9	-5	0	3	11	20	28
-35	-53	-45	-36	-28	-24	-20	-15	-11	-7	-3	0	5	13	21	30
-30	-51	-43	-35	-26	-22	-18	-14	-10	-5	-1	2	6	15	23	31
-25	-50	-41	-33	-25	-20	-16	-12	-8	-4	0	4	8	16	25	33
-20	-48	-40	-31	-23	-19	-15	-10	-6	-2	1	5	10	18	26	35
-15	-46	-38	-30	-21	-17	-13	-9	-5	0	3	7	11	20	28	36
-12	-45	-37	-29	-20	-16	-12	-8	-4	0	4	8	12	21	29	37
-10	-45	-36	-28	-20	-15	-11	-7	-3	0	5	9	13	21	30	38
-8	-44	-36	-27	-19	-15	-11	-6	-2	1	5	9	14	22	30	39
-6	-43	-35	-27	-18	-14	-10	-6	-2	2	6	10	14	23	31	39
-4	-43	-34	-26	-18	-13	-9	-5	-1	2	7	11	15	23	32	40
-2	-42	-34	-25	-17	-13	-9	-4	0	3	7	11	16	24	32	41
0	-41	-33	-25	-16	-12	-8	-4	0	4	8	12	16	25	33	41
2	-41	-32	-24	-16	-11	-7	-3	0	4	9	13	17	25	34	42
4	-40	-32	-23	-15	-11	-7	-2	1	5	9	13	18	26	34	43
6	-39	-31	-23	-14	-10	-6	-2	2	6	10	14	18	27	35	43
8	-39	-30	-22	-14	-9	-5	-1	2	6	11	15	19	27	36	44
10	-38	-30	-21	-13	-9	-5	0	3	7	11	15	20	28	36	45
12	-37	-29	-21	-12	-8	-4	0	4	8	12	16	20	29	37	45
15	-36	-28	-20	-11	-7	-3	0	5	9	13	17	21	30	38	46
20	-35	-26	-18	-10	-5	-1	2	6	10	15	19	23	31	40	48
25	-33	-25	-16	-8	-4	0	4	8	12	16	20	25	33	41	50
30	-31	-23	-15	-6	-2	1	5	10	14	18	22	26	35	43	51
35	-30	-21	-13	-5	0	3	7	11	15	20	24	28	36	45	53
40	-28	-20	-11	-3	0	5	9	13	17	21	25	30	38	46	55
45	-26	-18	-10	-1	2	6	10	15	19	23	27	31	40	48	56
50	-25	-16	-8	0	4	8	12	16	20	25	29	33	41	50	58
55	-23	-15	-6	1	5	10	14	18	22	26	30	35	43	51	60
60	-21	-13	-5	3	7	11	15	20	24	28	32	36	45	53	61
70	-18	-10	-1	6	10	15	19	23	27	31	35	40	48	56	65

Table 5-129

Separation, inch: 18

Leg Distance, inch: 35

SHIM REQUIRED FOR ALIGNMENT

Reverse Indicator Method

riveN IR ILS	DriveR TIR, MILS													
-50	-40	-30	-20	-15	-10	-5	0	5	10	15	20	30	40	50
-70	-81	-71	-62	-52	-47	-42	-37	-33	-28	-23	-18	-13	-3	5	15
-60	-76	-67	-57	-47	-42	-38	-33	-28	-23	-18	-13	-8	0	10	20
-55	-74	-64	-55	-45	-40	-35	-30	-25	-21	-16	-11	-6	3	12	22
-50	-72	-62	-52	-43	-38	-33	-28	-23	-18	-13	-9	-4	5	15	25
-45	-69	-60	-50	-40	-35	-30	-26	-21	-16	-11	-6	-1	7	17	27
-40	-67	-57	-48	-38	-33	-28	-23	-18	-14	-9	-4	0	10	20	29
-35	-65	-55	-45	-35	-31	-26	-21	-16	-11	-6	-1	2	12	22	32
-30	-62	-53	-43	-33	-28	-23	-19	-14	-9	-4	0	5	15	24	34
-25	-60	-50	-40	-31	-26	-21	-16	-11	-6	-2	2	7	17	27	36
-20	-58	-48	-38	-28	-24	-19	-14	-9	-4	0	5	10	19	29	39
-15	-55	-45	-36	-26	-21	-16	-11	-7	-2	2	7	12	22	31	41
-12	-54	-44	-34	-25	-20	-15	-10	-5	0	4	8	13	23	33	42
-10	-53	-43	-33	-24	-19	-14	-9	-4	0	5	9	14	24	34	43
-8	-52	-42	-32	-23	-18	-13	-8	-3	1	5	10	15	25	35	44
-6	-51	-41	-32	-22	-17	-12	-7	-2	2	6	11	16	26	36	45
-4	-50	-40	-31	-21	-16	-11	-6	-1	2	7	12	17	27	37	46
-2	-49	-39	-30	-20	-15	-10	-5	0	3	8	13	18	28	37	47
0	-48	-38	-29	-19	-14	-9	-4	0	4	9	14	19	29	38	48
2	-47	-37	-28	-18	-13	-8	-3	0	5	10	15	20	30	39	49
4	-46	-37	-27	-17	-12	-7	-2	1	6	11	16	21	31	40	50
6	-45	-36	-26	-16	-11	-6	-2	2	7	12	17	22	32	41	51
8	-44	-35	-25	-15	-10	-5	-1	3	8	13	18	23	32	42	52
10	-43	-34	-24	-14	-9	-5	0	4	9	14	19	24	33	43	53
12	-42	-33	-23	-13	-8	-4	0	5	10	15	20	25	34	44	54
15	-41	-31	-22	-12	-7	-2	2	7	11	16	21	26	36	45	55
20	-39	-29	-19	-10	-5	0	4	9	14	19	24	28	38	48	58
25	-36	-27	-17	-7	-2	2	6	11	16	21	26	31	40	50	60
30	-34	-24	-15	-5	0	4	9	14	19	23	28	33	43	53	62
35	-32	-22	-12	-2	1	6	11	16	21	26	31	35	45	55	65
40	-29	-20	-10	0	4	9	14	18	23	28	33	38	48	57	67
45	-27	-17	-7	1	6	11	16	21	26	30	35	40	50	60	69
50	-25	-15	-5	4	9	13	18	23	28	33	38	43	52	62	72
55	-22	-12	-3	6	11	16	21	25	30	35	40	45	55	64	74
60	-20	-10	0	8	13	18	23	28	33	38	42	47	57	67	76
70	-15	-5	3	13	18	23	28	33	37	42	47	52	62	71	81

Table 5-130

Separation, inch: 18

Leg Distance, inch: 40

SHIM REQUIRED FOR ALIGNMENT

Reverse Indicator Method

DriveN TIR MILS	DriveR TIR, MILS -50	-40	-30	-20	-15	-10	-5	0	5	10	15	20	30	40	50
-70	-98	-87	-76	-65	-59	-53	-48	-42	-37	-31	-26	-20	-9	1	12
-60	-92	-81	-70	-58	-53	-47	-42	-36	-31	-25	-20	-14	-3	7	18
-55	-89	-78	-66	-55	-50	-44	-39	-33	-28	-22	-16	-11	0	10	21
-50	-86	-75	-63	-52	-47	-41	-36	-30	-25	-19	-13	-8	2	13	25
-45	-83	-71	-60	-49	-44	-38	-33	-27	-21	-16	-10	-5	5	16	28
-40	-80	-68	-57	-46	-41	-35	-30	-24	-18	-13	-7	-2	8	20	31
-35	-76	-65	-54	-43	-38	-32	-26	-21	-15	-10	-4	0	11	23	34
-30	-73	-62	-51	-40	-35	-29	-23	-18	-12	-7	-1	3	15	26	37
-25	-70	-59	-48	-37	-31	-26	-20	-15	-9	-4	1	6	18	29	40
-20	-67	-56	-45	-34	-28	-23	-17	-12	-6	-1	4	10	21	32	43
-15	-64	-53	-42	-31	-25	-20	-14	-9	-3	1	7	13	24	35	46
-12	-62	-51	-40	-29	-24	-18	-12	-7	-1	3	9	14	26	37	48
-10	-61	-50	-39	-28	-22	-17	-11	-6	0	5	10	16	27	38	49
-8	-60	-49	-38	-27	-21	-16	-10	-4	0	6	11	17	28	39	50
-6	-59	-48	-37	-25	-20	-14	-9	-3	1	7	13	18	29	40	51
-4	-58	-46	-35	-24	-19	-13	-8	-2	3	8	14	19	30	42	53
-2	-56	-45	-34	-23	-17	-12	-6	-1	4	9	15	21	32	43	54
0	-55	-44	-33	-22	-16	-11	-5	0	5	11	16	22	33	44	55
2	-54	-43	-32	-21	-15	-9	-4	1	6	12	17	23	34	45	56
4	-53	-42	-30	-19	-14	-8	-3	2	8	13	19	24	35	46	58
6	-51	-40	-29	-18	-13	-7	-1	3	9	14	20	25	37	48	59
8	-50	-39	-28	-17	-11	-6	0	4	10	16	21	27	38	49	60
10	-49	-38	-27	-16	-10	-5	0	6	11	17	22	28	39	50	61
12	-48	-37	-26	-14	-9	-3	1	7	12	18	24	29	40	51	62
15	-46	-35	-24	-13	-7	-1	3	9	14	20	25	31	42	53	64
20	-43	-32	-21	-10	-4	1	6	12	17	23	28	34	45	56	67
25	-40	-29	-18	-6	-1	4	9	15	20	26	31	37	48	59	70
30	-37	-26	-15	-3	1	7	12	18	23	29	35	40	51	62	73
35	-34	-23	-11	0	4	10	15	21	26	32	38	43	54	65	76
40	-31	-20	-8	2	7	13	18	24	30	35	41	46	57	68	80
45	-28	-16	-5	5	10	16	21	27	33	38	44	49	60	71	83
50	-25	-13	-2	8	13	19	25	30	36	41	47	52	63	75	86
55	-21	-10	0	11	16	22	28	33	39	44	50	55	66	78	89
60	-18	-7	3	14	20	25	31	36	42	47	53	58	70	81	92
70	-12	-1	9	20	26	31	37	42	48	53	59	65	76	87	98

Table 5-131

Separation, inch: 18

Leg Distance, inch: 50

SHIM REQUIRED FOR ALIGNMENT

Reverse Indicator Method

riveN IR ILS	DriveR TIR, MILS -50	-40	-30	-20	-15	-10	-5	0	5	10	15	20	30	40	50
-70	-131	-117	-103	-90	-83	-76	-69	-62	-55	-48	-41	-34	-20	-6	7
-60	-122	-108	-95	-81	-74	-67	-60	-53	-46	-39	-32	-25	-11	2	16
-55	-118	-104	-90	-76	-69	-62	-55	-48	-41	-35	-28	-21	-7	6	20
-50	-113	-100	-86	-72	-65	-58	-51	-44	-37	-30	-23	-16	-2	11	25
-45	-109	-95	-81	-67	-60	-53	-46	-40	-33	-26	-19	-12	1	15	29
-40	-105	-91	-77	-63	-56	-49	-42	-35	-28	-21	-14	-7	6	20	33
-35	-100	-86	-72	-58	-51	-45	-38	-31	-24	-17	-10	-3	10	24	38
-30	-96	-82	-68	-54	-47	-40	-33	-26	-19	-12	-5	1	15	28	42
-25	-91	-77	-63	-50	-43	-36	-29	-22	-15	-8	-1	5	19	33	47
-20	-87	-73	-59	-45	-38	-31	-24	-17	-10	-3	3	10	23	37	51
-15	-82	-68	-55	-41	-34	-27	-20	-13	-6	0	7	14	28	42	56
-12	-80	-66	-52	-38	-31	-24	-17	-10	-3	3	10	17	31	44	58
-10	-78	-64	-50	-36	-29	-22	-15	-8	-1	5	11	18	32	46	60
-8	-76	-62	-48	-34	-27	-21	-14	-7	0	6	13	20	34	48	62
-6	-74	-60	-47	-33	-26	-19	-12	-5	1	8	15	22	36	50	64
-4	-73	-59	-45	-31	-24	-17	-10	-3	3	10	17	24	38	52	65
-2	-71	-57	-43	-29	-22	-15	-8	-1	5	12	19	26	39	53	67
0	-69	-55	-41	-27	-20	-13	-6	0	6	13	20	27	41	55	69
2	-67	-53	-39	-26	-19	-12	-5	1	8	15	22	29	43	57	71
4	-65	-52	-38	-24	-17	-10	-3	3	10	17	24	31	45	59	73
6	-64	-50	-36	-22	-15	-8	-1	5	12	19	26	33	47	60	74
8	-62	-48	-34	-20	-13	-6	0	7	14	21	27	34	48	62	76
10	-60	-46	-32	-18	-11	-5	1	8	15	22	29	36	50	64	78
12	-58	-44	-31	-17	-10	-3	3	10	17	24	31	38	52	66	80
15	-56	-42	-28	-14	-7	0	6	13	20	27	34	41	55	68	82
20	-51	-37	-23	-10	-3	3	10	17	24	31	38	45	59	73	87
25	-47	-33	-19	-5	1	8	15	22	29	36	43	50	63	77	91
30	-42	-28	-15	-1	5	12	19	26	33	40	47	54	68	82	96
35	-38	-24	-10	3	10	17	24	31	38	45	51	58	72	86	100
40	-33	-20	-6	7	14	21	28	35	42	49	56	63	77	91	105
45	-29	-15	-1	12	19	26	33	40	46	53	60	67	81	95	109
50	-25	-11	2	16	23	30	37	44	51	58	65	72	86	100	113
55	-20	-6	7	21	28	35	41	48	55	62	69	76	90	104	118
60	-16	-2	11	25	32	39	46	53	60	67	74	81	95	108	122
70	-7	6	20	34	41	48	55	62	69	76	83	90	103	117	131

Table 5-132

Separation, inch: 18

Leg Distance, inch: 60

SHIM REQUIRED FOR ALIGNMENT

Reverse Indicator Method

DriveN TIR MILS	DriveR TIR, MILS -50	-40	-30	-20	-15	-10	-5	0	5	10	15	20	30	40	50
-70	-165	-148	-131	-115	-106	-98	-90	-81	-73	-65	-56	-48	-31	-15	1
-60	-153	-136	-120	-103	-95	-86	-78	-70	-61	-53	-45	-36	-20	-3	13
-55	-147	-130	-114	-97	-89	-80	-72	-64	-55	-47	-39	-30	-14	2	19
-50	-141	-125	-108	-91	-83	-75	-66	-58	-50	-41	-33	-25	-8	8	25
-45	-135	-119	-102	-85	-77	-69	-60	-52	-44	-35	-27	-19	-2	14	30
-40	-130	-113	-96	-80	-71	-63	-55	-46	-38	-30	-21	-13	3	20	36
-35	-124	-107	-90	-74	-65	-57	-49	-40	-32	-24	-15	-7	9	25	42
-30	-118	-101	-85	-68	-60	-51	-43	-35	-26	-18	-10	-1	15	31	48
-25	-112	-95	-79	-62	-54	-45	-37	-29	-20	-12	-4	4	20	37	54
-20	-106	-90	-73	-56	-48	-40	-31	-23	-15	-6	1	10	26	43	60
-15	-100	-84	-67	-50	-42	-34	-25	-17	-9	0	7	15	32	49	65
-12	-97	-80	-64	-47	-39	-30	-22	-14	-5	2	11	19	36	52	69
-10	-95	-78	-61	-45	-36	-28	-20	-11	-3	5	13	21	38	55	71
-8	-92	-76	-59	-42	-34	-26	-17	-9	-1	7	15	24	40	57	74
-6	-90	-73	-57	-40	-32	-23	-15	-7	1	9	18	26	43	59	76
-4	-88	-71	-54	-38	-29	-21	-13	-4	3	12	20	28	45	62	78
-2	-85	-69	-52	-35	-27	-19	-10	-2	6	14	22	31	47	64	81
0	-83	-66	-50	-33	-25	-16	-8	0	8	16	25	33	50	66	83
2	-81	-64	-47	-31	-22	-14	-6	2	10	19	27	35	52	69	85
4	-78	-62	-45	-28	-20	-12	-3	4	13	21	29	38	54	71	88
6	-76	-59	-43	-26	-18	-9	-1	7	15	23	32	40	57	73	90
8	-74	-57	-40	-24	-15	-7	1	9	17	26	34	42	59	76	92
10	-71	-55	-38	-21	-13	-5	3	11	20	28	36	45	61	78	95
12	-69	-52	-36	-19	-11	-2	5	14	22	30	39	47	64	80	97
15	-65	-49	-32	-15	-7	0	9	17	25	34	42	50	67	84	100
20	-60	-43	-26	-10	-1	6	15	23	31	40	48	56	73	90	106
25	-54	-37	-20	-4	4	12	20	29	37	45	54	62	79	95	112
30	-48	-31	-15	1	10	18	26	35	43	51	60	68	85	101	118
35	-42	-25	-9	7	15	24	32	40	49	57	65	74	90	107	124
40	-36	-20	-3	13	21	30	38	46	55	63	71	80	96	113	130
45	-30	-14	2	19	27	35	44	52	60	69	77	85	102	119	135
50	-25	-8	8	25	33	41	50	58	66	75	83	91	108	125	141
55	-19	-2	14	30	39	47	55	64	72	80	89	97	114	130	147
60	-13	3	20	36	45	53	61	70	78	86	95	103	120	136	153
70	-1	15	31	48	56	65	73	81	90	98	106	115	131	148	165

Table 5-133

Separation, inch: 18

Leg Distance, inch: 70

SHIM REQUIRED FOR ALIGNMENT

Reverse Indicator Method

DriveN TIR MILS	DriveR TIR, MILS -50	-40	-30	-20	-15	-10	-5	0	5	10	15	20	30	40	50
-70	-198	-178	-159	-140	-130	-120	-110	-101	-91	-81	-71	-62	-42	-23	-3
-60	-183	-164	-145	-125	-115	-106	-96	-86	-76	-67	-57	-47	-28	-8	10
-55	-176	-157	-137	-118	-108	-98	-89	-79	-69	-60	-50	-40	-21	-1	17
-50	-169	-150	-130	-111	-101	-91	-81	-72	-62	-52	-43	-33	-13	5	25
-45	-162	-142	-123	-103	-94	-84	-74	-65	-55	-45	-35	-26	-6	12	32
-40	-155	-135	-116	-96	-86	-77	-67	-57	-48	-38	-28	-18	0	20	39
-35	-147	-128	-108	-89	-79	-70	-60	-50	-40	-31	-21	-11	7	27	46
-30	-140	-121	-101	-82	-72	-62	-53	-43	-33	-23	-14	-4	15	34	53
-25	-133	-113	-94	-75	-65	-55	-45	-36	-26	-16	-6	2	22	41	61
-20	-126	-106	-87	-67	-58	-48	-38	-28	-19	-9	0	10	29	48	68
-15	-118	-99	-80	-60	-50	-41	-31	-21	-11	-2	7	17	36	56	75
-12	-114	-95	-75	-56	-46	-36	-27	-17	-7	2	11	21	41	60	79
-10	-111	-92	-72	-53	-43	-33	-24	-14	-4	5	14	24	43	63	82
-8	-108	-89	-69	-50	-40	-31	-21	-11	-1	7	17	27	46	66	85
-6	-105	-86	-67	-47	-37	-28	-18	-8	1	10	20	30	49	69	88
-4	-103	-83	-64	-44	-34	-25	-15	-5	3	13	23	33	52	72	91
-2	-100	-80	-61	-41	-32	-22	-12	-2	6	16	26	36	55	74	94
0	-97	-77	-58	-38	-29	-19	-9	0	9	19	29	38	58	77	97
2	-94	-74	-55	-36	-26	-16	-6	2	12	22	32	41	61	80	100
4	-91	-72	-52	-33	-23	-13	-3	5	15	25	34	44	64	83	103
6	-88	-69	-49	-30	-20	-10	-1	8	18	28	37	47	67	86	105
8	-85	-66	-46	-27	-17	-7	1	11	21	31	40	50	69	89	108
10	-82	-63	-43	-24	-14	-5	4	14	24	33	43	53	72	92	111
12	-79	-60	-41	-21	-11	-2	7	17	27	36	46	56	75	95	114
15	-75	-56	-36	-17	-7	2	11	21	31	41	50	60	80	99	118
20	-68	-48	-29	-10	0	9	19	28	38	48	58	67	87	106	126
25	-61	-41	-22	-2	6	16	26	36	45	55	65	75	94	113	133
30	-53	-34	-15	4	14	23	33	43	53	62	72	82	101	121	140
35	-46	-27	-7	11	21	31	40	50	60	70	79	89	108	128	147
40	-39	-20	0	18	28	38	48	57	67	77	86	96	116	135	155
45	-32	-12	6	26	35	45	55	65	74	84	94	103	123	142	162
50	-25	-5	13	33	43	52	62	72	81	91	101	111	130	150	169
55	-17	1	21	40	50	60	69	79	89	98	108	118	137	157	176
60	-10	8	28	47	57	67	76	86	96	106	115	125	145	164	183
70	3	23	42	62	71	81	91	101	110	120	130	140	159	178	198

Table 5-134

Separation, inch: 18

Leg Distance, inch: 80

SHIM REQUIRED FOR ALIGNMENT

Reverse Indicator Method

DriveN TIR MILS	DriveR TIR, MILS -50	-40	-30	-20	-15	-10	-5	0	5	10	15	20	30	40	50
-70	-231	-209	-187	-165	-153	-142	-131	-120	-109	-98	-87	-76	-53	-31	-9
-60	-214	-192	-170	-147	-136	-125	-114	-103	-92	-81	-70	-58	-36	-14	7
-55	-205	-183	-161	-139	-128	-116	-105	-94	-83	-72	-61	-50	-28	-5	16
-50	-197	-175	-152	-130	-119	-108	-97	-86	-75	-63	-52	-41	-19	2	25
-45	-188	-166	-144	-121	-110	-99	-88	-77	-66	-55	-44	-33	-10	11	33
-40	-180	-157	-135	-113	-102	-91	-80	-68	-57	-46	-35	-24	-2	20	42
-35	-171	-149	-126	-104	-93	-82	-71	-60	-49	-38	-26	-15	6	28	50
-30	-162	-140	-118	-96	-85	-73	-62	-51	-40	-29	-18	-7	15	37	59
-25	-154	-131	-109	-87	-76	-65	-54	-43	-31	-20	-9	1	23	45	68
-20	-145	-123	-101	-78	-67	-56	-45	-34	-23	-12	-1	10	32	54	76
-15	-136	-114	-92	-70	-59	-48	-36	-25	-14	-3	7	18	40	63	85
-12	-131	-109	-87	-65	-54	-42	-31	-20	-9	1	12	23	46	68	90
-10	-128	-106	-83	-61	-50	-39	-28	-17	-6	5	16	27	49	71	93
-8	-124	-102	-80	-58	-47	-36	-24	-13	-2	8	19	30	52	75	97
-6	-121	-99	-77	-54	-43	-32	-21	-10	0	11	23	34	56	78	100
-4	-118	-95	-73	-51	-40	-29	-18	-6	4	15	26	37	59	82	104
-2	-114	-92	-70	-47	-36	-25	-14	-3	7	18	29	41	63	85	107
0	-111	-88	-66	-44	-33	-22	-11	0	11	22	33	44	66	88	111
2	-107	-85	-63	-41	-29	-18	-7	3	14	25	36	47	70	92	114
4	-104	-82	-59	-37	-26	-15	-4	6	18	29	40	51	73	95	118
6	-100	-78	-56	-34	-23	-11	0	10	21	32	43	54	77	99	121
8	-97	-75	-52	-30	-19	-8	2	13	24	36	47	58	80	102	124
10	-93	-71	-49	-27	-16	-5	6	17	28	39	50	61	83	106	128
12	-90	-68	-46	-23	-12	-1	9	20	31	42	54	65	87	109	131
15	-85	-63	-40	-18	-7	3	14	25	36	48	59	70	92	114	136
20	-76	-54	-32	-10	1	12	23	34	45	56	67	78	101	123	145
25	-68	-45	-23	-1	9	20	31	43	54	65	76	87	109	131	154
30	-59	-37	-15	7	18	29	40	51	62	73	85	96	118	140	162
35	-50	-28	-6	15	26	38	49	60	71	82	93	104	126	149	171
40	-42	-20	2	24	35	46	57	68	80	91	102	113	135	157	180
45	-33	-11	10	33	44	55	66	77	88	99	110	121	144	166	188
50	-25	-2	19	41	52	63	75	86	97	108	119	130	152	175	197
55	-16	5	28	50	61	72	83	94	105	116	128	139	161	183	205
60	-7	14	36	58	70	81	92	103	114	125	136	147	170	192	214
70	9	31	53	76	87	98	109	120	131	142	153	165	187	209	231

Table 5-135

Separation, inch: 18

Leg Distance, inch: 90

SHIM REQUIRED FOR ALIGNMENT

Reverse Indicator Method

DriveN TIR MILS	DriveR TIR, MILS -50	-40	-30	-20	-15	-10	-5	0	5	10	15	20	30	40	50
-70	-265	-240	-215	-190	-177	-165	-152	-140	-127	-115	-102	-90	-65	-40	-15
-60	-245	-220	-195	-170	-157	-145	-132	-120	-107	-95	-82	-70	-45	-20	5
-55	-235	-210	-185	-160	-147	-135	-122	-110	-97	-85	-72	-60	-35	-10	15
-50	-225	-200	-175	-150	-137	-125	-112	-100	-87	-75	-62	-50	-25	0	25
-45	-215	-190	-165	-140	-127	-115	-102	-90	-77	-65	-52	-40	-15	10	35
-40	-205	-180	-155	-130	-117	-105	-92	-80	-67	-55	-42	-30	-5	20	45
-35	-195	-170	-145	-120	-107	-95	-82	-70	-57	-45	-32	-20	5	30	55
-30	-185	-160	-135	-110	-97	-85	-72	-60	-47	-35	-22	-10	15	40	65
-25	-175	-150	-125	-100	-87	-75	-62	-50	-37	-25	-12	0	25	50	75
-20	-165	-140	-115	-90	-77	-65	-52	-40	-27	-15	-2	10	35	60	85
-15	-155	-130	-105	-80	-67	-55	-42	-30	-17	-5	7	20	45	70	95
-12	-149	-124	-99	-74	-61	-49	-36	-24	-11	1	13	26	51	76	101
-10	-145	-120	-95	-70	-57	-45	-32	-20	-7	5	17	30	55	80	105
-8	-141	-116	-91	-66	-53	-41	-28	-16	-3	9	21	34	59	84	109
-6	-137	-112	-87	-62	-49	-37	-24	-12	0	13	25	38	63	88	113
-4	-133	-108	-83	-58	-45	-33	-20	-8	4	17	29	42	67	92	117
-2	-129	-104	-79	-54	-41	-29	-16	-4	8	21	33	46	71	96	121
0	-125	-100	-75	-50	-37	-25	-12	0	12	25	37	50	75	100	125
2	-121	-96	-71	-46	-33	-21	-8	4	16	29	41	54	79	104	129
4	-117	-92	-67	-42	-29	-17	-4	8	20	33	45	58	83	108	133
6	-113	-88	-63	-38	-25	-13	0	12	24	37	49	62	87	112	137
8	-109	-84	-59	-34	-21	-9	3	16	28	41	53	66	91	116	141
10	-105	-80	-55	-30	-17	-5	7	20	32	45	57	70	95	120	145
12	-101	-76	-51	-26	-13	-1	11	24	36	49	61	74	99	124	149
15	-95	-70	-45	-20	-7	5	17	30	42	55	67	80	105	130	155
20	-85	-60	-35	-10	2	15	27	40	52	65	77	90	115	140	165
25	-75	-50	-25	0	12	25	37	50	62	75	87	100	125	150	175
30	-65	-40	-15	10	22	35	47	60	72	85	97	110	135	160	185
35	-55	-30	-5	20	32	45	57	70	82	95	107	120	145	170	195
40	-45	-20	5	30	42	55	67	80	92	105	117	130	155	180	205
45	-35	-10	15	40	52	65	77	90	102	115	127	140	165	190	215
50	-25	0	25	50	62	75	87	100	112	125	137	150	175	200	225
55	-15	10	35	60	72	85	97	110	122	135	147	160	185	210	235
60	-5	20	45	70	82	95	107	120	132	145	157	170	195	220	245
70	15	40	65	90	102	115	127	140	152	165	177	190	215	240	265

Table 5-136

Separation, inch: 18

Leg Distance, inch: 100

SHIM REQUIRED FOR ALIGNMENT

Reverse Indicator Method

DriveN TIR MILS	DriveR TIR, MILS -50	-40	-30	-20	-15	-10	-5	0	5	10	15	20	30	40	50
-70	-298	-270	-242	-215	-201	-187	-173	-159	-145	-131	-117	-103	-76	-48	-20
-60	-275	-247	-220	-192	-178	-164	-150	-136	-122	-108	-95	-81	-53	-25	2
-55	-264	-236	-208	-180	-166	-153	-139	-125	-111	-97	-83	-69	-41	-14	13
-50	-252	-225	-197	-169	-155	-141	-127	-113	-100	-86	-72	-58	-30	-2	25
-45	-241	-213	-185	-158	-144	-130	-116	-102	-88	-74	-60	-46	-19	8	36
-40	-230	-202	-174	-146	-132	-118	-105	-91	-77	-63	-49	-35	-7	20	47
-35	-218	-190	-163	-135	-121	-107	-93	-79	-65	-51	-38	-24	3	31	59
-30	-207	-179	-151	-123	-110	-96	-82	-68	-54	-40	-26	-12	15	42	70
-25	-195	-168	-140	-112	-98	-84	-70	-56	-43	-29	-15	-1	26	54	81
-20	-184	-156	-128	-101	-87	-73	-59	-45	-31	-17	-3	10	37	65	93
-15	-173	-145	-117	-89	-75	-61	-48	-34	-20	-6	7	21	49	76	104
-12	-166	-138	-110	-82	-69	-55	-41	-27	-13	0	14	28	56	83	111
-10	-161	-133	-106	-78	-64	-50	-36	-22	-8	5	18	32	60	88	116
-8	-157	-129	-101	-73	-59	-46	-32	-18	-4	9	23	37	65	92	120
-6	-152	-124	-97	-69	-55	-41	-27	-13	0	14	28	41	69	97	125
-4	-148	-120	-92	-64	-50	-36	-23	-9	4	18	32	46	74	102	129
-2	-143	-115	-87	-60	-46	-32	-18	-4	9	23	37	51	78	106	134
0	-138	-111	-83	-55	-41	-27	-13	0	13	27	41	55	83	111	138
2	-134	-106	-78	-51	-37	-23	-9	4	18	32	46	60	87	115	143
4	-129	-102	-74	-46	-32	-18	-4	9	23	36	50	64	92	120	148
6	-125	-97	-69	-41	-28	-14	0	13	27	41	55	69	97	124	152
8	-120	-92	-65	-37	-23	-9	4	18	32	46	59	73	101	129	157
10	-116	-88	-60	-32	-18	-5	8	22	36	50	64	78	106	133	161
12	-111	-83	-56	-28	-14	0	13	27	41	55	69	82	110	138	166
15	-104	-76	-49	-21	-7	6	20	34	48	61	75	89	117	145	173
20	-93	-65	-37	-10	3	17	31	45	59	73	87	101	128	156	184
25	-81	-54	-26	1	15	29	43	56	70	84	98	112	140	168	195
30	-70	-42	-15	12	26	40	54	68	82	96	110	123	151	179	207
35	-59	-31	-3	24	38	51	65	79	93	107	121	135	163	190	218
40	-47	-20	7	35	49	63	77	91	105	118	132	146	174	202	230
45	-36	-8	19	46	60	74	88	102	116	130	144	158	185	213	241
50	-25	2	30	58	72	86	100	113	127	141	155	169	197	225	252
55	-13	14	41	69	83	97	111	125	139	153	166	180	208	236	264
60	-2	25	53	81	95	108	122	136	150	164	178	192	220	247	275
70	20	48	76	103	117	131	145	159	173	187	201	215	242	270	298

Table 5-137

Separation, inch: 20

Leg Distance, inch: 25

SHIM REQUIRED FOR ALIGNMENT

Reverse Indicator Method

DriveN TIR MILS	DriveR TIR, MILS -50	-40	-30	-20	-15	-10	-5	0	5	10	15	20	30	40	50
-70	-40	-33	-27	-21	-18	-15	-11	-8	-5	-2	0	3	10	16	22
-60	-38	-32	-26	-20	-16	-13	-10	-7	-4	-1	1	5	11	17	23
-55	-38	-31	-25	-19	-16	-13	-10	-6	-3	0	2	5	11	18	24
-50	-37	-31	-25	-18	-15	-12	-9	-6	-3	0	3	6	12	18	25
-45	-36	-30	-24	-18	-15	-11	-8	-5	-2	0	3	6	13	19	25
-40	-36	-30	-23	-17	-14	-11	-8	-5	-1	1	4	7	13	20	26
-35	-35	-29	-23	-16	-13	-10	-7	-4	-1	1	5	8	14	20	26
-30	-35	-28	-22	-16	-13	-10	-6	-3	0	2	5	8	15	21	27
-25	-34	-28	-21	-15	-12	-9	-6	-3	0	3	6	9	15	21	28
-20	-33	-27	-21	-15	-11	-8	-5	-2	0	3	6	10	16	22	28
-15	-33	-26	-20	-14	-11	-8	-5	-1	1	4	7	10	16	23	29
-12	-32	-26	-20	-14	-10	-7	-4	-1	1	4	7	11	17	23	29
-10	-32	-26	-20	-13	-10	-7	-4	-1	1	5	8	11	17	23	30
-8	-32	-26	-19	-13	-10	-7	-4	-1	2	5	8	11	17	24	30
-6	-32	-25	-19	-13	-10	-7	-3	0	2	5	8	11	18	24	30
-4	-31	-25	-19	-13	-9	-6	-3	0	2	5	8	12	18	24	30
-2	-31	-25	-19	-12	-9	-6	-3	0	2	6	9	12	18	24	31
0	-31	-25	-18	-12	-9	-6	-3	0	3	6	9	12	18	25	31
2	-31	-24	-18	-12	-9	-6	-2	0	3	6	9	12	19	25	31
4	-30	-24	-18	-12	-8	-5	-2	0	3	6	9	13	19	25	31
6	-30	-24	-18	-11	-8	-5	-2	0	3	7	10	13	19	25	32
8	-30	-24	-17	-11	-8	-5	-2	1	4	7	10	13	19	26	32
10	-30	-23	-17	-11	-8	-5	-1	1	4	7	10	13	20	26	32
12	-29	-23	-17	-11	-7	-4	-1	1	4	7	10	14	20	26	32
15	-29	-23	-16	-10	-7	-4	-1	1	5	8	11	14	20	26	33
20	-28	-22	-16	-10	-6	-3	0	2	5	8	11	15	21	27	33
25	-28	-21	-15	-9	-6	-3	0	3	6	9	12	15	21	28	34
30	-27	-21	-15	-8	-5	-2	0	3	6	10	13	16	22	28	35
35	-26	-20	-14	-8	-5	-1	1	4	7	10	13	16	23	29	35
40	-26	-20	-13	-7	-4	-1	1	5	8	11	14	17	23	30	36
45	-25	-19	-13	-6	-3	0	2	5	8	11	15	18	24	30	36
50	-25	-18	-12	-6	-3	0	3	6	9	12	15	18	25	31	37
55	-24	-18	-11	-5	-2	0	3	6	10	13	16	19	25	31	38
60	-23	-17	-11	-5	-1	1	4	7	10	13	16	20	26	32	38
70	-22	-16	-10	-3	0	2	5	8	11	15	18	21	27	33	40

Table 5-138

Separation, inch: 20

Leg Distance, inch: 30

SHIM REQUIRED FOR ALIGNMENT

Reverse Indicator Method

DriveN TIR MILS	DriveR TIR, MILS -50	-40	-30	-20	-15	-10	-5	0	5	10	15	20	30	40	50
-70	-55	-47	-40	-32	-28	-25	-21	-17	-13	-10	-6	-2	5	12	20
-60	-52	-45	-37	-30	-26	-22	-18	-15	-11	-7	-3	0	7	15	22
-55	-51	-43	-36	-28	-25	-21	-17	-13	-10	-6	-2	1	8	16	23
-50	-50	-42	-35	-27	-23	-20	-16	-12	-8	-5	-1	2	10	17	25
-45	-48	-41	-33	-26	-22	-18	-15	-11	-7	-3	0	3	11	18	26
-40	-47	-40	-32	-25	-21	-17	-13	-10	-6	-2	1	5	12	20	27
-35	-46	-38	-31	-23	-20	-16	-12	-8	-5	-1	2	6	13	21	28
-30	-45	-37	-30	-22	-18	-15	-11	-7	-3	0	3	7	15	22	30
-25	-43	-36	-28	-21	-17	-13	-10	-6	-2	1	5	8	16	23	31
-20	-42	-35	-27	-20	-16	-12	-8	-5	-1	2	6	10	17	25	32
-15	-41	-33	-26	-18	-15	-11	-7	-3	0	3	7	11	18	26	33
-12	-40	-33	-25	-18	-14	-10	-6	-3	0	4	8	12	19	27	34
-10	-40	-32	-25	-17	-13	-10	-6	-2	1	5	8	12	20	27	35
-8	-39	-32	-24	-17	-13	-9	-5	-2	1	5	9	13	20	28	35
-6	-39	-31	-24	-16	-12	-9	-5	-1	2	6	9	13	21	28	36
-4	-38	-31	-23	-16	-12	-8	-4	-1	2	6	10	14	21	29	36
-2	-38	-30	-23	-15	-11	-8	-4	0	3	7	10	14	22	29	37
0	-37	-30	-22	-15	-11	-7	-3	0	3	7	11	15	22	30	37
2	-37	-29	-22	-14	-10	-7	-3	0	4	8	11	15	23	30	38
4	-36	-29	-21	-14	-10	-6	-2	1	4	8	12	16	23	31	38
6	-36	-28	-21	-13	-9	-6	-2	1	5	9	12	16	24	31	39
8	-35	-28	-20	-13	-9	-5	-1	2	5	9	13	17	24	32	39
10	-35	-27	-20	-12	-8	-5	-1	2	6	10	13	17	25	32	40
12	-34	-27	-19	-12	-8	-4	0	3	6	10	14	18	25	33	40
15	-33	-26	-18	-11	-7	-3	0	3	7	11	15	18	26	33	41
20	-32	-25	-17	-10	-6	-2	1	5	8	12	16	20	27	35	42
25	-31	-23	-16	-8	-5	-1	2	6	10	13	17	21	28	36	43
30	-30	-22	-15	-7	-3	0	3	7	11	15	18	22	30	37	45
35	-28	-21	-13	-6	-2	1	5	8	12	16	20	23	31	38	46
40	-27	-20	-12	-5	-1	2	6	10	13	17	21	25	32	40	47
45	-26	-18	-11	-3	0	3	7	11	15	18	22	26	33	41	48
50	-25	-17	-10	-2	1	5	8	12	16	20	23	27	35	42	50
55	-23	-16	-8	-1	2	6	10	13	17	21	25	28	36	43	51
60	-22	-15	-7	0	3	7	11	15	18	22	26	30	37	45	52
70	-20	-12	-5	2	6	10	13	17	21	25	28	32	40	47	55

Table 5-139

Separation, inch: 20

Leg Distance, inch: 35

SHIM REQUIRED FOR ALIGNMENT

Reverse Indicator Method

DriveN TIR MILS	DriveR TIR, MILS -50	-40	-30	-20	-15	-10	-5	0	5	10	15	20	30	40	50
-70	-70	-61	-52	-43	-39	-35	-30	-26	-21	-17	-13	-8	0	8	17
-60	-66	-57	-48	-40	-35	-31	-26	-22	-18	-13	-9	-5	3	12	21
-55	-64	-55	-46	-38	-33	-29	-25	-20	-16	-11	-7	-3	5	14	23
-50	-62	-53	-45	-36	-31	-27	-23	-18	-14	-10	-5	-1	7	16	25
-45	-60	-51	-43	-34	-30	-25	-21	-16	-12	-8	-3	0	9	18	26
-40	-58	-50	-41	-32	-28	-23	-19	-15	-10	-6	-1	2	11	20	28
-35	-56	-48	-39	-30	-26	-21	-17	-13	-8	-4	0	4	13	21	30
-30	-55	-46	-37	-28	-24	-20	-15	-11	-6	-2	1	6	15	23	32
-25	-53	-44	-35	-26	-22	-18	-13	-9	-5	0	3	8	16	25	34
-20	-51	-42	-33	-25	-20	-16	-11	-7	-3	1	5	10	18	27	36
-15	-49	-40	-31	-23	-18	-14	-10	-5	-1	3	7	11	20	29	38
-12	-48	-39	-30	-22	-17	-13	-8	-4	0	4	8	13	21	30	39
-10	-47	-38	-30	-21	-16	-12	-8	-3	0	5	9	13	22	31	40
-8	-46	-38	-29	-20	-16	-11	-7	-3	1	5	10	14	23	32	40
-6	-46	-37	-28	-19	-15	-11	-6	-2	2	6	10	15	24	32	41
-4	-45	-36	-27	-19	-14	-10	-5	-1	2	7	11	16	24	33	42
-2	-44	-35	-27	-18	-13	-9	-5	0	3	8	12	16	25	34	43
0	-43	-35	-26	-17	-13	-8	-4	0	4	8	13	17	26	35	43
2	-43	-34	-25	-16	-12	-8	-3	0	5	9	13	18	27	35	44
4	-42	-33	-24	-16	-11	-7	-2	1	5	10	14	19	27	36	45
6	-41	-32	-24	-15	-10	-6	-2	2	6	11	15	19	28	37	46
8	-40	-32	-23	-14	-10	-5	-1	3	7	11	16	20	29	38	46
10	-40	-31	-22	-13	-9	-5	0	3	8	12	16	21	30	38	47
12	-39	-30	-21	-13	-8	-4	0	4	8	13	17	22	30	39	48
15	-38	-29	-20	-11	-7	-3	1	5	10	14	18	23	31	40	49
20	-36	-27	-18	-10	-5	-1	3	7	11	16	20	25	33	42	51
25	-34	-25	-16	-8	-3	0	5	9	13	18	22	26	35	44	53
30	-32	-23	-15	-6	-1	2	6	11	15	20	24	28	37	46	55
35	-30	-21	-13	-4	0	4	8	13	17	21	26	30	39	48	56
40	-28	-20	-11	-2	1	6	10	15	19	23	28	32	41	50	58
45	-26	-18	-9	0	3	8	12	16	21	25	30	34	43	51	60
50	-25	-16	-7	1	5	10	14	18	23	27	31	36	45	53	62
55	-23	-14	-5	3	7	11	16	20	25	29	33	38	46	55	64
60	-21	-12	-3	5	9	13	18	22	26	31	35	40	48	57	66
70	-17	-8	0	8	13	17	21	26	30	35	39	43	52	61	70

Table 5-140

Separation, inch: 20

Leg Distance, inch: 40

SHIM REQUIRED FOR ALIGNMENT

Reverse Indicator Method

DriveN TIR MILS	DriveR TIR, MILS -50	-40	-30	-20	-15	-10	-5	0	5	10	15	20	30	40	50
-70	-85	-75	-65	-55	-50	-45	-40	-35	-30	-25	-20	-15	-5	5	15
-60	-80	-70	-60	-50	-45	-40	-35	-30	-25	-20	-15	-10	0	10	20
-55	-77	-67	-57	-47	-42	-37	-32	-27	-22	-17	-12	-7	2	12	22
-50	-75	-65	-55	-45	-40	-35	-30	-25	-20	-15	-10	-5	5	15	25
-45	-72	-62	-52	-42	-37	-32	-27	-22	-17	-12	-7	-2	7	17	27
-40	-70	-60	-50	-40	-35	-30	-25	-20	-15	-10	-5	0	10	20	30
-35	-67	-57	-47	-37	-32	-27	-22	-17	-12	-7	-2	2	12	22	32
-30	-65	-55	-45	-35	-30	-25	-20	-15	-10	-5	0	5	15	25	35
-25	-62	-52	-42	-32	-27	-22	-17	-12	-7	-2	2	7	17	27	37
-20	-60	-50	-40	-30	-25	-20	-15	-10	-5	0	5	10	20	30	40
-15	-57	-47	-37	-27	-22	-17	-12	-7	-2	2	7	12	22	32	42
-12	-56	-46	-36	-26	-21	-16	-11	-6	-1	4	9	14	24	34	44
-10	-55	-45	-35	-25	-20	-15	-10	-5	0	5	10	15	25	35	45
-8	-54	-44	-34	-24	-19	-14	-9	-4	1	6	11	16	26	36	46
-6	-53	-43	-33	-23	-18	-13	-8	-3	2	7	12	17	27	37	47
-4	-52	-42	-32	-22	-17	-12	-7	-2	3	8	13	18	28	38	48
-2	-51	-41	-31	-21	-16	-11	-6	-1	4	9	14	19	29	39	49
0	-50	-40	-30	-20	-15	-10	-5	0	5	10	15	20	30	40	50
2	-49	-39	-29	-19	-14	-9	-4	1	6	11	16	21	31	41	51
4	-48	-38	-28	-18	-13	-8	-3	2	7	12	17	22	32	42	52
6	-47	-37	-27	-17	-12	-7	-2	3	8	13	18	23	33	43	53
8	-46	-36	-26	-16	-11	-6	-1	4	9	14	19	24	34	44	54
10	-45	-35	-25	-15	-10	-5	0	5	10	15	20	25	35	45	55
12	-44	-34	-24	-14	-9	-4	1	6	11	16	21	26	36	46	56
15	-42	-32	-22	-12	-7	-2	2	7	12	17	22	27	37	47	57
20	-40	-30	-20	-10	-5	0	5	10	15	20	25	30	40	50	60
25	-37	-27	-17	-7	-2	2	7	12	17	22	27	32	42	52	62
30	-35	-25	-15	-5	0	5	10	15	20	25	30	35	45	55	65
35	-32	-22	-12	-2	2	7	12	17	22	27	32	37	47	57	67
40	-30	-20	-10	0	5	10	15	20	25	30	35	40	50	60	70
45	-27	-17	-7	2	7	12	17	22	27	32	37	42	52	62	72
50	-25	-15	-5	5	10	15	20	25	30	35	40	45	55	65	75
55	-22	-12	-2	7	12	17	22	27	32	37	42	47	57	67	77
60	-20	-10	0	10	15	20	25	30	35	40	45	50	60	70	80
70	-15	-5	5	15	20	25	30	35	40	45	50	55	65	75	85

Table 5-141

Separation, inch: 20

Leg Distance, inch: 50

SHIM REQUIRED FOR ALIGNMENT

Reverse Indicator Method

DriveN TIR MILS	DriveR TIR, MILS -50	-40	-30	-20	-15	-10	-5	0	5	10	15	20	30	40	50
-70	-115	-102	-90	-77	-71	-65	-58	-52	-46	-40	-33	-27	-15	-2	10
-60	-107	-95	-82	-70	-63	-57	-51	-45	-38	-32	-26	-20	-7	5	17
-55	-103	-91	-78	-66	-60	-53	-47	-41	-35	-28	-22	-16	-3	8	21
-50	-100	-87	-75	-62	-56	-50	-43	-37	-31	-25	-18	-12	0	12	25
-45	-96	-83	-71	-58	-52	-46	-40	-33	-27	-21	-15	-8	3	16	28
-40	-92	-80	-67	-55	-48	-42	-36	-30	-23	-17	-11	-5	7	20	32
-35	-88	-76	-63	-51	-45	-38	-32	-26	-20	-13	-7	-1	11	23	36
-30	-85	-72	-60	-47	-41	-35	-28	-22	-16	-10	-3	2	15	27	40
-25	-81	-68	-56	-43	-37	-31	-25	-18	-12	-6	0	6	18	31	43
-20	-77	-65	-52	-40	-33	-27	-21	-15	-8	-2	3	10	22	35	47
-15	-73	-61	-48	-36	-30	-23	-17	-11	-5	1	7	13	26	38	51
-12	-71	-59	-46	-34	-27	-21	-15	-9	-2	3	9	16	28	41	53
-10	-70	-57	-45	-32	-26	-20	-13	-7	-1	5	11	17	30	42	55
-8	-68	-56	-43	-31	-24	-18	-12	-6	0	6	12	19	31	44	56
-6	-67	-54	-42	-29	-23	-17	-10	-4	1	8	14	20	33	45	58
-4	-65	-53	-40	-28	-21	-15	-9	-3	3	9	15	22	34	47	59
-2	-64	-51	-39	-26	-20	-14	-7	-1	4	11	17	23	36	48	61
0	-62	-50	-37	-25	-18	-12	-6	0	6	12	18	25	37	50	62
2	-61	-48	-36	-23	-17	-11	-4	1	7	14	20	26	39	51	64
4	-59	-47	-34	-22	-15	-9	-3	3	9	15	21	28	40	53	65
6	-58	-45	-33	-20	-14	-8	-1	4	10	17	23	29	42	54	67
8	-56	-44	-31	-19	-12	-6	0	6	12	18	24	31	43	56	68
10	-55	-42	-30	-17	-11	-5	1	7	13	20	26	32	45	57	70
12	-53	-41	-28	-16	-9	-3	2	9	15	21	27	34	46	59	71
15	-51	-38	-26	-13	-7	-1	5	11	17	23	30	36	48	61	73
20	-47	-35	-22	-10	-3	2	8	15	21	27	33	40	52	65	77
25	-43	-31	-18	-6	0	6	12	18	25	31	37	43	56	68	81
30	-40	-27	-15	-2	3	10	16	22	28	35	41	47	60	72	85
35	-36	-23	-11	1	7	13	20	26	32	38	45	51	63	76	88
40	-32	-20	-7	5	11	17	23	30	36	42	48	55	67	80	92
45	-28	-16	-3	8	15	21	27	33	40	46	52	58	71	83	96
50	-25	-12	0	12	18	25	31	37	43	50	56	62	75	87	100
55	-21	-8	3	16	22	28	35	41	47	53	60	66	78	91	103
60	-17	-5	7	20	26	32	38	45	51	57	63	70	82	95	107
70	-10	2	15	27	33	40	46	52	58	65	71	77	90	102	115

Table 5-142

Separation, inch: 20

Leg Distance, inch: 60

SHIM REQUIRED FOR ALIGNMENT

Reverse Indicator Method

DriveN TIR MILS	DriveR TIR, MILS -50	-40	-30	-20	-15	-10	-5	0	5	10	15	20	30	40	50
-70	-145	-130	-115	-100	-92	-85	-77	-70	-62	-55	-47	-40	-25	-10	5
-60	-135	-120	-105	-90	-82	-75	-67	-60	-52	-45	-37	-30	-15	0	15
-55	-130	-115	-100	-85	-77	-70	-62	-55	-47	-40	-32	-25	-10	5	20
-50	-125	-110	-95	-80	-72	-65	-57	-50	-42	-35	-27	-20	-5	10	25
-45	-120	-105	-90	-75	-67	-60	-52	-45	-37	-30	-22	-15	0	15	30
-40	-115	-100	-85	-70	-62	-55	-47	-40	-32	-25	-17	-10	5	20	35
-35	-110	-95	-80	-65	-57	-50	-42	-35	-27	-20	-12	-5	10	25	40
-30	-105	-90	-75	-60	-52	-45	-37	-30	-22	-15	-7	0	15	30	45
-25	-100	-85	-70	-55	-47	-40	-32	-25	-17	-10	-2	5	20	35	50
-20	-95	-80	-65	-50	-42	-35	-27	-20	-12	-5	2	10	25	40	55
-15	-90	-75	-60	-45	-37	-30	-22	-15	-7	0	7	15	30	45	60
-12	-87	-72	-57	-42	-34	-27	-19	-12	-4	3	10	18	33	48	63
-10	-85	-70	-55	-40	-32	-25	-17	-10	-2	5	12	20	35	50	65
-8	-83	-68	-53	-38	-30	-23	-15	-8	0	7	14	22	37	52	67
-6	-81	-66	-51	-36	-28	-21	-13	-6	1	9	16	24	39	54	69
-4	-79	-64	-49	-34	-26	-19	-11	-4	3	11	18	26	41	56	71
-2	-77	-62	-47	-32	-24	-17	-9	-2	5	13	20	28	43	58	73
0	-75	-60	-45	-30	-22	-15	-7	0	7	15	22	30	45	60	75
2	-73	-58	-43	-28	-20	-13	-5	2	9	17	24	32	47	62	77
4	-71	-56	-41	-26	-18	-11	-3	4	11	19	26	34	49	64	79
6	-69	-54	-39	-24	-16	-9	-1	6	13	21	28	36	51	66	81
8	-67	-52	-37	-22	-14	-7	0	8	15	23	30	38	53	68	83
10	-65	-50	-35	-20	-12	-5	2	10	17	25	32	40	55	70	85
12	-63	-48	-33	-18	-10	-3	4	12	19	27	34	42	57	72	87
15	-60	-45	-30	-15	-7	0	7	15	22	30	37	45	60	75	90
20	-55	-40	-25	-10	-2	5	12	20	27	35	42	50	65	80	95
25	-50	-35	-20	-5	2	10	17	25	32	40	47	55	70	85	100
30	-45	-30	-15	0	7	15	22	30	37	45	52	60	75	90	105
35	-40	-25	-10	5	12	20	27	35	42	50	57	65	80	95	110
40	-35	-20	-5	10	17	25	32	40	47	55	62	70	85	100	115
45	-30	-15	0	15	22	30	37	45	52	60	67	75	90	105	120
50	-25	-10	5	20	27	35	42	50	57	65	72	80	95	110	125
55	-20	-5	10	25	32	40	47	55	62	70	77	85	100	115	130
60	-15	0	15	30	37	45	52	60	67	75	82	90	105	120	135
70	-5	10	25	40	47	55	62	70	77	85	92	100	115	130	145

Table 5-143

Separation, inch: 20

Leg Distance, inch: 70

SHIM REQUIRED FOR ALIGNMENT

Reverse Indicator Method

riveN IR ILS	DriveR TIR, MILS														
	-50	-40	-30	-20	-15	-10	-5	0	5	10	15	20	30	40	50
-70	-175	-157	-140	-122	-113	-105	-96	-87	-78	-70	-61	-52	-35	-17	0
-60	-162	-145	-127	-110	-101	-92	-83	-75	-66	-57	-48	-40	-22	-5	12
-55	-156	-138	-121	-103	-95	-86	-77	-68	-60	-51	-42	-33	-16	1	18
-50	-150	-132	-115	-97	-88	-80	-71	-62	-53	-45	-36	-27	-10	7	25
-45	-143	-126	-108	-91	-82	-73	-65	-56	-47	-38	-30	-21	-3	13	31
-40	-137	-120	-102	-85	-76	-67	-58	-50	-41	-32	-23	-15	2	20	37
-35	-131	-113	-96	-78	-70	-61	-52	-43	-35	-26	-17	-8	8	26	43
-30	-125	-107	-90	-72	-63	-55	-46	-37	-28	-20	-11	-2	15	32	50
-25	-118	-101	-83	-66	-57	-48	-40	-31	-22	-13	-5	3	21	38	56
-20	-112	-95	-77	-60	-51	-42	-33	-25	-16	-7	1	10	27	45	62
-15	-106	-88	-71	-53	-45	-36	-27	-18	-10	-1	7	16	33	51	68
-12	-102	-85	-67	-50	-41	-32	-23	-15	-6	2	11	20	37	55	72
-10	-100	-82	-65	-47	-38	-30	-21	-12	-3	5	13	22	40	57	75
-8	-97	-80	-62	-45	-36	-27	-18	-10	-1	7	16	25	42	60	77
-6	-95	-77	-60	-42	-33	-25	-16	-7	1	10	18	27	45	62	80
-4	-92	-75	-57	-40	-31	-22	-13	-5	3	12	21	30	47	65	82
-2	-90	-72	-55	-37	-28	-20	-11	-2	6	15	23	32	50	67	85
0	-87	-70	-52	-35	-26	-17	-8	0	8	17	26	35	52	70	87
2	-85	-67	-50	-32	-23	-15	-6	2	11	20	28	37	55	72	90
4	-82	-65	-47	-30	-21	-12	-3	5	13	22	31	40	57	75	92
6	-80	-62	-45	-27	-18	-10	-1	7	16	25	33	42	60	77	95
8	-77	-60	-42	-25	-16	-7	1	10	18	27	36	45	62	80	97
10	-75	-57	-40	-22	-13	-5	3	12	21	30	38	47	65	82	100
12	-72	-55	-37	-20	-11	-2	6	15	23	32	41	50	67	85	102
15	-68	-51	-33	-16	-7	1	10	18	27	36	45	53	71	88	106
20	-62	-45	-27	-10	-1	7	16	25	33	42	51	60	77	95	112
25	-56	-38	-21	-3	5	13	22	31	40	48	57	66	83	101	118
30	-50	-32	-15	2	11	20	28	37	46	55	63	72	90	107	125
35	-43	-26	-8	8	17	26	35	43	52	61	70	78	96	113	131
40	-37	-20	-2	15	23	32	41	50	58	67	76	85	102	120	137
45	-31	-13	3	21	30	38	47	56	65	73	82	91	108	126	143
50	-25	-7	10	27	36	45	53	62	71	80	88	97	115	132	150
55	-18	-1	16	33	42	51	60	68	77	86	95	103	121	138	156
60	-12	5	22	40	48	57	66	75	83	92	101	110	127	145	162
70	0	17	35	52	61	70	78	87	96	105	113	122	140	157	175

Table 5-144

Separation, inch: 20

Leg Distance, inch: 80

SHIM REQUIRED FOR ALIGNMENT

Reverse Indicator Method

DriveN TIR MILS	DriveR TIR, MILS -50	-40	-30	-20	-15	-10	-5	0	5	10	15	20	30	40	50
-70	-205	-185	-165	-145	-135	-125	-115	-105	-95	-85	-75	-65	-45	-25	-5
-60	-190	-170	-150	-130	-120	-110	-100	-90	-80	-70	-60	-50	-30	-10	10
-55	-182	-162	-142	-122	-112	-102	-92	-82	-72	-62	-52	-42	-22	-2	17
-50	-175	-155	-135	-115	-105	-95	-85	-75	-65	-55	-45	-35	-15	5	25
-45	-167	-147	-127	-107	-97	-87	-77	-67	-57	-47	-37	-27	-7	12	32
-40	-160	-140	-120	-100	-90	-80	-70	-60	-50	-40	-30	-20	0	20	40
-35	-152	-132	-112	-92	-82	-72	-62	-52	-42	-32	-22	-12	7	27	47
-30	-145	-125	-105	-85	-75	-65	-55	-45	-35	-25	-15	-5	15	35	55
-25	-137	-117	-97	-77	-67	-57	-47	-37	-27	-17	-7	2	22	42	62
-20	-130	-110	-90	-70	-60	-50	-40	-30	-20	-10	0	10	30	50	70
-15	-122	-102	-82	-62	-52	-42	-32	-22	-12	-2	7	17	37	57	77
-12	-118	-98	-78	-58	-48	-38	-28	-18	-8	2	12	22	42	62	82
-10	-115	-95	-75	-55	-45	-35	-25	-15	-5	5	15	25	45	65	85
-8	-112	-92	-72	-52	-42	-32	-22	-12	-2	8	18	28	48	68	88
-6	-109	-89	-69	-49	-39	-29	-19	-9	1	11	21	31	51	71	91
-4	-106	-86	-66	-46	-36	-26	-16	-6	4	14	24	34	54	74	94
-2	-103	-83	-63	-43	-33	-23	-13	-3	7	17	27	37	57	77	97
0	-100	-80	-60	-40	-30	-20	-10	0	10	20	30	40	60	80	100
2	-97	-77	-57	-37	-27	-17	-7	3	13	23	33	43	63	83	103
4	-94	-74	-54	-34	-24	-14	-4	6	16	26	36	46	66	86	106
6	-91	-71	-51	-31	-21	-11	-1	9	19	29	39	49	69	89	109
8	-88	-68	-48	-28	-18	-8	2	12	22	32	42	52	72	92	112
10	-85	-65	-45	-25	-15	-5	5	15	25	35	45	55	75	95	115
12	-82	-62	-42	-22	-12	-2	8	18	28	38	48	58	78	98	118
15	-77	-57	-37	-17	-7	2	12	22	32	42	52	62	82	102	122
20	-70	-50	-30	-10	0	10	20	30	40	50	60	70	90	110	130
25	-62	-42	-22	-2	7	17	27	37	47	57	67	77	97	117	137
30	-55	-35	-15	5	15	25	35	45	55	65	75	85	105	125	145
35	-47	-27	-7	12	22	32	42	52	62	72	82	92	112	132	152
40	-40	-20	0	20	30	40	50	60	70	80	90	100	120	140	160
45	-32	-12	7	27	37	47	57	67	77	87	97	107	127	147	167
50	-25	-5	15	35	45	55	65	75	85	95	105	115	135	155	175
55	-17	2	22	42	52	62	72	82	92	102	112	122	142	162	182
60	-10	10	30	50	60	70	80	90	100	110	120	130	150	170	190
70	5	25	45	65	75	85	95	105	115	125	135	145	165	185	205

Table 5-145

Separation, inch: 20

Leg Distance, inch: 90

SHIM REQUIRED FOR ALIGNMENT

Reverse Indicator Method

riveN IR ILS	DriveR TIR, MILS -50	-40	-30	-20	-15	-10	-5	0	5	10	15	20	30	40	50
-70	-235	-212	-190	-167	-156	-145	-133	-122	-111	-100	-88	-77	-55	-32	-10
-60	-217	-195	-172	-150	-138	-127	-116	-105	-93	-82	-71	-60	-37	-15	7
-55	-208	-186	-163	-141	-130	-118	-107	-96	-85	-73	-62	-51	-28	-6	16
-50	-200	-177	-155	-132	-121	-110	-98	-87	-76	-65	-53	-42	-20	2	25
-45	-191	-168	-146	-123	-112	-101	-90	-78	-67	-56	-45	-33	-11	11	33
-40	-182	-160	-137	-115	-103	-92	-81	-70	-58	-47	-36	-25	-2	20	42
-35	-173	-151	-128	-106	-95	-83	-72	-61	-50	-38	-27	-16	6	28	51
-30	-165	-142	-120	-97	-86	-75	-63	-52	-41	-30	-18	-7	15	37	60
-25	-156	-133	-111	-88	-77	-66	-55	-43	-32	-21	-10	1	23	46	68
-20	-147	-125	-102	-80	-68	-57	-46	-35	-23	-12	-1	10	32	55	77
-15	-138	-116	-93	-71	-60	-48	-37	-26	-15	-3	7	18	41	63	86
-12	-133	-111	-88	-66	-54	-43	-32	-21	-9	1	12	24	46	69	91
-10	-130	-107	-85	-62	-51	-40	-28	-17	-6	5	16	27	50	72	95
-8	-126	-104	-81	-59	-47	-36	-25	-14	-2	8	19	31	53	76	98
-6	-123	-100	-78	-55	-44	-33	-21	-10	0	12	23	34	57	79	102
-4	-119	-97	-74	-52	-40	-29	-18	-7	4	15	26	38	60	83	105
-2	-116	-93	-71	-48	-37	-26	-14	-3	7	19	30	41	64	86	109
0	-112	-90	-67	-45	-33	-22	-11	0	11	22	33	45	67	90	112
2	-109	-86	-64	-41	-30	-19	-7	3	14	26	37	48	71	93	116
4	-105	-83	-60	-38	-26	-15	-4	7	18	29	40	52	74	97	119
6	-102	-79	-57	-34	-23	-12	0	10	21	33	44	55	78	100	123
8	-98	-76	-53	-31	-19	-8	2	14	25	36	47	59	81	104	126
10	-95	-72	-50	-27	-16	-5	6	17	28	40	51	62	85	107	130
12	-91	-69	-46	-24	-12	-1	9	21	32	43	54	66	88	111	133
15	-86	-63	-41	-18	-7	3	15	26	37	48	60	71	93	116	138
20	-77	-55	-32	-10	1	12	23	35	46	57	68	80	102	125	147
25	-68	-46	-23	-1	10	21	32	43	55	66	77	88	111	133	156
30	-60	-37	-15	7	18	30	41	52	63	75	86	97	120	142	165
35	-51	-28	-6	16	27	38	50	61	72	83	95	106	128	151	173
40	-42	-20	2	25	36	47	58	70	81	92	103	115	137	160	182
45	-33	-11	11	33	45	56	67	78	90	101	112	123	146	168	191
50	-25	-2	20	42	53	65	76	87	98	110	121	132	155	177	200
55	-16	6	28	51	62	73	85	96	107	118	130	141	163	186	208
60	-7	15	37	60	71	82	93	105	116	127	138	150	172	195	217
70	10	32	55	77	88	100	111	122	133	145	156	167	190	212	235

Table 5-146

Separation, inch: 20

Leg Distance, inch: 100

SHIM REQUIRED FOR ALIGNMENT

Reverse Indicator Method

DriveN TIR MILS	DriveR TIR, MILS -50	-40	-30	-20	-15	-10	-5	0	5	10	15	20	30	40	50
-70	-265	-240	-215	-190	-177	-165	-152	-140	-127	-115	-102	-90	-65	-40	-15
-60	-245	-220	-195	-170	-157	-145	-132	-120	-107	-95	-82	-70	-45	-20	5
-55	-235	-210	-185	-160	-147	-135	-122	-110	-97	-85	-72	-60	-35	-10	15
-50	-225	-200	-175	-150	-137	-125	-112	-100	-87	-75	-62	-50	-25	0	25
-45	-215	-190	-165	-140	-127	-115	-102	-90	-77	-65	-52	-40	-15	10	35
-40	-205	-180	-155	-130	-117	-105	-92	-80	-67	-55	-42	-30	-5	20	45
-35	-195	-170	-145	-120	-107	-95	-82	-70	-57	-45	-32	-20	5	30	55
-30	-185	-160	-135	-110	-97	-85	-72	-60	-47	-35	-22	-10	15	40	65
-25	-175	-150	-125	-100	-87	-75	-62	-50	-37	-25	-12	0	25	50	75
-20	-165	-140	-115	-90	-77	-65	-52	-40	-27	-15	-2	10	35	60	85
-15	-155	-130	-105	-80	-67	-55	-42	-30	-17	-5	7	20	45	70	95
-12	-149	-124	-99	-74	-61	-49	-36	-24	-11	1	13	26	51	76	101
-10	-145	-120	-95	-70	-57	-45	-32	-20	-7	5	17	30	55	80	105
-8	-141	-116	-91	-66	-53	-41	-28	-16	-3	9	21	34	59	84	109
-6	-137	-112	-87	-62	-49	-37	-24	-12	0	13	25	38	63	88	113
-4	-133	-108	-83	-58	-45	-33	-20	-8	4	17	29	42	67	92	117
-2	-129	-104	-79	-54	-41	-29	-16	-4	8	21	33	46	71	96	121
0	-125	-100	-75	-50	-37	-25	-12	0	12	25	37	50	75	100	125
2	-121	-96	-71	-46	-33	-21	-8	4	16	29	41	54	79	104	129
4	-117	-92	-67	-42	-29	-17	-4	8	20	33	45	58	83	108	133
6	-113	-88	-63	-38	-25	-13	0	12	24	37	49	62	87	112	137
8	-109	-84	-59	-34	-21	-9	3	16	28	41	53	66	91	116	141
10	-105	-80	-55	-30	-17	-5	7	20	32	45	57	70	95	120	145
12	-101	-76	-51	-26	-13	-1	11	24	36	49	61	74	99	124	149
15	-95	-70	-45	-20	-7	5	17	30	42	55	67	80	105	130	155
20	-85	-60	-35	-10	2	15	27	40	52	65	77	90	115	140	165
25	-75	-50	-25	0	12	25	37	50	62	75	87	100	125	150	175
30	-65	-40	-15	10	22	35	47	60	72	85	97	110	135	160	185
35	-55	-30	-5	20	32	45	57	70	82	95	107	120	145	170	195
40	-45	-20	5	30	42	55	67	80	92	105	117	130	155	180	205
45	-35	-10	15	40	52	65	77	90	102	115	127	140	165	190	215
50	-25	0	25	50	62	75	87	100	112	125	137	150	175	200	225
55	-15	10	35	60	72	85	97	110	122	135	147	160	185	210	235
60	-5	20	45	70	82	95	107	120	132	145	157	170	195	220	245
70	15	40	65	90	102	115	127	140	152	165	177	190	215	240	265

References

Because of the importance of alignment, much has been written on the subject. The following references are not necessarily in agreement, but do make interesting reading.

1. Dodd, V. R., "Shaft Alignment Monitoring Cuts Costs," *The Oil and Gas Journal,* Sept. 25, 1972, pp. 91–96.
2. Flexibox Limited, "Coupling Alignment Calculator," 1976.
3. *High Performance Couplings,* Koppers Company, Inc., Baltimore, Maryland
4. "Horizontal Pump Installation Manual," United Centrifugal Pump Company, San Jose, CA.
5. Jackson, C., "Alignment of Rotating Equipment," NPRA Refinery and Petrochemical Plant Maintenance Conference, Feb. 1, 1974, Paper MC-74-1.
6. Jackson, C., "Techniques for Alignment of Rotating Equipment," *Hydrocarbon Processing,* January, 1976, pp. 81–86.
7. Murray, M. G., "Machinery Shaft/Coupling Alignment Strategy Cuts Manhours and Downtime," *Oil and Gas Journal,* July 11, 1983, pp. 55–61.

8. *Surveillance and Maintenance of Rotating Equipment.* The Petroleum Publishing Company, Tulsa, OK, 1978.

9. Thomas Flexible Coupling Catalog 861, Rexnord Coupling Division, Warren, Pa.